Plant Disease Management Under Natural Farming

NIPA® GENX ELECTRONIC RESOURCES & SOLUTIONS P. LTD.
New Delhi-110 034

About the Author

Dr. Sanjeev Kumar is an Assistant Professor in the Department of Plant Pathology, Office of Dean,Faculty of Agriculture, Jawaharlal Nehru Krishi Vishwavidyalaya, Jabalpur, Madhya Pradesh. He is a very sincere and enthusiastic teacher. He has taught with dedication to the satisfaction of students in many undergraduate, graduate and doctoral programs and has been actively involved in all educational activities. Dr. Kumar has mentored 23 MSc students and 2 of his PhD students and is also actively involved in their research and extension activities. As a researcher, he has been involved in his eight major research projects sponsored by ICAR, NATP, TSP and JICA, Japan. He has published 75 research and review articles, 25 book chapters, 105 popular articles, 1 Technical Bulletin and 3 Technical Folders. He is an active member of more than six national and international societies. He is a member of Indian Phytopathological Society, IARI New Delhi, Indian Society of Mycology and Phytopathology, Udaipur, Rajasthan, Society for Advancement of Biocontrol, Banglulu. He has participated in many conferences, symposiums, workshops and presented papers. He also attended the World Soybean Research Conference in Durban, South Africa.

Dr. Kumar has authored nine books entitled Plant Pathogens and Principles of Plant Pathology, Diseases of Horticultural Crops: Identification and Management, Diseases of Field Crop and their Integrated Management, Pesticides and Plant Protection Appliances, Diseases of Field and Horticultural Crop and Their Management-II, Fundamentals of Plant Pathology, Integrated Disease Management; Principles & Practices,Chemicals & Botanicals in Plant Diease Managment published by New India. Publishing Agency, Pithampura, New Delhi and Diseases of Field and Horticultural Crop and their Management-I published by, Brillion Publishing, Desh Bandhu Gupta Road, Karol Bagh, New Delhi-110005, and his three practical manuals. His research interests include pulse pathology and biological control.

Plant Disease Management Under Natural Farming

Sanjeev Kumar
Assistant Professor/Scientist
Department of Plant Pathology
Office of Dean, Faculty of Agriculture
Jawaharlal Nehru Krishi Vishwa Vidyalaya
Jabalpur-482004, Madhya Pradesh, India

NIPA® GENX ELECTRONIC RESOURCES & SOLUTIONS P. LTD.
New Delhi-110 034

NIPA® GENX ELECTRONIC RESOURCES & SOLUTIONS P. LTD.

101,103, Vikas Surya Plaza, CU Block
L.S.C. Market, Pitam Pura, New Delhi-110 034
Ph : +91-11-43860225, Mob.: +91 9717133558, 9540816132
E-mail: newindiapublishingagency@gmail.com
Website: www.nipaersources.com

Print ISBN: 978-93-58878-56-1

ebook ISBN: 978-93-58875-26-3

Composed and Designed by NIPA®.

Preface

Education is vital for each individual in the country as it is not only for creating awareness of a subject matter but also for broadening scope of work. As per NEP-2020, the education should provide all opportunities to children and youth for being skilled and be more flexible for better life opportunities. The country adopted the Green Revolution model—an input- and chemical-intensive model—in the 1960s to increase food production. The focus was on development of new varieties and increased use of synthetic fertilizers and irrigation for increasing food production. This approach succeeded in increasing food production but failed the test of sustainability.

We envisage & reemphasize the concept of "Vasudhaiva Kutumbakam" which means “The world is one family” and emphasizes on integration of all human developmental activities with nature and lead to harmony of all components on this earth system. Proper education in society towards commitment for nature can bring a new paradigm shift in farming throughout the world. New ecological innovations and business models are emerging throughout world to redesign the global food system which is more sustainable with zero harm to nature. In this background, the degree course on Natural Farming has been envisaged. These alternative forms of agriculture have been the part of on-going B.Sc (Hons.) Agriculture degree courses, PG & Ph. D. courses implemented by different SAUs/ Institutions under Agriculture Education Division of ICAR, New Delhi.

The curriculum for the Undergraduate Program in Natural Farming has been developed lately and is being implemented nationwide in all agriculture universities The Plant Pathology undergraduate syllabus has been reorganized and new courses have been introduced. The curriculum has been developed with objective to teach students the important plant diseases and plant pathogenic organisms, impart practical knowledge on crop loss assessment and disease management methods under natural farming, impart practical knowledge on the preparation of inputs for the management of plant diseases under natural farming. There is no textbook accessible for teachers or students because the courses are new.This ‘Plant Disease Management under Natural Farming ’ book has been carefully written for the undergraduate class and is fully compliant with the latest syllabus. The purpose of producing this text is to provide basic and fresh facts to enable students to introduce the scientific foundations and principles of plant pathology in this course. Great care was taken to present facts appropriate to the student's level and grade the content accordingly. The text of the book is clear, well presented and arranged according to the curriculum. Chapters are designed to lead to comprehensive learning, with key concepts to help develop students' investigative skills. All of the syllabus's key subjects and commonly asked questions are included in the sample question papers at the end of each chapter, which make them comprehensive review tools. Regular use of them enhances knowledge, honed thinking skills, and removes doubt, leading to a deeper understanding of the subject matter.

I make no claims of originality in the creation of this book and have had the help of numerous books, magazines, bulletins, and electronic resources. We would like to express our sincere gratitude to the pioneers past and present in the field of plant pathology.

The book entitled Plant Disease Management Under Natural Farming written by me is dedicated to my parents, who sacrificed everything to give me the most and best education possible, to my wife, 'Archana' whose love and support have been the most precious things to me throughout our life together, and who helped me in many facets of preparation of this and of previous books and finally, to "Saumya" and "Adyan", our youngest children, who, someday, when they read their names in the book, may be reassured of "Papa" love for them, and may feel proud of their father.

In this endeavor, omissions and errors will inevitably be found, and I accept responsibility for factual errors. Teachers and students are invited to comment on the book and make suggestions for further improvement.

January, 2025 **Sanjeev Kumar**
Jabalpur

Contents

1

Natural Farming

Definition

"Natural Farming in the Indian context is a holistic agricultural production system in tandem with the laws of nature to provide food for all living beings ensuring production without harming the panchmahaboota or panchtatva (Prithvi, Agni, Jal, Vaayu and Aakash).

Principles of Natural Farming

Based on philosophy and application factors, cause and effect relationship, natural farming has following four principles.

- Principle of improved natural resource efficiency.
- Principle of self sustainability/no dependency.
- Principle of resilience.
- Principle of social equity/responsibility.

Philosophy of Natural Farming

The basic principle of natural farming is to maintain both the soil and ourselves healthy while growing nutritious food in harmony with the natural world. Farmers are seen as facilitators to sustain the natural equilibrium because they are a part of the web of life. As previously said, the primary goal of natural farming is to utilize as few external inputs as possible on farms, only those that are naturally degenerative. In natural farming, natural microbial inoculums such as jeevamruth and beejamruth are added to the soil to hasten the growth of the soil microflora.

Thus, natural faming suggests

- No external inputs.
- Local seeds (using regional cultivars).
- Microbial treatment of seeds.
- Microbial inoculants for the health of soil.
- Cover crops for mulching and biomass productionv.
- Incorporation for the highest level of beneficial microbial activity.
- Mixed cropping.
- Integrating cattle and using native breeds for cow dung and urine as vital inputs for various practices.

- Preservation of moisture and water, etc.v

Organic and Natural Farming Systems

The primary difference between natural farming and organic farming lies in their approach to external inputs. Natural farming completely avoid using synthetic and external inputs, while organic farming incorporates approved organic inputs as part of its practices (https://times of agriculture.in/natural-farming-vs-organic-farming . Using ecological processes, biodiversity, and cycles that are tailored to the local environment, organic agriculture is a production system that maintains the health of soils and ecosystems. Better tasting and more nutritious food is produced through organic farming, which also lowers pesticide and chemical residue in the soil, increases biodiversity, uses less energy, and is more sustainable over the long run. The lack of organic biomass, the gap between supply and demand, crop management time, high input prices, the absence of specialized infrastructure, including transportation, and the absence of a guaranteed marketing outlet for organic produce are some of the difficulties associated with organic farming.

Among the principles of organic farming are the avoidance of tillage, fertilizer, pesticides, herbicides, weeding, and pruning. The philosophy behind this approach is that the only way to attain superior quality and production is to preserve the natural order of things.Instead of using medication to treat the disease, this approach aims to prevent it. The following is a summary of how organic and natural farming differs:

Organic Farming	Natural Farming
Farmlands receive organic fertilizers and manures from external sources, such as compost, vermin compost, and cow dung manure.	The soil is not supplemented with chemical or organic fertilizers from external sources.
It makes use of standard agricultural techniques like tilling, weeding, mixing manures, and plowing.	It is done without ploughing, and cultivation are carried out the way it would be in natural ecosystems.
Because organic farming requires the use of bulk manures, it is still costly and has an effect on the environment.	It is a remarkably low cost farming technique that fully integrates with the surrounding biodiversity of the area.
Because organic farming entails interfering with natural processes, it has a modest negative impact on the environment.	Natural farming methods are in harmony with regional biodiversity processes and have no negative impact on the environment.

Scope of Natural Farming (NF)

1. **Natural Farming as a social movement:** Through community involvement, local bodies in Andhra Pradesh and Karnataka are promoting community-based natural farming, which has promising future prospects. Natural farming methods are currently being promoted on a global scale as an agro-ecological project in the food chain of agriculture.

2. **Natural Farming as an innovative approach;** NF as a cutting-edge strategy for nutrition and food security in sustainable food systems.
3. **Natural Farming as a science:** Agro-ecosystem approach and utilizing natural cycles and inputs based on long-term observations are the foundations of several farming techniques, including Organic Farming, Regenerative Agriculture, Rishi Krishi, Permaculture, Biodynamic Farming, Yogic Kheti, No-till farming, etc.
4. **Natural Farming as a set of practice:** As a body of practice, NF: Small and marginal farmers, as well as producers of low-input agriculture, have developed a variety of techniques to lower the chance of crop failure and cut production costs. Over time, they have developed ecological practices that are feasible and inexpensive using the resources at their disposal.
5. **Natural Farming for convergence with organic agriculture:** NF techniques have a wide range of commercial potential and convergence because they are primarily included in organic farming standards.
6. **Natural Farming for trade off advantages in carbon trading and climate mitigation:** NF farmers can obtain trade-off benefits and contribute to the reduction of ecological footprints by registering with carbon trading agencies.
7. **Natural Farming for knowledge generation and sharing for achieving SDG by 2025:** Seven of the 17 SDGs established by UNO are closely related to NF and, as a result, are receiving a lot of international attention and financing opportunities.
8. **Natural Farming for risk management under low input resource conditions for small land holders:** Enhancing water holding capacity, promoting biodiversity, sequestering carbon, protecting soil and water, and improving ecological services are all made possible by NF. Ecological, food, and nutritional security are national priorities that can be met by community-based natural farming, which can benefit small land owners as well.

Advantages of Natural Farming

1. Natural farming uses materials that are found locally and naturally as inputs, making it environmentally friendly.
2. Natural farming is the most effective approach to achieve abundant harvests and superior quality crops because it honors life as it is.
3. Natural farming yields superior quality vegetables with enhanced flavor.
4. Natural agricultural products are more nutritious and don't include any chemical residues.
5. Farmers lower the initial cost of manufacturing by preparing farm inputs.
6. It enhances the ecosystem of the soil, increasing soil productivity.

Component of Natural Farming

1. Natural Farming System
2. Crop/Breed/Strain Improvement
3. Soil Health Management
4. Livestock Management
5. Plant Protection
6. Weather Forecasting
7. Biodiversity
8. Agronomic Practices
9. Bio Inputs
10. Crop Management
11. Water Management
12. Farm Machinery
13. Renewable Energy Sources (Wind, Water, Solar)
14. Post Harvest Management
15. Bio Waste Resource Management
16. Nutrition and Health
17. Certification
18. Community Mobilization
19. Marketing
20. Economics
21. Agro Eco Tourism

Sub Component of Plant Protection

1. Entomology: Insects and Pest; Natural enemies, Trap crops, Pest repellent plants, Pollinators; Beneficial insects.
2. Pathology: Diseases Management.
3. Nematology: Nematode Management.
4. Microbiology: PGPR (Plant Growth Promoting Rhizobacteri(a).
5. Cultural Methods.
6. Traditional Knowledge Base.

Basic Pillars of Natural Farming

1. Jivamrita

It provides the soil nutrients and also functions as a catalyst to increase the activity of microorganisms. Additionally, it boosts earthworm activity. It keeps bacterial and fungal pathogens away from crops. A barrel containing 200 liters of water, 10 kg of fresh local cow dung, 5 to 10 liters of aged cow urine, 2 kilogram of jaggery (a form of brown sugar found locally), 2 kg of pulse flour, and a handful of soil from the farm's bund must all be used to prepare Jivamrita. Jeevamritha can be applied once the ingredients are thoroughly mixed and left to ferment in the shade for 48 hours. For an acre of land, 200 liters of Jeevamritha are plenty. The aerobic and anaerobic bacteria in the cow dung and urine grow during the 48-hour fermentation process as they consume organic components (such jaggery and pulse flour). A small amount of unaltered soil serves as an inoculant for indigenous microorganisms and organisms. Additionally, Jeevamritha helps in the prevention of bacterial and fungal plant diseases. Jeevamritha should be sprayed on crops twice a month either as a 10% foliar spray or in irrigation water. The preparation is used in the field by spraying it on or combining it with irrigation water, and it is kept for a maximum of 15 days in storage.

2. Bijamrita

This product is a seed treatment that guards young roots against soil-borne and seed-borne diseases, as well as fungi. Water (20 l), cow dung (5 kg), urine (5 l), lime (50 g), and a small amount of soil make up its basic composition. Lime, soil, and local cow dung—which is regarded as a natural fungicide—as well as cow urine, which is a liquid antibacterial—must be combined properly. Usually, the excrement is wrapped in a cloth and left in urine for around twelve hours. Any crop's seeds must be coated with beejamritha, mixed by hand, well dried, and then used for planting. Leguminous seeds need only be briefly immersed, then allowed to air dry.

3. Acchadana

By lowering evaporation through the application of soil mulch, straw mulch, etc., achadana preserves soil moisture. In order to provide the soil and crops with all the necessary ingredients, live mulch is necessary for the development of numerous cropping patterns in the same area.

4. Whapasa Moisture

The term "whapasa" refers to the mixing of 50% water vapor and 50% air in the space between two soil particles. The majority of the moisture and some of the nutrients that soil organisms and roots need are found in the microclimate of the soil. In order to meet the crops' moisture requirements, irrigation frequency should be decreased and it should be done in alternating furrows at noon. This will result in a large decrease in the need for irrigation under natural farming. Other common practices used in natural farming include crop rotation, intercropping, fallowing (allowing the soil to rest and rejuvenate), organic compost and bio-inoculants, biological and natural pest control techniques, indigenous resistant plant varieties, crop rotation, and bio-inoculants.

Model Practice Question

A. Objective Questions

a.Multiple Choice Questions

1. A holistic agricultural production system in tandem with the laws of nature to provide food for all living beings ensuring production without harming the panchtatva
   (a) Organic farming (b) Natural farming
   (c) Digital farming (d) Mixed farming
2. The principles of natural farming.
   (a) Principle of improved natural resource efficiency
   (b) Principle of self sustainability/no dependency
   (c) Principle of resilience
   (d) All of the above
3. Which among the following provides the soil nutrients and keeps bacterial and fungal pathogens away from crops.
   (a) Jivamrita (b) Bijamrita
   (c) Acchadana (d) Whapasa moisture

4 The product is used as seed treatment that guards young roots against soil-borne and seed-borne diseases, as well as fungi
   (a) Jivamrita (b) Bijamrita
   (c) Acchadana (d) Whapasa moisture

5. The term refers to the mixing of 50% water vapor and 50% air in the space between two soil particles
   (a) Jivamrita (b) Bijamrita
   (c) Acchadana (d) Whapasa moisture

6.The concept of "Vasudhaiva Kutumbakam" which means
   (a) The world is one family (b) The world is one village
   (c) The world is one district (d) The world is one State

Answer

Q. No.	Answer	Q. No.	Answer
1	(b) Natural Farming	4	(b) Bijamrita
2	(a) All of the above	5	(d) Whapasa moisture
3	(a) Jivamrita	6	(a) The world is one family

b.True/False

1. The primary difference between natural farming and organic farming lies in their approach to external inputs.
2. Natural farming completely eschews synthetic and external inputs.

3. Organic farming incorporates approved organic inputs as part of its practices.
4. Natural farming is a remarkably low cost farming technique.
5. Natural farming methods are currently being promoted on a global scale as an agro-ecological project in the food chain of agriculture.
6. Natural farming techniques have a wide range of commercial potential and convergence because they are primarily included in organic farming standards.
7. Organic farming is still expensive since it needs the usage of bulk manures.
8. Natural farming farmers can obtain trade-off benefits and contribute to the reduction of ecological footprints by registering with carbon trading agencies.
9. Natural farming is the most effective approach to achieve abundant harvests and superior quality crops because it honors life as it is.
10. Natural farming yields superior quality vegetables with enhanced flavor.

Answer

Q. No.	Answer	Q. No.	Answer
1	True	6	True
2	True	7	True
3	True	8	True
4	True	9	True
5	True	10	True

2. Descriptive Questions

a. Short answers

1. Define Natural farming.
2. Write down the principles of natural farming.
3. Describe the philosophy of natural farming
4. Write down the advantages of natural farming
5. How does natural farming differ from conventional farming?

b. Long answers

1. Discuss in details about the scope of natural farming
2. What is natural farming? Explain the component of natural farming
3. Write down the difference in natural and organic farming.
4. Explain in details about the basic pillars of natural farming (Jivamrita, Bijamrita, Acchadana and Whapasa moisture)
5. What is the key technique in natural farming?

2

Plant Pathology: Definition, Objectives, Scope and Importance

Plant Pathology

The term Plant Pathology is derived from Greek words- *Pathos* (suffering) + *Logos* (study) i.e. the study of the suffering plant. Plant Pathology has two phases broadly-

- **Science-** *understanding of the disease* i.e. the theoretical consideration of the suffering plants and *how do pathogens invade, how do plants defend themselves, what causes symptoms etc.*
- **Art-** *the application of the science to the field problems* i.e. Diagnosis and Control.

Plant pathology or phytopathology is the branch of agricultural, botanical or biological science which deals with the cause, etiology, resulting in losses and management methods of plant diseases.

Plant pathology is for plants mainly what medicine is for humans and veterinary medicine is for animals. Plant pathology is an integrative science and profession that uses and combines fundamental knowledge of botany, mycology, bacteriology, virology, nematology, plant anatomy, plant physiology, genetics and molecular biology and genetic engineering, biochemistry, horticulture, agriculture, tissue culture, soil science, forestry, chemistry, physics, meteorology, and many other scientific fields. Plant pathology has benefited from advances in each of these sciences, and many advances have been made in other sciences to solve plant pathological problems.

As a science, plant pathology seeks to improve our knowledge of plant diseases. At the same time, plant pathology seeks to develop methods, devices, and materials that can be used to prevent plant diseases or controlled. Uncontrolled crop diseases can lead to food shortages, higher food prices, or lower food quality. Affected plant products may be toxic and unsuitable for consumption. Some plant diseases may wipe out entire plant species, many of which affect beauty and landscape of our environment. Although plant disease control results in better quality food and a more aesthetically pleasing environment, consumers must pay for the cost of materials, equipment, and labor used to control plant diseases, and sometimes other less obvious costs of such as contamination of the environment.

Major Objectives

- To study biotic, mesobiotic and abiotic causes of diseases or disorders.
- To study the mechanisms of disease development by pathogens.
- To study the plant (host)-pathogen interaction in relation to environment.
- To develop methods of management of plant diseases.

Scope of Plant Pathology

- Field surveys to determine disease prevalence and incidence.
- Recordings of newly emerging diseases of economic importance, including details and extent of outbreaks.
- Assessment of losses due to various economically important diseases.
- Studies of the etiology, symptoms, predisposition, and recurrence of such diseases.
- Identification of suitable and economical methods to managme economically important crop diseases.
- Support for breeding disease-resistant varieties.
- Training of advisors and subject matter experts to bridge the gap between pathologists and farmers for better crop production.

Importance of Plant Pathology in Agriculture

- Plant pathology has advanced the knowledge to protect the crop from losses due to diseases.
- Most of the diseases with known disease cycle can now be avoided by the modification of cultural practices.
- Many economically significant diseases can now be controlled thanks to our growing understanding of how diseases spread.
- Crop improvement and varietal resistance have been achieved against many diseases through the joint effort of breeder and plant pathologist.
- The science of plant pathology has contributed disease free certified seed production.
- Plant pathology has made it possible to limit the spread of plant diseases from one place to another and from one country to another through appropriate measures and quarantine laws.
- Knowledge of plant pathology allows various preventive measures to be taken to successfully treat disease. These measures are seed treatments, soil treatments and crop rotation.
- Diseases in cold storage can be avoided by taking phytopathological measures, following various disease recommendations to protect stored fruits and vegetables.
- Plant pathology has enabled toxicants to be recognized, removed and utilized through toxicant production, competition and parasitism. Organisms that exert such lethal or harmful effects on other organisms are called antagonists.

- Plant pathology has advanced the knowledge to protect the crop from losses due to diseases.
- Most of the diseases with known disease cycle can now be avoided by the modification of cultural practices.
- With the knowledge of mode of disease spread, many diseases of economic importance can now be managed .
- Crop improvement and varietal resistance have been achieved against many diseases through the joint effort of breeder and plant pathologist.
- The science of plant pathology has contributed disease free certified seed production.
- Plant pathology has made it possible to limit the spread of plant diseases from one place to another and from one country to another through appropriate measures and quarantine laws.
- Knowledge of plant pathology allows various preventive measures to be taken to successfully treat disease. These measures are seed treatments, soil treatments and crop rotation.
- Diseases in cold storage can be avoided by taking phytopathological measures, following various disease recommendations to protect stored fruits and vegetables.
- Plant pathology has enabled toxicants to be recognized, removed and utilized through toxicant production, competition and parasitism. Organisms that exert such lethal or harmful effects on other organisms are called antagonists.

Challenges for plant pathology

The challenges for plant pathology are

- To reduce food losses.
- To improve food quality.
- To safeguarde our environment.

Model Practice Questions

A.Objective Questions

a. Multiple Choice Questions

1. The study of algae

(a) Mycology (b) Phycology

(c) Bacteriology (d) Virology

2. The study of the nature, cause and management of plant diseases.

(a) Plant Pathology (b) Plant Breeding

(c) Plant Taxonomy (d) Plant Physiology

3. The famous Irish famine of 1845 was caused due to the outbreak of

(a) Helminthosporiose of rice (b) Stem rust of wheat

(c) Late blight of potato (d) Downy mildew of grapes

4. In 2005 which pathological scientist got Borlaug award
   (a) Rattan Lal (b) VL Chopra
   (c) CD Mayee (d) S Nagarajan
5. Renowned scientist Dr. N.E. Borlaug belongs to–
   (a) Agronomy (b) Genetics
   (c) Entomology (d) Plant Pathology
6. The term plant pathology derived from
   (a) Greek word (b) American word
   (c) French word (d) Latin word
7. The study of the suffering plant is called
   (a) Plant Pathology (b) Plant Epidemiology
   (c) Plant Taxonomy (d) Plant Physiology
8. Plant Pathology is
   (a) Science (b) Art
   (c) Commerce (d) All
9. The study of epidemics and of the factors that influence them is called.
   (a) Plant Pathology (b) Plant Breeding
   (c) Plant Taxonomy (d) Plant Epidemiology
10. The study of fungi
   (a) Mycology (b) Phycology
   (c) Bacteriology (d) Virology

Answer

Q. No.	Answer	Q. No.	Answer
1	(b) Phycology	6	(a) Greek word
2	(a) Plant Pathology	7	(a) Plant Pathology
3	(c) Late blight of potato	8	(d) All
4	(d) S Nagarajan	9	(a) Plant Epidemiology
5	(d) Plant Pathology	10	(a) Mycology

b.True /False

1. Plant pathology has advanced the knowledge to protect the crop from losses due to diseases.
2. Understanding of the plant disease is called science.
3. The science of plant pathology has contributed disease free certified seed production.
4. Crop improvement and varietal resistance have been achieved against many diseases through the joint effort of breeder and plant pathologist.

5. All plant diseases are infectious..
6. Most of the diseases with known disease cycle can now be avoided by the modification of cultural practices.
7. All plant diseases are infectious..
8. Disease requires a susceptible host, a conducive environment, and a virulent pathogen.
9. Irish famine happen in Ireland due late blight disease of potato.
10. Bengal famine happen in West Bengal due Helminthosporium leaf blight disease of rice.

Q. No.	Answer	Q. No.	Answer
1	True	6	True
2	True	7	False
3	True	8	True
4	True	9	True
5	False	10	True

A. Subjective Questions

(a) Short Answer Questions

1. What is plant pathology?
2. Why is plant pathology important?
3. What are the major causes of plant diseases?
4. How does plant pathology contribute to agriculture?
5. What role do plant pathologists play in crop protection?
6. How does plant pathology help in sustainable farming?
7. How does climate change affect plant diseases?
8. What is the scope of plant pathology in research?

(b) Long Answer Questions

1. Plant pathologists proclaim that plant pathology is a discipline which encompasses both "art" and "science". What do you understand by the "art" of plant pathology?
2. What are the objectives and scope of Plant Pathology?.
3. Write the importance of plant pathology in Agriculture.
4. What is Plant Pathology ? Discuss its importance in human welfare.
5. Define Plant Pathology. Write down the objective and scope of plant pathology in agriculture.
6. Define Plant Protection. Explain its responsibilities to Agriculture?

3

Plant Diseases Definitions, Importance, and Concepts

Disease is one of those terms that are very difficult to define. It is realized that disease (literally *dis-ease*) implies *lack of 'comfort'* and therefore, involves *deviation from normal functioning*. From time to time several definitions which have been proposed, in fact descriptive but not simultaneously exclusive. The definitions for the term disease are:

- Disease is a malfunctioning process that is caused by continuous irritation, which results in some suffering producing symptoms.
- Disease is an alteration in one or more of the ordered sequential series of physiological processes culminating in a loss of coordination of energy utilization in a plant as a result of the continuous irritation from the presence or absence of some factor or agent.
- Disease is any morphological or physiological abnormality in a plant or any of its parts caused by continuous irritation and results in economic losses to human beings.
- Any deviation from normal growth or structure of plants that is sufficiently pronounced and permanent to produce visible symptoms or to impair quality and economic value.
- A plant is said to be diseased when there is a harmful deviation from normal functioning of physiological process .

All these definitions indicate that disease

- Is not a pathogen but it is *caused by a pathogen.*
- It is not symptom but *results in symptoms.*
- It is not a condition as the condition results from disease and is not synonymous with it.
- It is *not an injury.*
- Disease cannot be infectious, it is actually the pathogen which is infectious.
- Disease result from continuous irritation.
- Disease is a malfunctioning process and this result in some suffering and therefore disease is a pathological process.

Disease *vs* Injury

- Disease occurs whenever a vital process in the host plant is continuously disrupted.
- Disease is particularly severe when the regulated intake and use of energy is impaired.
- Injury is sometimes confused with disease, and injurious agents are sometimes confused with pathogens. However, this confusion can be avoided.
- Injuries occur when plants are temporarily disturbed. E.g.. By mowers or grasshoppers.
- In contrast, disease occurs when plants are exposed to long-term injury or damage (or, as some authors prefer, "continuous irritation") by pathogens. Examples include wilt fungus and yellow spot fungus.

The Concept of Disease in Plants

It is difficult to pinpoint exactly when a plant is diseased because it is not known whether plants feel pain or discomfort and because, in any case, plants do not speak or otherwise communicate with us.

Healthy Plant

- A plant is healthy, or normal, when it can carry out its physiological functions to the best of its genetic potential.
- The meristematic cells of a healthy plant divide and differentiate as required, and different types of specialized cells absorb water and nutrients from the soil; translocate these to all plant parts; carry on photosynthesis, translocate, metabolize, or store the photosynthetic products; and produce seed or other reproductive organs for survival and multiplication.

Diseased Plant

- When the ability of the cells of a plant or plant part to carry out one or more of these essential functions is interfered with by either a biotic or abiotic agents, the activities of the cells are disrupted, altered, or inhibited, the cells malfunction or die, and the plant becomes diseased.
- At first, the infection is localized to one or a few cells and is invisible. Soon, the reaction becomes more widespread and affected plant parts develop changes visible to the naked eye. These visible changes are symptoms of the disease.
- Visible or measurable detrimental changes in plants in response to biological infection or adverse environmental factors are measures of the degree of plant disease.
- Plant diseases can therefore be defined as a series of invisible responses of plant cells and tissues to biotic or abiotic agents. This can adversely affect plant morphology, function, or integrity, leading to partial plant failure or death.

- The type of cells and tissues affected determines the type of physiology that is disrupted first.
- Infection of roots might cause roots to rot and cause them to not able to soak up water and nutrients from the soil.
- Infection of xylem vessels, as takes place in vascular wilts and in a few cankers, interferes with the translocation of water and minerals to the crown of the plant.
- Infection of the foliage, as takes place in leaf spots, blights, rusts, mildews, mosaics, and so on, interferes with photosynthesis.
- Infection of phloem cells inside the veins of leaves and within side the bark of stems and shoots, as takes place in cankers and in diseases because of viruses, mollicutes, and protozoa, interferes with the downward translocation of photosynthetic products.
- Infection of flowers and fruits result interferes with reproduction.
- Although infected cells in most diseases are weakened or die, in a few diseases, e.g., infected cells are prompted to divide much faster hyperplasia or to make bigger a excellent deal extraordinary amorphous overgrowths (tumors) or abnormal organs.
- Pathogenic microorganisms which can cause disease/ailment are commonly called pathogens, commonly cause ailment in plants via way of means of disturbing the metabolism of plant cells via enzymes, toxins, growth regulators, and different materials they secrete and via way of means of absorbing foodstuffs from the host cells for his or her personal use.
- Some pathogens may additionally reason diseases via way of means of developing and multiplying within the xylem or phloem vessels of plants, thereby blockading the upward transportation of water or the downward motion of sugars, respectively, via those tissues.
- Environmental elements reason disease in plants whilst abiotic elements, which include temperature, moisture, mineral, nutrients, and pollutants, arise at stages above or beneath a certain range tolerated by plants.

Importance of Plant Diseases

- Crop diseases can occur throughout the crop production chain and remain one of the greatest threats to the sustainable development of society. The result is a 13% to 22% reduction in annual yield or costing billions of dollars in economic costs in staple of rice wheat, corn and potatoes plus additional costs for education and development of management strategies.
- Plant diseases caused by infectious pathogens have had a profound impact on human society and nature throughout human history through damage to food production, economic development, ecosystem resilience and natural landscapes.

- Famine and malnutrition in Ireland caused by *Phytophthora infestans* (Mont.) de Bary, the pathogen of potato blight, and his *Bipolaris oryzae* (Breda de Haan), the pathogen of rice brown spot famine in Bengal, led to millions of deaths and uprooted families and social structures.
- The pandemic of Chestnut wilt caused by *Cryphonectria parasitica* (Murrill) Barr and Dutch elm caused by *Ophiostoma novo-ulmi* (Buism.) Nann. destroyed much of the primary and secondary forestry industry in North America and Europe, causing an ecological disaster in the region.
- Plant diseases make plants toxic to humans and animals. Many plant pathogens produce mycotoxins that directly or indirectly threaten human and animal health.
- These biological and economic losses are responsible, at least in part, for the latest estimates of about 800 million people worldwide suffering from hunger or malnutrition.
- Plant diseases can also limit the variety of local crops and industries and reduce the quality and quantity of plant products.
- The costs of controlling crop diseases are also direct disease losses.

Model Practice Questions

a. Objective Questions

1. 'Fungi and Diseases in plants' is written by
   (a) K.C. Mehta (b) B.B. Mundkur
   (c) E.J. Butler (d) R.N. Vasudeva
2. Which among the following are true for disease
   (a) It is not a pathogen but it is caused by a pathogen.
   (b) It is not symptom but results in symptoms.
   (c) It is not an injury.
   (d) All
3. Bengal famine of 1942-43 was due to
   (a) Blast of rice (b) Brown spot of rice
   (c) Bacterial blight of rice (d) Rice tungro
4. Irish famine of 1945 was due to
   (a) Early blight of potato (b) Brown spot of rice
   (c) Bacterial blight of rice (d) Late bligtht of potato
5. Rust disease belongs to
   (a) Basidiomycotina (b) Ascomycotina
   (c) Oomycetes (d) Deuteromycotina

6. Deviation from normal functioning of plant is called
   (a) Disease (b) Disorder
   (c) Pathology (d) Pathogen
7. Disease in plants is caused by
   (a) Biotic agents (b) Mesobiotic agents
   (c) Abiotic agents (d) All
8. Disorder in plants is caused by
   (a) Biotic agents (b) Mesobiotic agents
   (c) Abiotic agents (d) All
9. Diseases is
   (a) Continnuous irritation (b) Discontinuous irritation
   (c) Both (d) None
10. Injury is
   (a) Continnuous irritation (b) Discontinuous irritation
   (c) Both (d) None

Answer

Q. No.	Answer	Q. No.	Answer
1	(b) B.B. Mundkur	6	(a) Disease
2	(d)All	7	(a) All
3	(b) Brown spot of rice	8	(c) Abiotic agents
4	(d) Late bligtht of potato	9	(a) Continnuous irritation
5	(b) Basidiomycotina	10	(b) Discontinuous irritation

b.True /False

1. Disease results from continuous irritation.
2. Disease is a pathological process.
3. Plant diseases may limit the kinds of plants and industries in the area.
4. Injuries occur when plants are temporarily disturbed.
5. Coffee rust happen in Srilanka in seventees.
6. Disease as not a pathogen.
7. Disease is a symptom.
8. Disease is not a condition as the condition results from disease and is not synonymous with it.
9. Disease is not an injury.
10. Disease cannot be infectious, it is actually the pathogen which is infectious.

Q. No.	Answer	Q. No.	Answer
1	True	6	True
2	False	7	True
3	True	8	True
4	True	9	True
5	True	10	True

B. Descriptive Questions

a. Short Answer

1. Which pathogens do not satisfy Koch's postulates and why?
2. Describe the socio-economic impacts of plant diseases.
3. Write down the difference between disease *and* injury.
4. Define disease and its importance.
5. Mention the importance of plant disease in agriculture.

b. Long Answer

1. What is disease? Mention importance of plant disease in agriculture with suitable classical examples.
2. Differentiate between the healthy (normal) and the diseased plant.
3. Disdinguish between i.Science and art of plant pathology II.Disease and injury III. Disease and disorder IV. Soil invaders and Soil inhabitant.
4. Mention two important milestones in the development of the subject of Plant Pathology.
5. Mention five plant disease outbreaks with significant effects in more recent times on human civilization.
6. Expalin old and new concept of disease in plants

4

Factors Affecting Disease Development: Disease Triangle and Tetrahedron

The Disease Triangle and Disease Pyramid

One of the first ideas that students learn in an introductory plant pathology course is the disease triangle, which they may also meet in more advanced studies. The triangle that represents the interplay of the three elements—the susceptible host, the virulent pathogen, and the favorable environments—is known as the "DISEASE TRAINGLE." The Disease Triangle has been expanded upon by certain plant pathologists by the addition of one or more factors. Three other parameters—human activity, vectors, and time—have been proposed. By adding a single parameter, a three-dimensional disease triangle can become a four-dimensional disease pyramid or disease telaheredon.

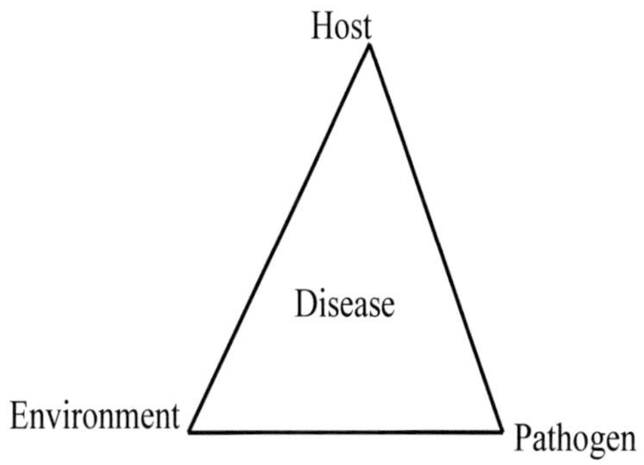

A very virulent pathogen must interact with a susceptible host in a setting that promotes the spread of disease in order for an epidemic to arise. The host's susceptibility and the pathogen's activity can both be impacted by the environment. Both the host and the pathogen have the ability to change the environment. Forecasting disease outbreaks and taking action to lessen the spread of disease are made possible by an understanding of these variables and how they interact with one another for a given disease in a given area.

Factors affecting disease development

There are three major factors that affect disease development

1. Pathogen factors
2. Host factors
3. Environmental factors

1. **Pathogen factors**
    - Presence of pathogen
    - Pathogenicity (virulence and aggressiveness)
    - Adaptability
    - Dispersal efficiency
    - Survival efficiency
    - Reproductive fitness
2. **Host factors**
    - Susceptibility
    - Growth stage and form
    - Population density and structure
    - General health
3. **Environmental factors**
    - Temperature
    - Rainfall
    - Dew
    - Leaf wetness period
    - Soil temperature
    - Soil moisture
    - Soil fertility
    - Air pollution

1. Pathogen factors

(a) **Presence of pathogen:** The pathogen's presence or absence is the primary determinant of whether disease develops at all. The primary determinant of whether or not disease develops in the late blight outbreak of potato in north India is the presence or absence of the pathogen.

b) **Pathogenicty:** The pathogenicity of the predominant population of the pathogen is often what determines how much disease occurs. Both the pathogen's virulence and its aggressiveness are included in the term "pathogenicity."Aspects of dispersal and survival fitness are also included. Monitoring the pathogenicity of the dominant pathogen population is a common practice in disease control, serving as a roadmap for the development of resistant crop types.

c) **Adaptibility:** The ability of a pathogen to overcome resistance in recently released varieties or adapt to altered environmental conditions is largely dependent on how adaptable it is. It is determined by the pathogen population's genetic adaptability and rate of reproduction. *O. theobromae* on cocoa has a far lower reproductive capacity and is therefore expected to adapt more slowly than *Puccinia graminis*, which has a high reproductive efficiency and will likely adapt fairly quickly.

d) **Dispersal efficiency:** A pathogen's capacity to spread quickly over great distances determines its capacity to unleash devastating disease outbreaks. Rust urediniospores may travel great distances in a matter of days, which enables epidemics to spread quickly across a vast area. Conversely, soil-borne pathogens are limited in their ability to spread and can only result in localized outbreaks of disease.

e) **Survival efficiency:** A pathogen's capacity to spread disease over multiple seasons is based on how long it can live. The pathogens that cause cruciferous club roots, *Plasmodiophora brassicae*, produce spores that allow them to persist in the soil for several years, decreasing the efficacy of crop rotation as a disease management strategy. *Sclerotium rolfsii* and Sclerotinia spp.'s sclerotia are very good at allowing these pathogens to endure the summer and winter, which produces more inoculums to start epidemics during the next growing season.

2. Host factors

(a) **Susceptibility:** The primary element influencing the development of a disease is the presence of individuals within the host population who are susceptible to the specific infection. The host plant population must be primarily vulnerable to assault by the pathotypes of the pathogen in the area for an epidemic of disease to transpire. Planting cultivars resistant to the prevalent population of the pathogen of concern is the most crucial approach to managing diseases.

b) **Growth stage and form:** The host's growth stage and morphology have a significant impact on the likelihood of disease. The degree of canopy closure is also determined by the growth stage, and this has an impact on the microclimate that exists within the canopy. Many diseases, like damping off, are more prevalent in seedlings, whereas many other diseases are specific to fully grown plants.

c) **Population density and structure:** The host's population density and structure will have a significant impact on how quickly disease spreads throughout the population. Dense plantings are far more susceptible to plant diseases than sparse plants. Since clones of a single plant or a variety of them are frequently planted across a wide region, modern intensive agriculture creates a perfect environment for the development of plant diseases. Few or no individuals of non-host species or cultivars impede the spread of inoculums. When sensitive host types were planted over sizable areas or when a new pathogen was introduced to such plantations, epidemics of certain diseases have happened. Similar circumstances resulted from the introduction of coffee rust to Ceylon in the 1870s and the late blight of potatoes to Ireland in the 1840s. Destructive outbreaks were created by both diseases when they came into contact with large monocultures of their respective hosts, coffee and potatoes.

d) **General health:** The host's past health frequently plays a significant role in predicting the onset of disease. While biotrophs are the origin of necrotrophic diseases, the latter are frequently more harmful to slowly growing hosts than to strong ones.

3. Environmental factors

The entire biosphere, which is where the epidemiologic processes converge, is included in the environment. For ease of understanding, the entire environment is separated into micro, meso, and macro environments.

The area where epidemiologic activities at the cell and organ levels take place is known as the microenvironment. The majority of the infection cycle, which includes both deposition and takeoff. The term "phyllosphere" refers to the leaf's microenvironment, which consists of a laminar air layer that can be up to 1 mm thick. The real leaf surface, or phylloplane, is a two-dimensional space, while the phyllosphere is a three-dimensional space. There are two types of rhizoplanes and a three-dimensional rhizosphere in relation to the roots (Hiltner, 1904).

The crop creates the mesoenvironment.It is created by the plants and affects both the plants and the crop's epidemiologic processes at the same time. The air layer extending from the crop surface to the troposphere is known as the macroenvironment.There are quite significant epidemiologic processes that take place in the macroenvironment. Three main environmental elements influence the development of disease:

(a) **Temperature:** Temperature affects *Latent period,Generation time* and *Infectious period.* The latent period denotes the interval between the onset of disease symptoms and infection, while the generation time represents the interval between infection and sporulation. The amount of time that the pathogen is still generating propagules is known as the infectious phase. Higher temperatures hasten the disease cycle, which accelerates the spread of epidemics. Epidemics typically spread more slowly in cooler weather. As a result, disease incidence and severity might not rise to the critical point required to result in a large loss of agricultural yield.

b) **Moisture:** The most significant environmental factor impacting the spread of diseases brought on by bacteria, nematodes, and fungi is moisture. We refer to dew, rainfall, relative humidity, and leaf wetness as "moisture." It's also crucial to consider how raindrop splashing and flowing water affect the spread of disease propagules. Rainfall and the duration of the dew period are less reliable indicators of disease than leaf moisture. The proporation of pathogen propagules that infect the host is frequently determined by the combination of leaf moisture and ambient temperature. Moisture has a significant impact on the spread of infections. Almost all bacteria and most fungi can be released and spread by free water or the impact of rainfall. This is a very helpful adaptation for a pathogen, as it increases the likelihood of favorable conditions for infection and germination after the propagules are distributed. For instance, coelomycetes release moist spores that are carried and distributed by raindrop splashing. The power of rainfall hitting surfaces releases a large number of dry spores. The primary environmental factors affecting soil-borne diseases are those found in the soil. Wetness is frequently a crucial element. Certain diseases, like as pythiaceous fungi that cause damping off, need a period of saturated soil to enable root infection and the development of surviving propagules. For release and motility, pyhecious fungus zoospores need free water.

c) **Soil fertility** Both soil and soil-borne disease growth are influenced by soil fertility. Certain facultative infections increase disease when plants grow poorly in nutrient-deficient environments. However, strong, well-fertilized plants are generally more susceptible to biotrophic diseases such powdery mildews and rusts than are underdeveloped plants. For certain soil-borne diseases, the amount of organic matter in the soil has a significant role. For instance, sandy or gravelly soils deficient in organic matter may experience eucalyptus dieback. Avocados have successfully managed *P. cinnamomi* by significantly raising the organic matter content. Greater populations of hostile microorganisms are found in soils with high organic matter content, which shortens the time plant diseases may survive and hence lowers the incidence of disease.

d) **Pollution** Plant disease has been shown to be influenced by the airborne environment. Acid rain, a form of atmospheric pollution, can directly harm plants in areas where industrial expansion has resulted in elevated levels of aerial pollution. These pollutants include nitrogen and sulfur oxides.

Model Practice Questions

A.Objective Questions

a. Multiple Choice Questions

1. The triangle that represents the interplay of the three elements—the susceptible host, the virulent disease, and the favorable environments—is known as

(a) Disease Traingle (b) Disease Tetrahedron

(c) Disease Pyramid (d) All

2. A pathogen's capacity to spread quickly over great distances determines its
   (a) Dispersal efficiency (b) Adaptibility
   (c) Pathogenicty (d) Survival efficiency
3. A pathogen's capacity to spread disease over multiple seasons is based on
   (a) Dispersal efficiency (b) Adaptibility
   (c) Pathogenicty (d) Survival efficiency
4. Temperature affects plant pathogen
   (a) Latent period (b) Generation time
   (c)Infectious period (d) All
5. The interval between the onset of disease symptoms and infection
   (a) Latent period (b) Generation time
   (c) Infectious period (d) Gestation period
6. The interval between infection and sporulation
   (a) Latent period (b) Generation time
   (c) Infectious period (d) Gestation period
7. The amount of time that the pathogen is still generating propagules is known as
   (a) Latent period (b) Generation time
   (c) Infectious period (d) Gestation period
8. Those diseases which occur at very irregular intervals and locations and in relatively few instances
   (a) Epidemic (b) Endemic
   (c) Pandemic (d) Sporadic
9. The diseases which appear on the plant while managing major diseases are called
   (a) Iatrogenic diseases (b) High sugar disease
   (c) Low sugar disease (d) All
10. The establishment of host parasitic relationship is called
   (a) Inoculation (b) Infection
   (c) Incubation (d) Dissemination

Answer

Q. No.	Answer	Q. No.	Answer
1	(a) Disease triangle	6	(b) Generation time
2	(a) Dispersal efficiency	7	(c) Infectious period
3	(d) Survival efficiency	8	(d) Sporadic
4	(d) All	9	(a) Iatrogenic diseases
5	(a) Latent period	10	(b) Infection

b. True/False

1. A four-dimensional disease pyramid or disease telaheredon can be created by adding a single parameter to a three-dimensional disease triangle.

2. The term "pathogenicity" encompasses both the pathogen's virulence and aggressiveness.
3. Monitoring the pathogenicity of the dominant pathogen population is a common practice in disease control.
4. Soil-borne pathogens are limited in their ability to spread and can only result in localized outbreaks of disease.
5. Planting cultivars resistant to the prevalent population of the pathogen of concern is the most crucial approach to managing diseases.
6. Destructive outbreaks of the coffee rust disease occurred in Sri Lanka as a result of a massive coffee monoculture.
7. Rainfall and the duration of the dew period are less reliable indicators of disease than leaf moisture.
8. Strong, well-fertilized plants are generally more susceptible to biotrophic diseases.
9. Sandy or gravelly soils deficient in organic matter may experience eucalyptus dieback.
10. Epidemics typically spread more slowly in cooler weather.

Answer

Q. No.	Answer	Q. No.	Answer
1	True	6	False
2	False	7	True
3	True	8	True
4	True	9	False
5	True	10	True

B. Descriptive questions

a. Short Answer Questions

1. Classify disease on the basis of type of infection.
2. Group diseases according to their causes.
3. Classify disease on based on their occurrence.
4. Based on the production of inoculums, categorize diseases.
5. Classify disease based on type of perpetuation and spread.
6. Discuss in details about itragenic diseases

b. Long answer questions

1. What are the major categories of plant diseases based on causative agents, and how are they classified?
2. Explain the difference between biotic and abiotic plant diseases with examples.
3. How are plant diseases classified based on the affected plant part, and

4. What are the implications for disease management?
5. Discuss the classification of plant diseases based on their symptoms and provide examples.
6. What are the methods used in the classification of plant diseases based on their mode of transmission?
7. How are plant diseases classified based on their life cycle, and why is this important for disease management?
8. Describe the classification of plant diseases according to their impact on agricultural productivity and economic importance.
9. What are the challenges in classifying plant diseases, and how can advances in molecular biology help overcome these challenges?
10. How do environmental factors influence the classification of plant diseases, and what are the key environmental factors involved?

5

Classification of Plant Diseases

Crop plants are susceptible to several diseases. For easier and better comprehension, plant diseases must therefore be categorized into more manageable and focused groups according to a number of criteria. Let's talk about each in turn.

I. Depending on the kind of infection

There are two categories of disease based on the type of infection.

i. **Localized disease**:which only damage a portion of the plant like leaf spots and anthracnoses.

ii. **Systemic diseases**,which impact the entire plant such as mosaic and leaf curl.

II Considering how frequently they occur

The four classes that the disease falls into are based on how often it occurs:

1. **Endemic disease**: A disease is considered endemic if it is more or less continuously prevalent in a certain nation or region of the world in one or more forms, or more or less continuously present year to year in a moderate to severe form. Ex: Darjelling district potato wart disease
2. **Epiphytotic or pandemic disease**: The word "epidemic" refers to "among the people" and, in fact, refers to human diseases that virulently affect a significant portion of the population. In order to convey a similar meaning while discussing plant diseases, the term "epiphytotic" was created. A disease that is widespread but periodic is called an epiphytotic disease. While it may always be there in the area, it only sometimes takes on a severe form. For example, the potato blight that caused the Irish famine
3. **Sporadic diseases:** Sporadic diseases are those that manifest relatively infrequently and locations and in relatively few instances. For example, potato spindle tuber
4. **Pandemic diseases:** When a disease is prevalent throughout the country, continent or the world it is known as a pandemic disease. current illustration, corona illness.

III. Based on the cause

Based on the cause the disease can be divided into two parts

1. **Infectious or Biotic disease**: These are pathogens brought on by biotic or mesobiotic agents under the favourable environment . For instance, bacterial, viral, or fungal diseases.

2. **Non-infectious or Abiotic disease**: These are the diseases that are not linked to any biotic or mesobiotic agents, stay non-contagious, and cannot spread from one sick plant to a healthy one. For instance. disease brought on by the sun, hailstorm, wind, rain, and frost.

IV. Based on the Production of Inoculum

Based on the production of inoculum, the disease can also be divided into two parts;

1. Single cycle disease (Simple Interest Disease):

Simple interest disease is the name given to the progression of disease that has a mathematical analogy to the simple interest of money. Within an epidemic, there is only one generation of the disease. Such diseases develop from a common source of inoculums i.e the capital i.e inoculum is constant. Example. Loose smut of wheat

2. Multiple cycle Disease (Compound Interest Disease):

Compound interest disease is the term given to a disease when its progression is mathematically comparable to compound interest on money. There are several or many generations of the pathogen in one life cycle of the crop. The capital ie inoculum is increased by the amount of interest. Example. Late blight of potato.

V. On the Basis of Type of Perpetuation and Spread

On the basis of type of perpetuation and spread, the disease can be divided into three groups

1. **Soil-borne diseases:** The causative agents proliferate and disperse via soil. For instance. Fusarium wilt and dampening off.
2. **Seed-borne diseases:** The disease is propagated and sustained with the use of seed or seed materials. Agents that cause disease can spread either outwardly or internally through seeds.Examples of internally seed borne are loose smut of wheat and blast of rice, which are examples of outwardly borne seeds.
3. **Air-borne diseases:** The causative agents of these diseases are carried by the wind (air). For instance. Groundnut has early and late leaf spots.

VI. Based on Symptoms

The disease can be categorized into rusts, smuts, wilts, blights, cankers, mildews, rots, damping-off, die-back, scabs, etc. based on its symptoms.

VII. Based on the host plant

Diseases such as those of apples, wheat, roses, coconuts, coffee, and cotton are categorized according to the host they affect.

Model Practice Questions

A.Objective Questions

a.Multiple Choice Questions

1. When disease affecting only a part of the plant
   (a) Localized disease (b) Systemic disease
   (c) Aerial disease (d) Endemic disease
2. When disease affecting the entire plant
   (a) Localized disease (b) Systemic disease
   (c) Aerial disease (d) Endemic disease
3. The term 'epidemic' is derived from a Greek word meaning
   (a) Among the people (b) Among the plant
   (c) Among the animal (d) Among the insects
4. The term 'epiphytotic ' is derived from a Greek word meaning
   (a) Among the people (b) Among the plant
   (c) Among the animal (d) Among the insects
5. When a disease is prevalent throughout the country, continent or the world it is known as
   (a) Epidemic (b) Endemic
   (c) Pandemic (b) Sporadic
6. Those diseases which occur at very irregular intervals and locations and in relatively few instances
   (a) Epidemic (b) Endemic
   (c) Pandemic (d) Sporadic
7. 'Low sugar' disease is
   (a) Rust (b) Downy mildew
   (c) Powdery mildew (d) Leaf spot
8 'High sugar' disease is
   (a) Rust (b) Downy mildew
   (c) Powdery mildew (d) All the above
9. The diseases which appear on the plant while managing major diseases are called
   (a) Iatrogenic diseases (b) High sugar disease
   (c) Low sugar disease (d) All
10. The establishment of host parasitic relationship is called
   (a) Inoculation (b) Infection
   (c) Incubation (d) Dissemination

Answer

Q. No.	Answer	Q. No.	Answer
1	(a) Localized disease	6	(d) Sporadic
2	(b) Systemic disease	7	(d) Leaf spot
3	(a) Among the people	8	(d) All the above
4	(b) Among the plant	9	(a) Iatrogenic diseases
5	(c) Pandemic	10	(b) Infection

b.True/False

1. Damping off caused by fungi is a soil borne disease.
2. Loose smut of wheat is a internally seed borne in nature
3. The word endemic means prevalent in, and confined to, a particular country or district and is applied to disease.
4. An epiphytotic disease is one which occurs widely but periodically in plant
5. The term 'endemic' is derived from a Greek word meaning 'among the people.
6. Loose smut of wheat is a cmpound interest disease.
7. Infectious diseases which are incited by biotic or mesobiotic agents.
8. Non-infectious diseases cannot be transmitted from one diseased plant to another healthy plant.
9. Late blight of potato is a simple interest disease.
10. Blast of rice is a externally seed-borne in nature.

Answer

Q. No.	Answer	Q. No.	Answer
1	True	6	False
2	False	7	True
3	True	8	True
4	True	9	False
5	True	10	True

B. Descriptive questions

a. Short Answer Questions

1. Classify disease on the basis of type of infection.
2. Group diseases according to their causes.
3. Classify disease on based on their occurrence.
4. Based on the production of inoculums, categorize diseases.
5. Classify disease based on type of perpetuation and spread.
6. Discuss in details about itragenic diseases

b. Long answer questions

1. What are the major categories of plant diseases based on causative agents, and how are they classified?

2. Explain the difference between biotic and abiotic plant diseases with examples.
3. How are plant diseases classified based on the affected plant part, and 4.what are the implications for disease management?
5. Discuss the classification of plant diseases based on their symptoms and provide examples.
5. What are the methods used in the classification of plant diseases based on their mode of transmission?
6. How are plant diseases classified based on their life cycle, and why is this important for disease management?
7. Describe the classification of plant diseases according to their impact on agricultural productivity and economic importance.
8. What are the challenges in classifying plant diseases, and how can advances in molecular biology help overcome these challenges?
9. How do environmental factors influence the classification of plant diseases, and what are the key environmental factors involved?
10. What is the significance of integrated pest and disease management (IPDM) in the classification and control of plant diseases?

6

Important Plant Pathogenic Organisms

Plant diseases are classified on the basis of type of pathogenic or non-pathogenic causes of the disease. The classification is based on the plant pathogenic organisms as follows.

1. Parasites

They include both biotic and mesobiotic agents. The diseases are incited by parasites under a set of suitable environment. Association of definite pathogen is essential with each disease.

i. Biotic agents: They are also called as animate causes. They are living organisms.

Biotic agents include

1. Prokaryotes

a. True bacteria or bacteria (Facultative parasites) e.g. Citrus canker.

b. Rickettsia-like bacteria (RL(B) e.g. Citrus greening, Pierce's disease of grape

c. Mollicutes or wall-less prokaryotes

  i. Mycoplasma-like organism (MLO) e.g. Sesame phyllody, Egg plant little leaf.

  ii. Spiroplasma e.g. Corn stunt, Citrus stubborn

2. Eukaryotes

a. **Protists** (Unicellular, coenocytic or multicellular with little or no differentiation of cells and tissues).

  i. Fungi e.g. wilt of cotton

  ii. Protozoa e.g. heart rot of coconut

  iii. Algae e.g. red rust of mango

b. **Plants** - Parasitic flowering plants or phanerogamic parasites - Broomrape of tobacco etc.

c. **Animals**- Nematodes –Root knot nematode etc.

d. **Mesobiotic agents**: They include viruses and viroids. They are infectious agents. They can be crystallized and are considered non-living. But their multiplication in the living plants ensures that they are living. Hence they are called as mesobiotic agents.

a. Viruses e.g. Yellow mosaic of blackgram

b. Viroids e.g. Spindle tuber of potato

2. Non-parasites or Abiotic agents

They are also called as non-infectious or physiological disorders. When no pathogen is found, cultured from or transmitted from a diseased plant, then the disease is said to be caused by a non-living or environmental factor. These diseases occur because of disturbances in the plant system by the improper environmental conditions in the air or soil or by mechanical influences. They are listed below.

i. Too low or too high temperature
ii. Lack or excess of soil moisture
iii. Lack or excess of light
iv. Lack of oxygen
v. Air pollution (Toxic gases in the atmosphere etc.)
vi. Mineral deficiencies or toxicities
vii. Soil acidity or alkalinity
viii. Toxicity of pesticides
ix. Improper agricultural practices.

Table: Comparison between different plant pathogens

Features	Fungi	Bacteria	Viruses	Viroids	Phytoplasma	Sprioplasma	RLO	Protozoa	Algae	Nematode
Visibility in micorscope	Yes	Yes	No (visible under Electronmicroscope)	No (visible under Electron microscope)	Yes	Yes	Yes	Yes	Yes	Yes
Prokaryote / Eukaryote	Eukaryote	Prokaryote	Kingdom -Virus	-	Prokaryote	Prokaryote	Prokaryote	Eukaryote	Eukaryote	Eukaryote
Cell Wall	Present	Present	Absent	Absent	Absent	Absent	Present	Present	Present	Present
Cell wall with mucopoly saccharides	No	Yes	Yes	No	No	No	Yes	No	No	No
Nucleic acid	Both DNA & RNA	Both DNA & RNA	Either DNA & RNA	RNA	Both DNA & RNA	Both DNA & RNA	Both DNA & RNA	Both DNA & RNA	Both DNA & RNA	Both DNA & RNA
Protein synthesis by their own enzyme	YES	YES	No	No	YES	YES		YES	YES	Yes
Culture in cell free medium	YES	YES	YES	No	No	No	Xylem inhibitory fastidious vascular bacteria can be cultured	YES	YES	Yes
Dependent on host nucleic acid for multiplication	No	No		No	No	No	No	No	No	No

Reproduction	Asexual &Sexual	Binary fission	Replication	Replication	Binary fission	Binary fission	Binary fission	Asexual &Sexual	Asexual &Sexual	Asexual &Sexual

Difference between disease and disorder

Features	Diasease	Disorder
Symptom expression	Appear progressively at definite stages.	Appear suddenly, nearly all at once in their fullest nature and intensity
Proportion of affected plants in an area	May vary in their extent of disease development	Tend to be affected with similar extent or in a similar way
Symptom variability	Symptoms variable in type, pattern and occurrence, although recognizable other wise	Symtom on individual plant or plant parts are regular or uniform in nature and pattern.
Sign	Present	Absent

Distribution of affected plants	Usually irregular or uneven distribution	Fairly regular or uniform in field or tightly clustered in affected area with no apparent spread pattern.

Difference between pathogen and parasite

Features	**Pathogen**	**Parasite**
Definition	Any agent that causes damage or disease	Organism existing in an intimate association with another living organisms and derives its nutrition from that host
Plant Pathogen relationship	Most (not all) pathogens are parasites	All parasites are not pathogens
Exapmle	*Ustilago tritici*	Root nodule bacteria (Rhizobium leguminoserum on the roots of pulses). Mycorhizal fungus parasitic on roots of trees
Sign	Present	Absent
Distribution of affected plants	Usually irregular or uneven distribution	Fairly regular or uniform in field or tightly clustered in affected area with no apparent spread pattern.

Multiple Choice Questions

A.Objective Questions

a. Multiple choice Questions

1. Rice blast pathogen perfect stage is–
   (a) *Pyricularia oryzae* (b) *Magnaporthe grisea*
   (c) *Helminthosporium solani* (d) *Rhizoctonia solani*
2. Black heart is a physiological disorder of
   (a) Tomato (b) Chilli
   (c) Potato (d) Cabbage
3. The scientific name of burrowing nematode is–
   (a) *Xiphinema* sp. (b) *Lougidorus* sp.
   (c) *Meloidogune* sp. (d) *Radopholus similis.*
4. Tea rust is caused by–
   (a) MLO (b) Virus
   (c) Bacteria (d) Algae
5. The major storage fungi that effects the food grain is–
   (a) Rhizobium (b) Mucor
   (c) Cercospora (d) Aspergillus
6. Application of potash increases
   (a) Resistance to water logging (b) Frost resistance in plants
   (c) Disease resistance in plants (d) None of these
7. Viruses contain–
   (a) RNA (b) DNA
   (c)Both RNA and DNA (d) Either RNA or DNA
8. The incidence of black scurf of potato is more in–
   (a) Sandy soil (b) Clay soil
   (c) Alluvial soil (d) Loam soil
9. The causal organism of bunchy top of banana is transmitted by–
   (a) *Peutalonia nigronervosa* (b) *Bemisia tabaci*
   (c) *Lipaphis erisimi* (d) Pollen
10. Mad cow disease is caused by
    (a) Virion (b) Pirion
    (c) Bacteria (d) MLO
11. Black tip of mango is caused due to the deficiency of
    (a) B (b) Zn
    (c) Cu (d) Mo
12. Heart rot of sugarbeet is caused due to the deficiency of
    (a) B (b) Zn
    (c) Cu (d) Mo

13. Khaira disease of rice is caused due to the deficiency of
    (a) B (b) Zn
    (c) Cu (d) Mo
14. Reclamation disease of oats is caused due to the deficiency of
    (a) B (b) Zn
    (c) Cu (d) Mo
15. White bud of corn is caused due to the deficiency of
    (a) B (b) Zn
    (c) Cu (d) Mo

Q. No.	Answer	Q. No.	Answer
1	(b) *Magnaporthe grisea*	6	(c) Disease resistance in plants
2	(c) Potato	7	(d) Either RNA or DNA
3	*Radopholus similis*	8	(a) Sandy soil
4	(d) Algae	9	(a) *Peutalonia nigronervosa*
5	(d) Aspergillus	10	(b) Pirion
11	(a) B	12	(b) Zn
13	(a) B	14	(c) Cu
15	(b) Zn	-	-

b.True /False

1. Sucidal germination takes place in striga.
2. Citrus greening caused by Fastidious bacteria.
3. 'Little leaf' in brinjal is caused by a Mycoplasma.
4. Wither tip of apple is caused due to deficiency of Molybdnum.
5. Pahala blight of Sugarcane is caused due to deficiency of Manganease.
6. The viruses which are usually helped or accompanied by smaller spherical particles of another serologically unrelated virus known as Satellite virus.
7. MLO disease transmitted by Leaf hopper.
8. Green netting of citrus is caused due to deficiency of Iron.
9. Whiptail disease of cauliflower is caused due to deficiency of Molybdnum.
10. Caulimovirus is an example of virus containing ds DNA.
11. Multiplication of virus occurs in Plasmodesmata.
12. Phytoplasmas contain DNA only.
13. NEPO viruses are transmitted by *Xiphinema index*.
14. Blossom end rot of tomato is caused due to deficiency of Calcium.
15. Organism existing in an intimate association with another living organisms and derives its nutrition from that host are called parasite.

Q. No.	Answer	Q. No.	Answer
1	True	6	True
2	False	7	True
3	True	8	True
4	True	9	True
5	True	10	True
11	True	12	False
13	False	14	True
15	True		

B. Descriptive Questions

a. Short Answer

1. How nutrient mobility determines the site of symptom development?
2. Which elements become deficients in acidic soils?
3. Why is iron not easily available to the plants?
4. Write notes on i. Microbial inoculants, ii.Iron efficient plants iii. Plant disorder caused by Zn. Disorder caused by Fe.
5. Write short notes on i. Disorder caused by B ii. Disorder caused by N iii. Disorder caused by Moiv. Disorder caused by Ca

b. Long Answer Questions

1. What are the biotic, abiotic and mesobiotic pathogens of plant diseases? Give in brief the classification of plant diseases according to major causal agents.
2. Disdinguish between biotic and abiotic pathogens .
3. Give two examples each of biotic, abiotic and mesobiotic pathogens.
4. Discuss about physiological disorders in plants.
5. Differentiate between pathogen and parasites.
6. Distinguish between:
   i. Iron efficient and Iron deficient plants
   ii. Freezing and Chilling injury
   iii. Primary and Secondary air pollutants

7

Different Groups: Fungi, Bacteria, Fastidious Vesicular Bacteria, Phytoplasmas, Spiroplasmas, Viruses, Viroids, Algae, Protozoa, Phanerogamic Parasites and Nematodes

Fungi

Definition

Fungi are eukaryotic, spore bearing, *achlorophyllus, heterotrophic* and thallophytic plants with varied forms and habitats, representing heterothallic or homothallic types of sexuality with characteristic sexual and /or asexual means of reproduction without tissue differentiation.

Main Characteristic for Fungi

The main characteristics of fungi are as follows:

Distribution	Cosmopolitan
Thallus	Unicellual, filamentous(myceli(a), or occasionally plasmodial
Cell wall	Made up of chitin (Cellulose only in class Oomycetes)
Chlorophyll	Lack chlorophyll
Nucleus	Eukaryotic
Life cycle	Simple to complex
Sexuality	Asexual and sexual
Habitat	Ubiquitous, saprobes, parasites or symbionts
Nutrition	Heterotrophic and absorptive
Storage product	Oil and glycogen

Types of Fungal Thalli

There are two types of thalli in fungi

1. **Homothallic Fungi**
   - Male and female sex organs or both the gametes are produced on the *same thallus*.
   - E.g. Powdery mildew of mung –*Sapharotheca fulginae.*

2. **Hetrothallic Fungi**
   - If male and female sex organs or both the gametes are produced on the *different thallus*
   - E.g. Rust Fungi.

Reproduction

Reproduction is the *formation of a progeny by either sexual or asexual means.* Spore is an unit of reproductions.

Types of reproduction

- Asexual reproduction
- Sexual reproduction

Asexual Reproduction:

This method of reproduction is characterized by *production of identical individuals without the union of the sex organs.*

Methods of Asexual Reproduction

1. By fragmentation
2. By arthrospores or oidia.
3. By budding
4. By binary fission
5. By production of spores

Asexual Reproduction

1. By fragmentation

- Breaking off of parts of the mycelium through external forces.
- Such pieces of mycelium under favorable conditions can start a new individual.

2. By arthrospores or oidia.

·Cells of the hyphae at the distal end round off and separate in basipetal succession.

· On germination, the arthrospores give rise to new fungus colonies.

3. By Budding

- Production of a small outgrowth from a parent cell.
- Bud increases in size while still attached to the parent cell.
- It eventually breaks off and forms a new individual .

· E.g. Rust, Smut fungi & yeast fungi.

4. Binary Fission

- Fission can occur through the simple splitting of a cell into two daughter cells by constriction.
- This is found among the bacteria generally, but some fungal yeasts may do this also.

5. Production of conidia

- Conidia produced on conidiophores in basipetal (oldest at the top and youngest at the bottom) and acropetal (oldest at the bottom and youngest at the top).

Sexual Reproduction

Sexual Reproduction:

Sexual reproduction in fungi involves the *union of two compatible nuclei.*

Methods of Sexual Reproduction:

a. Planogametic copulation
b. Gametangial contact
c. Gametangial copulation
d. Spermatization
e. Somatogamy
f. Heterokaryasis
g. Dikaryotization.

The process of sexual reproduction consists of three distinct phases.

1. **Plasmogamy:** It is *union of two protoplasts* brings the nuclei close together within the same cell.
2. **Karyogamy:** Actual fusion of *two haploid nuclei* brought together as a result of plasmogamy, Karyogamy immediately follows plasmogamy.
3. **Meiosis:** The *nuclear fusion is followed by meiosis.* Meiosis reduces the number of chromosomes to haploid.

Methods of Sexual Reproduction

a. Planogametic copulation (Gametogamy)

This involves the *fusion of two naked gametes one or both of which are motile.* Motile gametes are called **planogametes.** Depending on the size and motility of the fusing gametes, there are three types

- **Isogamy**: *Union of gametes that are similar in shape and size.* e.g. *Synchytricum, Olpidium.*
- **Anisogamy**: *Union of gametes that are morphologically similar but differ in size.* e.g. O. Blastocladiales. Genus- Allomyces.
- **Heterogamy**: *Union between a motile male gamete with a non motile female gamete* is known as heteroplanogametic copulation.

b. Gametangial Contact

In this method, *two gametangia of opposite sex come in contact, and one or more gamete nuclei migrate from the male to the female through especially developed fertilization tube serves as a passage for the male nuclei.* The zygote formed is called **Oospore**. e.g. Oomycetes, Ascomycetes.

c. Gametangial Copulation

Entire protoplast is transferred into the female gametangia a involes the fusion of two protoplast in a common cell. Sex gametes are indistinguishable or morphological identical. Copualtion occurs either by complete fusion of two protoplast. It is common in class Trichomycetes and zygomycetes.

d. Spermatization

It involves the formation of small spores or seed like structures or *spematia. Spermatia acts as male gamete which is uninucleate or spore like and carried out by wind, insects to the retentive hypha* (Female gametangium). A pore developed at the plant of contact and the contents of the spermatia passes into receptive hypha which serves as a female organ. e.g. Pycniospores and Receptive hypha in rusts.

e. Somatogamy

No gametes are involved. *Vegetative hypha itself acts as a male and female gamete and bring about sexual reproduction.* e. g. Smut Fungi.

f. Dikaryotization

It is accomplished through *migration of nuclei from one cell to another cell of vegetative hypha, often through mechanism of clamps.* The two nuclei remain in pair and divide as such and only fuse prior to the formation of spores. No special sex cells are produced. Clamp connections are formed during nuclear division. E.g. Class- Basidiomycetes.

g. Heterokaryosis:

The phenomenon of the *existence of genetically different kinds of nuclei in the same individual is calle heteokaryosis,* and the individuals that exhibit it are heterokaryotic. Heterkaryons may originate in a fungus thallus in four ways:

- By the introduction of genetically different nuclei into a homokaryon, a somatic cell in which all nuclei are similar.
- By the germination of a heterokaryotic spore, which will give rise to heterokaryotic soma.
- By mutation in a multinucleate, homokaryotic structure and the subsequent survival, multiplication, and spread of mutant nuclei among the wild type nuclei, and
- By fusion of some nuclei in a haploid homokaryon, and the subsequent survival, multiplication, and spread of the diploid nuclei among the haploid.

Major diseases caused by fungi

(i) Late blight of potato
(ii) Powdery mildew of pea
(iii) Downy mildew of pea
(iv) Anthracnose of mango
(v) Black stem rust of wheat
(vi) Blast of rice

Bacteria

Definition

Bacteria belong to proakaryota which encompasses organisms with a primitive type of nucleus lacking a clearly defined membrane .The bacteria are smaller than fungi and measure about 0.5 to 1.0 x 2.0 to 5.0µ.

Main characteristic for Plant Pathogenic Bacteria

1. A bacterial cell consists of a *cell wall* and acompound membrane enclosing protoplasm.
2. The cell wall of bacteria is composed of a *peptidoglycan.* It is composed of *acetyle-glucosamine* and *cetyle-muramic acid.*
3. The rigid peptidoglycan layer is located between the cytoplasmic membrane and an outer multiple tract layer.
4. The latter layer is composed of *lipoprotein* . It is common in gram negative bacteria.
5. Inside the protoplasm, nucleus, vacuoles, mesosomes, polysaccharides, lipids, mitochondric granules and spores are found.
6. Many bacteria possess other intracellular membrane systems such as mesosomes and chondrioids.
7. *Mesosomes* serve for compartmentalization and integration of biochemical system.
8. Bacteria do not have characteristic *nucleus.*
9. Bacteria contain *70 ribosomes* depending upon the bacteria. Ribosomes are the sites of protein synthesis.
10. Externally bacterial cells may show *capsules, pili, fimbrae* and *flagella.*
11. Some of the bacteria possess *flagella* which are useful for *motility* of the bacteria.
12. The bacteria belonging to Eubacteriales have *peritrichous flagellus* at all the sides.
13. Some bacteria transform themselves into small ovals or spheres which are highly resistant to adverse conditions. They are called as spores.
14. *Fibriae* are hair like structures that are observed as surface appendages on some bacteria. Fimbriae are common in plant pathogenic bacteria.
15. *Pili* are also hair like structures found in some bacteria. They serve as *adsorption organs* for bacteriophages. Pili mediate conjugation of bacteria.
16. Most of the plant pathogenic bacteria are rod shaped except *Streptomyce*s which are mostly filmentous.
17. Majority of plant pathogenic bacteria are aerobic except; *Erwinia*
18. Plant pathogenic bacteria required optimum growth temperature ranging between 27-30°C.

19. Most of the plant pathogenic bacteria reproduce by asexual process known as binary fission or fission.
20. Most of the plan pathogenic bacteria are gram negative except *Streptomyces, Clavibactor, Curtobacterium, Nocardia* and *Rhodococcus* which are gram positive in Gram's reaction.

Reproduction of Bacteria

1. Fission

- The most common mode of reproduction in bacteria.
- The bacterial cells divide into *two daughter cells.*
- It is *asexual* reproduction.

2. Conjugation

- Genetic material of one cell is transferred to another cell during conjugation.
- The two cells are genetically different.
- The donor cell transfers part of its genome to the recipient cell.
- Conjugation is one type of *genetic recombination.*
- Genetic recombination refers to *any process leading to the formation of a new individual which derives some of its gene from one parent and some from another, genetically different, parent.*

3. Transforamtion

- DNA from one type of bacterium is incorporated into the genetic makeup of another organism during transformation.
- In this DNA is absorbed through external source.
- This is a method employed in laboratory to bring about *recombination*: but it may be occurring in nature as well.

4. Transduction

- Virus that infects bacteria is called *bacteriophage.*
- Bacteriophage grows within a bacterial cell.
- The infected bacterial cell bursts and the bacteriophage particles are liberated.
- Some of the bacteriophage particles released carry some genetic material of the host bacterial cell.
- They enter into another bacterial cell and the genetic material brought into the cell by the virus is incorporated into that of the bacteria .
- By this process a new genotype of the bacterium arises and this process is called transduction.

Examples of disease caused by bacteria

(i) Wilts of tomato, banana, brinjal, potato
(ii) Wild fire of tobacco
(iii) Halo blight of beans

(iv) Black rot of cabbage
(v) Bacterial blight of rice
(vi) Citrus canker
(vii) Black arm of cotton

Fastidious Vascular Bacteria (Rickettesia Like Organisms)

Main characteristic for RLO

1. The fastidious vascular bacteria were formally known as *rickettesia like organisms* (RLOs).
2. They are usually rod shaped, aflagellate, bounded by a cell membrane and a cell wall measuring 1.0 -4.0x 0.2-0.5μm.
3. They are confined to *phloem or xylem of the host* plant but never to both.
4. Usually transmitted by *leaf hoppers* exception: Citrus greening bacteria (*Liberobacter asciaticum* is transmitted by citrus psylla (*Diaphorina citri*).
5. Al most all fastidious bacteria known so far are *gram negative* .Exception. Sugarcane ratoon stunting (*Clavibacter xyli pv.xyli*) is gram positive in nature.
6. There are two groups: one in which the fastidious bacteria occur only in phloem (phloem limited) and the other in which they occur in xylem (i.e xylem limited.
7. Xylem limited fastidious bacteria known so far have been successfully grown on nutrient media while phloem limited fastidious bacteria have not been culture so far.
8. They are sensitive to antibiotics such as *tetracycline* and *penicillin* and to high temperature.

Disease Caused by RLO

(i) Plecei's disease of grapes
(ii) Almond leaf scorch
(iii) Alfalfa dwarf
(iv) Phony peach
(v) Raton stunting
(vi) Plum leaf scald.

Phytoplasmas

Definition

Phytoplasmas are unicellular, ultramicroscopic, walless, prokaryotic, self-replicating, highly pleomorphic, filterable organisms or entities.

Main characteristic for Phytoplasmas

1. Phytoplasma *lack* rigid cell wall.
2. They are surrounded by a unit membrane, and consist of cytoplasma, ribosomes and nuclear material.

3. *Pleomorphic* or filamentous shape.
4. Size ranging from 200 to 800 nm and is less than 1 μm in diameter.
5. DNA is present in cytoplasm.
6. Genome is very small (689-1600kb)
7. Strictly host dependent and can survive and multiply only in the sap of phloem sieve tubes or insect haemolymph; also in their eggs.
8. Phytopalsma can only be acquire and transmitted by phloem –feeding insect vectors like, *Leafhoppers, Plant hoppers* and *Psllids*.
9. They *cannot* be grown on artificial media
10. Reproduce by *binary fission*.
11. They are sensitive to *tetracycline* but resistant to *penicillin.*
12. MLOs have no flagella, produce no spore and are *gram-ve*.

Disease caused by Phytoplasmas

(i) Pear decline
(ii) Grape yellows
(iii) Aster yellows of vegetables
(iv) Apple proliferation
(v) Little leaf of brinjal.
(vi) Coconut lethal yellowing
(vii) Elms yellows
(viii) Pigeonpea witche's broom
(ix) Rice yellow dwarf
(x) Stolbur
(xi) X.disease
(xii) Citrus greening
(xiii) Seasamum phyllody
(xiv) Grassy shoot of sugarcane
(xv) Little leaf of brinjal

Spiroplasma

Definition

Spiro plasma is a wall less mollicutes bounded by a triple layered unit membrane.

Main characteristic of Spiroplasma

1. Spiroplasma *lack* true cell wall, bounded by a single triple layered unit membrane.
2. They produce typical '*fried egg*' like colonies on agar medium.
3. They multiplied by *fission.*
4. They produce helical (100-240 nm in dia., 2-4μm inlength) form in liquid culture medium.

5. Spiroplasma can be *cultured* on nutrient media.
6. Resistant to *penicillin* but inhibited by *tetracyeline.*
7. Transmitted only through phloem –feeding insect vectors mostly through *leaf hoppers.*
8. Mostly found in the gut or haemolymph of insects, or in the phlowm of infected plants.

Disease Caused by Spiroplasma

(i) Corn stunt,
(ii) White bud of maize
(iii) Cirtus stubborn

Viruses

Definition of Virus

According to Luria and Darnell (1968), *Viruses are entities whose genome is a nucleic acid, either DNA or RNA, which reproduce inside living cells and use their synthetic machinery to direct the synthesis of specialized particles- the* ***virions,*** *which contain the viral genome and transfer it to other cells*

Nature of Virus

The physical and chemical nature of viruses eluded scientists until the 1930s. Stanley (1935) announced *crystallization* of Tobacco mosaic viruses, a feat for which he got a Nobel Prize. Stanley thought the crystals to be pure proteins. It was the pioneering work of F.C. Bawden and N.W. Pirie of England which showed that the viruses consists of proetein and nucleic acid. The chemical nature of viruses was settled. It consisted of nucleic acid and protein. The viruses chemically were proved to be nucleic acid. The mature particle of a plant virus is generally called **virion** and the whole infective particle is called as **Nuclcocapsid.**

Nucleic Acid of Viruses

1. Nucleic acid portion of the virus particle is called as Genome.
2. Genomes are organized as single nucleic acid molecules that are linear or circular.
3. They may have as few as four genes or as many as several hundred.
4. They may be double-stranded DNA, single-stranded DNA, double-stranded RNA, or single-stranded RNA.
5. Genome is actual infective component.
6. Majority of plant viruses contain RNA.
7. Some viruses Cauliflower mosaic virus, Maize streak virus, Mung bean yellow mosaic virus contain DNA.
8. Most plant viruses contain single strand of RNA.
9. In double stranded viruses (as RNA or ds the two strands are coiled around each other helically.

The Virus Protein

1. Protein coat that encloses the viral genome is called capsid.
2. Its structure may be rod-shaped, polyhedral, or complex.
3. Composed of many capsomeres, protein subunits made from only one or a few types of protein.
4. It helps viruses infect their host.
5. It protects nucleic acid (RNA or DNA) of virus.
6. It is made-up of different amino acid sequences in different viruses.
7. Protein coat protects viral nucleic acid from environment.
8. Viruses rely on living host cells for most of the enzymes necessary for their replication.
9. Viruses are in cable of growing for the synthesis of *Lipman system* is absent in viruses.

Different types of viruses

The following are the examples for different types of viruses

S.No	Types		Examples
1.	RNA viruses		
	a. Single stranded	With one segment	Tobacco mosaic viruses Lettuce necrotic yellow virus
		With two segments	Cowpea mosaic virus Tobacco ring spot virus
		With three segments	Cucumber mosaic virus
		With more number of segments	Potato virus x Tomato spotted wilt virus
	b. Double stranded		Wound tumour virus Fiji disease virus Reo virus
2.	DNA viruses		
	Double stranded		Cauliflower mosaic virus
	Single stranded		Maize streak virus, ϕX174

Examples of disease caused by viruses

1. Ring spot of papaya
2. Leaf roll of potato
3. Yellow mosaic of soybean
4. Leaf curl of papaya
5. Yellow mosaic of green gram
6. Bunchy top of banana

Viroids

Defnition

Viroid is circular, encapsulated, low molecular weight ($1.1\text{-}1.3x10^5$), self replicating, highly infectious ssRNA molecules .

Main characteristic for viroids

1. The viroid exists in *in vivo* as unencapsidated *RNA*.
2. They never contain any *protein coat*(capsid)
3. Genome of the viriod naked, *single stranded RNA* with 250-400 nucleotide, either linear or mostly circular.
4. Viroids are smaller in size than viruses i.e 50 nm or 1.1 to 1.3 X 103 molecular weight.
5. Despite its small size, the infectious RNA is replicated autonomously in susceptible cells; that is, no helper virus is required for multiplication.
6. Viroids are notable to synthesis protein and replicase enzyme required for replication.
7. Viroids replicate by direct RNA, copying in which all components required for viroid multiplication including RNA polymerase are provided by the host.
8. They cause diseases only in plants.
9. Viroids concentration and translocation is higher in growing parts of the plants (up to 0.2 mm from the apex).

Examples of disease caused by viroid

(i) Potato spindle tuber
(ii) Coconut cadang- cadang
(iii) Citrus exocortis
(iv) Chrysanthemum stunt
(v) Tomato bunchy top
(vi) Hot stunt viroid
(vii) Avocardo sunblotch
(viii) Peach latent mosic

Algae

Main characteristic for Algae

1. Algae are the *eukaryotic thallophytes* having chlorophyll as their primary photosynthetic pigments.
2. They are ubiquitous organisms abundantly present in aquatic environment.
3. They have wide range of shapes and size and the common shapes are rhizoidal, filamentous and mucilaginous.

4. The Chlorophycean (green algae) cause disease in plants.
5. Most of the algae are photoautotrophic, may reproduce either *sexually or asexually*.
6. Algae reproduce *asexually* by *binary fission*, fragmentation and formation of spores and *sexually* by formation of *zygotes*.
7. They spread through an air borne sporangia and produces 300 spores.
8. They enter in to the host through stomata and other natural openings.

Example

Red rust of tea, coffee, mango citrus, and guava (*Cephaleurous parasitica, C.minimum, C.coffeae)*

Protozoa

Main characteristic for protozoa

1. Protozoa are single celled, non photosynthetic, eukaryotic animal like organisms.
2. They are widely distributed in nature particularly in aquatic, environment or moist habitats.
3. Their size ranges from 2-3 micron.
4. They *lack rigid cell wall* and *do not contain chlorophyll.*
5. They make their movements with the help of *cilia, flagella or pseudopodia.*
6. Most of the protozoa are free living but some are symbiotic or parasitic causing diseases in plants, animals and humans.
7. Free living protozoa are commonly found in fresh water salt water, sand, soil and decaying organic matter.
8. Protozoa reproduce *sexually or asexually.*
9. Asexual reproduction is either by *binary fission* or by budding or by both.
10. Sexual reproduction is by fusion of two gametes.
11. They have characteristics ability to regenerate the lost parts of the body. For example, Genera: Amoeba, Paramecium etc.

Example

1. Phloem necrosis of coffee (*Phytomonas leptovasorum*)
2. Hart rot of coconut and oil palm (Phytomonas)
3. Sudden wilt of oil palm
4. Empty root of Cassava

Prions

Main characteristic for prions

1. *Proteinaceous* infectious particles that *lack nucleic acid.*
2. It is the smallest known protein often considered to be the cause various infectious diseases of the nervous system.

3. Prions are some times called slow viruses because of their slow effect.
4. Prions are transmissible particles .
5. No plant diseases are reported to be caused by prions.

Diseases caused by Prions

- Mad cow disease & Creutzfeldt- Jakob disease in humans
- Scrapie disease in sheep.

Phanerogamic Parasites

The parasitic flowering plants are also important pathogens. There are few seed plants, which are parasitic on living plants and are called parasitic higher plants or **Phanerogamic parasites.** These parasitic higher plants attack some valuable crops and trees causing considerable losses. More than 2500 species of higher plants are known to live parasitically on other plants.

Classification of flowering plant parasites:

i) Complete Parasite (Holoparasites)

a. Root: Rafflesia, Orobanchea (Broom rape)
b. Stem: Cuscuta (Dodder, Amarvel)

ii) Partial Parasites (Semi Parasites)

a. Root: Striga (Sandle wood, witch weed)
b. Stem: Loranthus (Banda or Deudrophae)

Nematode

Definition

Nematodes may be defined as *bilaterally symmetrical, triploblastic, pseudocoelomic unsegmented invertebrates with four hypodermal chords, a tri-radiate oesophagus,circum oesophageal nerve ring and a definite tail but lack circulatory muscles and specialized organs for locomotion and respiration.*

Main characteristic of plant parasitic nematodes

1. Plant parasitic nematodes are elongated, cylindrical, *bilaterally symmetrical*, unciliated worms with body tapering towards both the ends .
2. The body is covered by a transparent, tough, *resistant cuticle* that is secreted by a layer of hypodermal (epidermal) cells lying just beneath it.
3. Plant parasitic nematodes have a protrusible hypo-pharyngeal needle like knobbed *stylet (spear)* at the anterior end of the body, which is used to *suck* the cell cytoplasm from the host plants.
4. Nematodes possess a false body cavity or *pseudocoelom* i.e a cavity present between mesoderm and endoderm. Pseudocoelom is fluid filled and most part of it is occupied with intestine and components of reproductive system.
5. Oral aperture (mouth) is terminal and is surrounded by lips which contain sensory *cephalic papillae*. All other body openings like excretory pore,

vulva and anus (in females) and cloacal aperture (in males) are ventrally located.

6. Musculature consists of smooth longitudinal muscle cells located between the hypodermal chords and are directly connected to longitudinal nerves Circulatory muscles are absent.
7. *Digestive, reproductive, nervous* and *excretory systems* are *present.* Well defined *respiratory* and *circulatory* organs are *absent.*
8. They are usually *dioecious* with tubular reproductive organs. One or two gonads may be present.
9. They do not have any specialized structures for locomotion.
10. The life cycle of a plant parasitic nematode has six stages; egg, four juvenile stages and adult.
11. Nematodes belong to the animal kingdom, and phylum *Nematoda.* Most of the important parasitic genera belong to the order *Tylenchida* and few under Dorylaimida.

Examples of highly damaging plant parasitic nematodes

1	Root knot	*Meloidogyne* spp.
2	Cyst	*Heterodera* and *Globodera* spp.
3	Root lesion	*Pratylenchus* spp.
4	Spiral	*Helicotylenchus* spp.
5	Burrowing	*Radopholus similis*
6	Bulb and stem	*Ditylenchus dipsaci*
7	Reniform	*Rotylenchulus reniformis*
8	Dagger	*Xiphinema* spp.
9	Bud and leaf	*Aphelenchoides* spp.

Model Question Papers

A.Objective Types

a. Multiple Choice Questions

1. Sugarcane ratoon stunting is caused by

   (a) Phytoplasmas (b) Viroids
   (c) *Cavibacter xyli sub sp xyli* (d) Virus

2. Phytoplasmas contain

   (a) RNA only (b) DNA only
   (c) RNA+ DNA (d) RNA or DNA

3. Spiroplasma are mostly

   (a) Xylem inhibiting (b) Phloem inhibiting
   (c) Both (a) and (b) (d) Stomata inhibiting

4. Potato spindle tuber disease is caused by
   (a) Viroid (b) Phytoplasma
   (c) Spiroplasma (d) Virus
5. Bacteriophage is
   (a) Naked single stranded RNA particle (b) A Virus infecting bacteria
   (c) A virus infecting eukaryotic cell (d) Animal virus
6. Bacteriophage term was given by -----------------------
   (a) Mayer (b) D'Herelle
   (c) Hashimoto (d) Doi *etal*
7. Cuscuta is a-----------------------
   (a) Complete stem parasite (b) Complete root parasite
   (c) Semi stem parasite (d) Semi stem parasite
8. Phloem necrosis of coffee is caused by–
   (a) MLO (b) Virus
   (c) Protozoa (d) Algae
9. Spiroplasma is inhibited by an antibiotic –
   (a) Penicillin (b) Tetracycline
   (c) Streptocyclin (d) Aureofungin
10. The disease caused by Spiroplasma
   (a) Corn stunt (b) White bud of maize
   (c) Cirtus stubborn (d) All
11. Fastidious vascular bacteria are also called –
   (a) RLO (b) MLO
   (c) Viroids (d) Prion
12. RLO is usually transmitted by
   (a) Leaf hopper (b) White fly
   (c) Aphid (d) Thrips
13. The disease caused by Phytoplasma
   (a) Corn stunt (b) White bud of maize
   (c) Cirtus stubborn (d) Little leaf of brinjal
14. Fried egg' like colonies on agar medium is characteristic feature of
   (a) Phytoplasma (b) Spiroplasma
   (c) Viroid (d) Prion
15. Red rust of coffee is caused by
   (a) Phytoplasma (b) Spiroplasma
   (c) Viroid (d) Algae

Q. No.	Answer	Q. No.	Answer
1	(c) *Cavibacter xyli sub sp xyli*	9	(b) Tetracycline
2	(c) RNA+ DNA	10	(d) All

3	(b) Phloem inhibiting	11	(a) RLO
4	(a) Viroid	12	(a) Leaf hopper
5	(b) A Virus infecting bacteria	13	(d) Little leaf of brinjal
6	(b) D'Herelle	14	(a) Phytoplasma
7	(a) Complete stem parasite	15	(d) Algae
8	(c) Protozoa	16	

b.True/False

1. Orobanchea is a complete rot parasite
2. Cuscuta is a complete stem parasite
3. Striga is *partial root parasite* of *sugarcane, cereals, jawar, maize*, and *millets* in India.
4. Neamtode is triploblastic, bilaterally symmetrical, unsegmented, pseudocoelomate invertebrate.
5. Nematode lacks specialized organs for respiration and circulation.
6. The life cycle of a plant parasitic nematode has six stages.
7. The Mollicutes are a class of bacteria distinguished by the *absence of a cell wall.*
8. The genus *Spiroplasma* are helical and move by *twisting* .
9. Most of the important parasitic genera belong to the order *Tylenchida* .
10. When an ectoparasite feed at a particular site for a brief period and the move to the next site. It is called migratory ectoparasite.
11. Loranthus is common parasites of mango trees.
12. Witch weed posses' *chlorophyll bearing leaves.*
13. Endoparasites Considered to be the most destructive forms of plant parasitic nematodes.
14. *Meloidogyne is a* Sedentary endoparasites nematode.
15. *Tylenchulus semipenetrans* is a Sedentary semi-endoparasite nematode.
16. *Tylenchorhynchus is a* Migaratory semi-endoparasite.
17. *Heterodera,and Globodera* are considered to be the most advanced plant parasitic nematodes.
18. The best known genus in Mollicutes is *Mycoplasma.*
19. Phloem necrosis of coffee is caused by Protozoa.
20. Red rust of tea is caused by an algae.

Q. No	Answer	Q. No.	Answer
1	True	11	True
2	True	12	True
3	True	13	True
4	True	14	True
5	False	15	True

6	True	16	True
7	True	17	True
8	True	18	True
9	True	19	True
10	True	20	True

Descriptive Questions

a. Short answer

1. What features are used for identification of plant parasitic nematodes?
2. The nematodes placed in which two class and order.
3. Write any three modes of feeding in plant nematodes?
4. Differentiate between holo and hemiparasites of higher plants.
5. Write down the significant feateures of Phytoplasma.
6. Why is the designation RLO now abondaned?
7. Why are aphids unable to transmit mollicutes?
8. Why are Phytoplasma and Liberatobacter treated as Candidatus genera?
9. Why are Phytoplasmas considered as basically insect parasites?

b. Long Answer

1. Write notes on
   i. Classification of plant parasitic nematodes.
   ii. Under and above ground symptoms produced by nematodes.
   iii. Integrated Nematode Management.
2. Classify plant parasitic nematodes on the basis of their parasitism.
3. White down the symptoms and nature of damage caused by Heterodera, Meloidogyne and Anguina.
4. Distinguish between.
   i. Mycoplasma and Spiro plasma.
   ii. Orbanchae and Striga.
   iii. Host range and host preference of parasitic angiosperm.
5. Explain in brief about other plant pathogens: Mollicutes; Flagellant protozoa; FVB; Green algae and Parasitic higher plants.

8

Pathogenesis

A. Parasitism

An organism that lives within or upon another living organism from which it derives nourishment and in which it may cause various degrees of harm is called as **parasite** Among the parasite one can distinguish different degree of parasitism as

i) Obligate parasites
ii) Non obligate parasites
iii) Facultative saprophytes
iv) Facultative parasites

i) Obligate Parasites or Biotrophs

Those organisms which requires living hosts or tissues complete to their life cycle, are called as obligate parasites or An obligate parasite is an organism that can live only on living host. They can never be grown on dead, artificial food material. Examples; Rust, Mildews, Viruses.

ii) Non Obligate Parasites or Necrotrophs

Those organisms when kills the tissue in advance of penetration and then lives on it as saprophytically. E.g. *Sclerotium rolfsii, Ventruria, Claviceps.*

iii) Facultative Saprophyte

An organism that is ordinarily parasitic but under proper conditions may be saprophytic. e.g. *Smut, Sphacelotheca* sp.

iv) Facultative Parasites

An organism that is ordinarily saprophytic but under proper conditions may be parasitic. e.g. *Pythium, Phytophthora.*

B. Pathogenesis

Definition

Pathogenesis can be defined as the chain of events leading to the development of a disease. In another words it is also called disease cycle.

It tells us about

- Source of perennation of the pathogen.
- Mode of its dispersal from the source of survival and during its spread of disease.

- Help us to formulate effective control measures.
- The events constituting pathogenesis occurs in several well defined steps, one after the other, to complete the disease cyele.
- Pathogen perennates at some location during the absence of cultivated host.
- Its gets transported to the cultivated host through some agencies.
- The pathogen breaks the host barrier to establish infection.
- Effect the host physiology, damage the plant and the plant expresses symptoms.
- Finally the pathogen finds an exit from the host through its propagules.

Stages in the Pathogenesis or Disease cycle

The primary events in a disease cycle are

1. Inoculation
2. Penetration
3. Establishment of infection
4. Colonization (Invasion)
5. Growth and Reproduction of the pathogen
6. Dissemination of the pathogen
7. Survival of the pathogen

1. Inoculation

Inoculation is the coming in contact of a pathogen with a plant. Inoculum defined as the part of the pathogen, which on contact with a suitable host cause infection. E.g. of fungal inoculum; chlaymydospores, sclerotia, teleutospores, zoospores, conidiospores etc. In bacteria, mollicutes, the inoculum is whole individual. The propagule is a unit of inoculum that includes juvenile nematodes is the inoculum of nematode.

(a) Types of Inoculum

All inoculum that survives in the winter or summer and causes the org infection in the spring or autumn is called primary inoculum, and the it causes are called primary infections. An inoculum produced from primary infections is called secondary inoculum and it in turn causes secondary infections.

(b) Source of Inoculums

- Sowing and planting materials viz., seeds, tubers, bulbs, rhizomes.
- Diseased plants in the field.
- Debris left on the land or used in compost heaps.
- Host plants in a dormant stage when conditions are not favorable.
- Soil infected with certain soil borne pathogens.

Plant diseases are spread in various ways:

1. Spread through seeds and other planting materials such as tubers, rhizome etc.
2. Spread by natural agents such as wind, water and rain.
3. Spread by animal and insects.

2. Penetration

The penetration process has two phases

i) Pre penetration stage.
ii) Penetration.

i) Pre penetration stage

During the pre penetration stage the fungus hyphae attach with the host surface. The spores will germinate or spores to produce germ tube or infection threads. This stage may be abortive if environments do not favor growth of the hypha or germination of spores.

ii) Penetration

During penetration, the infection thread enters the host by any of the following methods

(a) Entry through natural openings like stomata & lenticels.
(b) Entry through rupture of the host surface due to development of organs like prop roots.
(c) Entry through wounds due to mechanical injury or insect injury.
(d) Direct penetration

The pathogen exerts its own effort to break the host barrier and directly enter through cuticle or epidermis without seeking the wounds or natural openings. Host barriers can be either structural barriers or chemical barriers. Enzymes produced by the pathogen will break down the epidermal cells and thus gain entry. Pathogens penetrate plant surfaces by direct penetration. Some fungi penetrate tissues in one way only, others bacteria enter plant mostly through wounds, less frequently through natural opening and never directly. Viruces, viroids, mollicutes and protozoa enter through wound made by vectors, although some viruses and viroids may also enter through wounds made by tools and other means. Nematodes enter plants by direct penetration and sometimes through natural openings. Penetration does not always lead to infection.

3. Infection

Infection is the process by which pathogen establish contact with the susceptible cells or tissues of the host and take nutrient from them. During infection pathogens grow or multiply or both within the plant tissues and invade and colonize the plant to a lesser or greater extent. Therefore, invasion of the plant tissues by the pathogen and growth and reproduction of the pathogen (colonization) in or on infected tissues are actual two concurrent sub stages of disease development within the stage of infection. Successful infections result in the appearance of symptom on the host plant. In most plant diseases, symptoms appear from a few days to few weeks after inoculation.

4. Colonization

Various pathogen invade hosts in different ways and to different extents. Many fungi spread into all the tissues of the plant organs they infect, either by growm directly through the cells as an intracellular mycelium or by growing between the cells as an intercellular mycelium. The fungi that cause vascular wilts invad the xylem vessels of plants. Bacteria invade tissues intercellularly, although where part of the cell wall dissolve, bacteria also grow intracellular. Bacteria causing vascular wilt, invade the xylem vessels. Viruses, viroids, mollicutes, fastidiou bacteria and protozoa invade tissues by moving from cell to cell intercelluarly Viruses and viroids invade all types of living plant cells, mollicutes and protozoa invade phloem sieve tubes and perhaps a few adjacent phloem parenchymatous cells. Most nematodes invade tissues intercellularly, but some can invade intracellularly-many nematodes do not invade cells or tissues at all but feed by piercing epidermal cells with their styles.

5. Growth and reproduction of the pathogen

Individual fungi and parasitic higher plants generally invade and infect tissues by growing into them from one initial point of inoculation. Most of these pathogens, whether producing a small spot, a large infected area, or a general necrosis of the plant continue to grow and branch out within the infected host indefinitely so that the same pathogen individual spread into more and more plant tissues until the spread at the infection is stopped or the plant is dead. All other pathogens like bacteria, virus, viroids, nematodes and protozoa, do not increase much, ifs all, in size with time, since, their size and shape remain relatively unchanged throghout their existance. These pathogens invade and infect new tissues within the plant by reproducing at a rapid rate and increasing their number tremendous in the infected tissues. The progeny may then be carried passively through:

(a) Plasmadesmata: virus and viroids
(b) Phloem: viruses, viroids, mollicutes protozoa, some fastidious bacteria
(c) Xylem: some bacteria nematodes

6. Dissemination of the pathogen

Following the infection, the pathogen will continue its growth, produce spores or reproductive units, which will find exit through host surface and spread to repeat the same process. Many pathogens spread through crops in a special and often spectacular way. Some cover tremendous distances at a remarkable speed, especially pathogens with airborne propagules, like rusts. Pathogens which infect subterranean plant organs exhibit a restricted pattern of dispersal. Then there are pathogens which depend for their dispersal of spores and propagules on rain or water, while others need a vector like man, insects and nematodes. Some parasites are transported with seed, thus depending on the hosts own dispersal processes.

There are many dispersal mechanisms. Some of these mechanisms are both elegant and highly efficient in that they are closely adapted to the biology of the host, like the synchronization of spore release in certain pathogens like Venturia and Claviceps. Other parasites produce millions of spores, saturating the environment with propagules in an apparently haphazard and wasteful way.

7. Survival of Pathogen

Pathogens must find some alternate source of their survival in the absence of their cultivated host; otherwise the infection chain will be remaining incomplete. Infection chain is the chain of events leading to the completion of pathogenesis.

The source of survival can be grouped into

(a) Infected host as reservoir of inoculums

  i. Cultivated host: Main host

  ii. Collateral host: Wild host of the same family

  iii. Alternate host: Wild host of other family

(b) Saprophytic survival outside the host. Soil and plant debris serve as the media for survival. E.g. Pythium, Rhizoctonia.

(c) Dormant organs of pathogen as a source of survival and primary inoculums; only fungi and nematodes have spores and cysts etc, while virus and bacteria have no resting stage.

Model Practice Question

A. Objective Questions

a. Multiple Choice Questions

1. The coming in contact of a pathogen with a plant.

   (a) Inoculation (b) Inoculum

   (c) Penetration (d) Infection

2. The organisms which kill host tissues in advance of penetration and then live saprophytically

   (a) Biotroph (b) Necrotroph

   (c) Hemibiotroph (d) All

3. Establishment of host parasitic relationship is called

   (a) Inoculation (b) Inoculum

   (c) Penetration (d) Infection

4. The part of the pathogen, which on contact with a suitable host cause infection

   (a) Inoculation (b) Inoculum

   (c) Penetration (d) Infection

5. Spread of pathogen inside the host is called
   (a) Saprophytes (b) Parasites
   (c) Symbionts (d) Colonization
6. Those fungi that feed on dead and decaying organic matter are called
   (a) Saprophytes (b) Parasites
   (c) Symbionts (d) All
7. Wild host of the different family
   (a) Cultivated host (b) Collateral host
   (c) Alternate host (d) Main host
8. Infection chain is the chain of events leading to the completion of
   (a) Pathogenesis. (b) Pathogenicty
   (c) Virulence (d) Aggressiveness
9. Rhizoctonia and Sclerotium is
   (a) Biotroph (b) Necrotroph
   (c) Hemibiotroph (d) All
10. Wild host of the same family
   (a) Cultivated host (b) Collateral host
   (c) Alternate host (d) Main host

Answer

Q. No.	Answer	Q. No.	Answer
1	(a) Inoculation	6	(a) Saprophytes
2	(b) Necrotroph	7	Alternate host
3	(d) Infection	8	(a) Pathogenesis.
4	(b) Inoculum	9	(b) Necrotroph
5	(d) Colonization	10	(b) Collateral host

b.True/False

1. All inoculum that survives in the winter or summer and causes the secondary infection in the spring or autumn is called primary inoculums.
2. Viruses, viroids, mollicutes, fastidiou bacteria and protozoa invade tissues by moving from cell to cell intercelluarly
3. An inoculum produced from primary infections is called secondary inoculum and it in turn causes secondary infections
4. The fungi that cause vascular wilts invade the phloem vessels of plants.
5. Penetration does not always lead to infection.
6. Bacteria causing vascular wilt, invade the xylem vessels.
7. Viruses and viroids invade all types of living plant cells.
8. Mollicutes and protozoa invade phloem sieve tubes and perhaps a few adjacent phloem parenchymatous cells.

9. Bacteria invade tissues intercellularly, although where part of the cell wall dissolve, bacteria also grow intracellular.
10. Successful infections result in the appearance of symptom on the host plant.

Answer

Q. No.	Answer	Q. No.	Answer
1	False	6	True
2	True	7	True
3	True	8	True
4	False	9	True
5	False	10	True

B.Descriptive Questions

a. Short answer questions

1. What is plant pathogenesis?
2. What are the primary types of plant pathogens?
3. How do fungi cause disease in plants?
4. What is the role of stomata in plant infection?
5. What is the concept of the "disease triangle" in plant pathology?
6. What are phytotoxins?
7. How do viruses infect plants?
8. What is a host plant's defense mechanism against pathogens?
9. What role do vectors play in plant pathogenesis?
10. What is the significance of the hypersensitive response in plant immunity?

b. Long answer questions

1. Explain the following: i. Infection ii. Latent infection iii. Invasion iv. Colonization v. Inoculum potential
2. Describe the general principles of plant infection? How infection is caused in some of the important plant pathogen.
3. What is infection? Describe in detail the mechanism of infection and establishment of any typical fungal pathogen.
4. What are the different stages in the development of a infectious plant disease. Describe in detail the pre penetration activities of fungal pathogens. Gives suitable examples.
5. Explain in details about growth and reproduction in plant pathogens.

9

Koch's Postulates

Koch's postulates are four criteria intended for establishing a causal relationship between a disease-causing microorganism and its disease. Robert Koch and Friedrich Loeffler developed the postulates in 1884, and Koch refined and published them in 1890. The postulates were utilized by Koch to establish the etiology of tuberculosis and anthrax, but they have since been extended to other diseases.

In the 19th century, Koch's postulates developed as broad criteria for identifying pathogens that could be isolated using the techniques that were at hand. Even though some infectious agents did not meet all of the postulates, it was known even in Koch's day that they were unquestionably the cause of disease. The early development of the field of virology may have been hampered by attempts in the late 19th century, when viruses could not be seen or isolated in culture, to strictly apply Koch's postulates to the diagnosis of viral diseases. Even if some infectious agents do not meet all of Koch's postulates, they are nevertheless recognized as the cause of disease today.

Koch's Postulates

- **Postulate 1**: The microorganism must be found in abundance in all organisms suffering from the disease, but should not be found in healthy organisms.
- **Postulate 2**: The microorganism must be isolated from a diseased organism and grown in pure culture.
- **Postulate 3**: The cultured microorganism should cause disease when introduced into a healthy organism.
- **Postulate 4**: The microorganism must be reisolated from the inoculated, diseased experimental host and identified as being identical to the original specific causative agent.

Advantage and Disadvantage of Koch's postulates

- Koch's rules are possible to implement.
- Although not always easy to carry out, with such pathogens as fungi, bacteria, parasitic higher plants, nematodes, most viruses and viroids, and the spiroplasmas.
- These organisms can be isolated and cultured, or can be purified, and they can then be introduced into the plant to see if they cause the disease.

- With the other pathogens, however, such as some viruses, phytoplasmas, fastidious phloem-inhabiting bacteria, protozoa, and even some plant pathogenic fungi that are obligate parasites of plants (such as the powdery mildew, downy mildew, and rust fungi), culture or purification of the pathogen is not yet possible and the pathogen often cannot be reintroduced into the plant to reproduce the disease.
- Despite the difficulties of carrying out Koch's postulates with some causal agents, they continue to be applied, sometimes with certain modifications, in all cases of disease.
- By attempting to carry out Koch's postulates in all newly discovered diseases, a great deal of work with potential saprophytes has been avoided.
- While, at the same time, doubt and criticism are reduced to a minimum while confidence in and use of the identification increase greatly and quickly.

10

Role of Enzymes, Toxins and Growth Regulators in Disease Development

Enzymes, toxins, growth regulators, and polysaccharides are much more common and perhaps more important in the development of plant diseases. They produce these substances when grown on certain substrates (inducible). Let's talk carefully one by one

1. Enzymes in plant diseases
2. Toxins in plant diseases
3. Growth regulator in plant diseases

1. Role of important enzymes in disease development

a. Cutinase

- Thee cutinase enzyme produce by pathogens degrades cutin layer.
- The amount of cutinase released by germinating spores of *Fusarium solani* f.sp.*pisi*, the pea rot was correlated with the degree of virulence of various isolates.
- The avirulent isolate did not produce cutinase.
- All these observations suggest the importance of cutinase in disease development.

b. Pectic enzymes (Pectinmethylesterase (PME) and Polygalacturonases)

- These enzymes degrade middle lamella and cell wall and facilitates free movement of pathogens from cell to cell.
- Pectic enzymes release peroxidases from cell walls. Peroxidase induces browning of vascular tissues, a typical wilt syndrome in many plants.
- Pectic enzymes release endogenous ethylene from host cells. These ethylene causes loss of turgor, chlorosis and necrosis of leaves.
- Pectic enzymes causes vascular plugging producing gels. Swelling of gels leads to occlusion of vascular lumina. It leads to wilting of plants.
- Pectic enzymes in some cases cause cell death. *Erwinia carotovora* produces endo polygalacturonate transeliminase and this purified enzyme induces leakage from potato cell membrane and cell death.

Table: Cell wall and membrane degrading enzymes produced by certain plant pathogens

Enzymes	Substrate	Pathogen
Cutinases	Cutin	*Fusarium solani* *Colletotrichum gloeosporioides*
Cellulases	Cellulose	*Erwinia carotovora, Ralstonia, Aschochyta pisi, Rhizoctonia solani*
Hemicellulases	Hemicellulose	*Sclerotium rolfsii, Erwinia carotovora, Penicillium expansum*
Proteinases	Protein	*Rhizoctonia solani, Sclerotium rolfsii,*
Ligninases	Lignin	*Botrytis cinerea, Heterobasidium annosus*
Esterases	Suberin	*Fusarium solani f. sp. pisi*

c. Cellulase

- Plant pathogens produces cellulotic enzymes.
- These enzymes become activated only after the action of pectic enzymes.
- Cellulases quickly degrade cell walls and facilitate the invasion of host cells by the pathogens.

d. Hemicllulases

- Hemicellulases degrade hemicellulosic substances present in plant cells.
- The most important hemicellulasic enzymes are endoglucanase, -xylosidase, endoxylanase, endomannanase,β-mannnosidase and α-galactosidase.
- *Sclerotium rolfsii* and *Fusarim* spp. Are known to produce these enzymes.

e.Galactanase and Arabansae

- These enzymes are also produced by pathogens.
- Degradation of galactan is accomplished by exo- and endo galactansases while araban is degraded by exo-arabanase.

f. Phospholipases

- Phospholipases decompose cell membrane and kill the cells.
- Only very few pathogens produce these enzymes.

g. Proteolytic enzymes

- Many pathogens produce these enzymes.
- They hydrolyze peptide bonds in proteins found in cell walls.

2. Role of toxins in disease development

Toxins

'Toxins' can be defined as low molecular weight, non enzymatic microbial products toxic to the higher plants. These are differs from other microbial products like 'mycotoxins' and 'antibiotics' which are toxic to the animals and microbes (except producers).

Toxins are classified as phytotoxins, vivotoxins and pathotoxins.

Phytotoxins: Toxins produced by pathogens, which are toxic to plants but not considered as of primary importance during pathogenesis, is called as phytotoxin. Non specific toxins are considered as phytotoxins.

Vivotoxins: The toxin produced *in vivo* (in infected tissues) which function in disease development but not as the initial inciting agent, is termed as vivotoxin.

Pathotoxins: Pathotoxins are the toxins which are major determinants of disease and of pathogenicity. Pathotoxin induces all the typical symptoms in reasonable concentration and is co-related with pathogenicity. Host specific toxin can be considered as pathotoxins.

In recent classification, toxins are divided into two categories.

1. **Host non specific toxin**: A toxin which may not affect many unrelated plant species in addition to main host of the pathogen producing toxin; it includes phytotoxin and vivotoxin.
2. **Host specific toxin**: A toxin which affects only the specific host of the pathogen; it includes pathotoxins.

Toxins in general, interact with cell membrane or organelles (chloroplast or mitochondria and alter their permeability. Major host specific and host non specific toxins are listed in table- .

Table: Host Specific and non host specific toxins

Host specific toxins			
Toxin	Pathogen	Host	Chemical nature
AK-toxin	*Alternaria kikuchiana*	Pear	Altenin (furanose ring)
HV- toxin	*Helminthosporium victorae*	Oats	Victoxinine
HC-toxin	*Helminthosporium carbonum*	Corn	Cyclic peptide
AM- toxin	*Alternaria Mali*	Apple	Alternariolide
AB-toxin	*Alternaria brassicae*	*Brassica* spp.	Destruxin B
PC-toxin	*Periconia circinata*	Sorghum	Proteinaceous
Host Non specific toxins			
Tentoxin	*Alternaria alternata*		Cyclic tetrapeptide
Tabtoxin	*Pseudomonas tabaci*	Tobacco	Aminoacid derivative
Alternaric acid	*Alternaria solani*	Potato	Hemiquinone derivative
Cerato-ulmin	*Ceratocystis ulmi*		Large M. carbohydrate
Pyricularin	*Pyricularia oryzae*		Nitrogen containing compound
Fusaric acid	*Fusarium spp.*		5-n-butypicolinic acid

Primary determinant: The toxin required for pathogenicity is called primary determinant. It is essential for colonization by the producing pathogens and disease development in the host. The primary determinant is otherwise called pathogenicity factor.

Secondary determinant: Thoe toxin required for virulence is called secondary determinant. Secondary determinant accounts for certain symptoms or contribute to virulence, but are not essential for colonization. The secondary determinant is otherwise called virulence factor.

Virulence is a quantitative term denoting relative disease inducing ability of a microorganism. Pathogenicity refers to the capacity of an organism to induce disease. A change in host range indicates a change in pathogenicity while a change in severity of disease indicates a change in virulence. But a question arises, when does pathogenesis begin and how much must the pathogen spread for colonization to be successful. Toxin is not required for penetration. Both toxin produced and no produced penetrate the host tissues. Initial colonization (upto24h) of both the isolates is also same. Only subsequent invasion is different. It is most likely that toxins play important role only as virulence factor rather than pathogenicity factor. In other words toxin is important in necrotrophic phase rather than in biotrophic phase. All the toxins isolated so far are from saprophytic pathogens. No toxin has been isolated from obligate parasites like powdery mildews, downy mildews and rusts.

3. Role of growth regulators in disease development

Plant growth is regulated by a small numbers of groups of naturally occurring compounds that act as hormones and are generally called growth regulators. The most important growth regulators are auxins, gibberellins, and cytokinins, but other compounds, such as ethylene and growth inhibitors. Alternation in concentration of growth regulators have been found to be associated with several plant diseases particularly in those resulting in abnormal plan/organ/ cell growth like, galls, tumors, knot, stunting, culing, hypertrophy, hyperplasia, etc. Increase in level of a growth regulator may be either owing to its induced production (by pathogen or host) or inhibited degradation. Decrease in concentration is due to the enhanced degradation by the host's or pathogen's enzymes. Biotrophs like powdery mildews employ these hormones to draw their nutrients from the host cells.

Alternations in the level of growth regulators seems to be a consequence of pathogenesis. However, at least in three diseases hormones have been demonstrated to be determinant of pathogenicity. These are crown gall caused by *Agrobacterium tumefaciens,* (cytokinin and auxin), oleander and olive knot caused by *Pseudomonas syringae sp. savastanoi* (auxin), fasciation disease caused by *Corynebacterium fasciens* (cytokinin). In all these three bacteria genes have been identified and cloned which are responsible for the hormonal production. Deletion of these genes result in conversion of virulent strains into avirulent.

Table: Certain Plant diseases involving altered concentration of growth regulators

Growth Regulators	Disease (Pathogens)
Auxins	**Increased concentration**: Crown gall (*Agrobacterium tumefaciens*), White stem rust (*Puccinina graminis tritici*), Downey mildew of rucifers (*Peronospora parasitica*, Tomato wilt (*Verticillium albo-atrum*), Banana wilt (*F. oxysporum f. sp. cubense*), Bacterial wilt of solanaceous crops (*Pseudomonas solanacearum*)
	Reduced concentration: Potato leaf roll(PLRV), Curly top of sugarbeet(SBCTV), Tobacco mosaic (TMV)
Gibberellins	**Increased concentration:** Bakanae disease of rice (*Gibberella fujikuri*), Creeping thistle rust (*Puccinia punctiformis*)
	Reduced concentration: Anther smut of sea campion(*Ustilago violacea*
Cytokinins	**Increased concentrations**: Club root of crucifers (*Plasmodiophora brassicae*), Crown gall (*Agrobacterium tumefaciens*), White rust of crucifers (*Albugo candida*, Root knot of tobacco (*Meloidogyne incognita*)
	Reduced concentrations: root knot of tomato (*M. incognita*), Verticillium wilt of tomato and cotton
Ethylene	**Increased concentration:** Tomato wilt (*Fusarium oxysporum f.sp. lycopersici*), Verticillium wilt of cotton (*Verticillium albo-atrum*), Wilt of tulip (*Fusarium oxysporum f. sp. tulipae*)

Model Practice Question

A. Objective Questions

a. Multiple Choice Questions

1. The change in severity of disease indicates a change in
   (a) Pathogenesis. (b) Pathogenicty
   (c) Virulence (d) Aggressiveness
2. The enzyme induces browning of vascular tissues, a typical wilt syndrome in many plants.
   (a) Peroxidase (b) Esterase
   (c) Zymase (d) Cutinase
3. Low molecular weight, non enzymatic microbial products toxic to the higher plants.
   (a) Enzymes (b) Toxins
   (c) Growth regulators (d) Polysaccharides
4. Toxins produced by pathogens, which are toxic to plants but not considered as of primary importance during pathogenesis,
   (a) Phytotoxin (b) Vivotoxin
   (c) Pathotoxin (d) Tabtoxin
5. The toxin produced *in vivo* (in infected tissues) which function in disease development but not as the initial inciting agent, is termed as

(a) Phytotoxin (b) Vivotoxin
(c) Pathotoxin (d) Tabtoxin

6. The toxins which are major determinants of disease and of pathogenicity
(a) Phytotoxin (b) Vivotoxin
(c) Pathotoxin (d) Tabtoxin

7. Which one is a host specific toxin
(a) Phytotoxin (b) Vivotoxin
(c) Pathotoxin (d) Tabtoxin

8. Tabtoxin is produced by
(a) *Pseudomonas tabaci* (b) *Alternaria kikuchiana*
(c) *Helminthosporium victorae* (d) *Ceratocystis ulmi*

9. Tentoxin is produced by
(a) *Alternaria alternata* (b) *Alternaria kikuchiana*
(c) *Helminthosporium victorae* (d) *Eratocystis ulmi*

10. The toxin required for pathogenicity is called
(a) Primary determinant (b) Secondary determinant
(c) Tertiary determinant (d) Primary indeterminant

11. The toxin required for virulence is called
(a) Primary determinant (b) Secondary determinant
(c) Tertiary determinant (d) Primary indeterminant

12. A quantitative term denoting relative disease inducing ability of a microorganism
(a) Pathogenesis. (b) Pathogenicty
(c) Virulence (d) Aggressiveness

13. A change in host range indicates a change in
(a) Pathogenesis (b) Pathogenicty
(c) Virulence (d) Aggressiveness

14. Which hormone concentration rise was responsible for the rise in rice occurrences of Bakanae disease (Gibberella fujikuri)?
(a) Auxins (b) Gibberellins
(c) Cytokinins (d) Ethylene

Answer

Q. No.	Answer	Q. No.	Answer
1	(c) Virulence	8	(a) *Pseudomonas tabaci*
2	(a) Peroxidase	9	(a) *Alternaria alternata*
3	(b) Toxins	10	Primary determinant
4	(a) Phytotoxin	11	(b) Secondary determinant
5	(b) Vivotoxin	12	(c) Virulence
6	(c) Pathotoxin	13	(b) Pathogenicty
7	(c) Pathotoxin	14	(b) Gibberellins

b.True/False

1. Thee cutinase enzyme produce by pathogens degrades cutin layer.
2. The secondary determinant is otherwise called virulence factor.
3. The avirulent isolate of pathogen did not produce cutinase.
4. Non specific toxins are considered as phytotoxins.
5. Phytotoxin and vivotoxin are host non specific toxin.
6. Toxins in general interact with cell membrane or organelles and alter their permeability.
7. The primary determinant is otherwise called pathogenicity factor.
8. Toxin is important in necrotrophic phase rather than in biotrophic phase.
9. Primary determinant is essential for colonization by the producing pathogens and disease development in the host.
10. Host specific toxin can be considered as pathotoxins.

Answer

Q. No.	Answer	Q. No.	Answer
1	False	6	True
2	True	7	True
3	True	8	True
4	True	9	True
5	True	10	True

B. Descriptive Questions

Short Answer

1. Name two cutinolytic and two cell wall degrading enzymes secreted by the pathogens.
2. What is tissue maceration what does it contribute in disease development?
3. Who first confirmed tabtoxin production in which plant disease? what was the casual organism of that disease?
4. Which fungal disease is related with the discovery of gibberellins? What was the pathogen of the disease?
5. What may be the usual causes for little or no water translocation in vascular wilt diseases?
6. Name some enzymes that are thought responsible for membrane damage in diseased plants.
7. What do you understand by 'extra haustorial membrane (EHM)'?
8. Name two disease with their causal agents in which the photosynthetic area is usually very much reduced.
9. Name two theories proposed to explain the increase in respiration rate in diseased plants.
10. Write about the role of enzymes in pathogenesis.

Long Answer

1. Write notes on:
   a. Pathotoxins and vivotoxins
   b. Lignolytic enzymes in disease development
   c. Criteria of degradative enzymes involvement in pathogenesis
   d. HC-toxins
2. Write in brief about the chemical weapons used by pathogens in pathogenesis.
3. Write an essay on the role of toxins in disease development.
4. Discuss the mechanism of respiratory increase in diseased plants.
5. Write notes on:
   a. Physical blockage of xylem vessels.
   b. Exceptions to the general pattern of reduction in photosynthetic activity in diseased plants.
   c. Protein metabolism in diseased plant tissue.
   d. Effect on transcription by pathogens.
   e. Effect on translation by pathogens.
   f. Give a concise account of growth physiology in diseased plants.

11

Defense Mechanism in Plants

When a host plant resists a disease, it means that it defeats the pathogen's onslaught and sustains little to no harm as a result. Most pathogen propagules that come into touch with resistant hosts are either prevented from developing or have their development and reproduction slowed down. Diverse plants have diverse strategies for self-defense. It's likely that every type of plant uses a different defense mechanism against every pathogen that threatens it. Plants create defense barriers in response to pathogen attack through a coordinated system of molecular, cellular, and tissue-based reactions.

Plants employ a variety of defense systems. These mechanisms can be categorized under two general headings for ease of reference:

1. **Pre-existing or passive defense** (Defense that has been a part of the plant from the start)
   a. Pre existing structural defense
   b. Pre-existing biochemical defense
2. **Post-infection or active defense** (Defense in reaction to the pathogen's onslaught)
   a. Post-infection structural defense
   b. Post-infection biochemical defense

1. Passive or preexisting defenses

a. **Pre existing structural defense**; Preexisting defense structures, also known as passive, static, or anti-infection structures, are the external or internal structural barriers that are in place prior to a pathogen attack. The following barriers, which appear in various host plants, are discussed starting from the periphery.

i. **Wax and cuticle:** Plant epidermal cells are covered in a cuticle that is made up of a wax, a pectin, and a cutinized layer. Cuticular wax lowers the risk of infection from a variety of pathogens.

ii. **Epidermal layer:** The thick, resilient walls of epidermal cells can either immediately stop or significantly hinder pathogen entrance. The polymers of cellulose, hemicelluloses, lignin, minerals, suberin, etc. are responsible for the toughness. For example, the majority of the outer walls of rice plant epidermal cells are lignified, making *Magnaporthe grisea*, which causes rice blast disease, difficult to penetrate. In contrast, the walls of motor cells are proteinaceous rather than lignified, making it easier for pathogens to enter the host.

iii. **Natural openings:** Many plant pathogenic fungi and bacteria can penetrate plants through natural openings like stomata, lenticels, hydathodes, nectarines, etc., however occasionally the structures of these openings impede pathogen access. For instance, the large cuticular lips that surround the stomata of mandarin oranges confer resistance against *Xanthomonas axenopodis pv. citri.* Likewise, the dimensions and internal organization of lenticels might have a protective function against the pathogens. The nectarthodes in the inflorescence of many plants are called hydathodes. They release sweet nectar that acts as a barrier to keep out pathogens that are unable to withstand high osmotic pressure. Chickpea' high level of hairiness on their leaves and pods confers resistance against *Aschochyta rabei.*

iv. **Internal structures:** The plant has a variety of pre-existing structures that shield it from pathogen entrance. For instance, some wheat cultivars have some degree of resistance to *Puccinina graminis f. sp. tritici* due to the presence of sclerenchymatous tissues in a relatively high proportion of the stem. Only young tissues are invaded by Pythium spp. As seedlings mature and develop secondary thickenings, they become resistant to the invaders.

Table: Pre-infectional morphological and structural defense mechanisms against plant pathogens

S. No	Defense barrier	Pathogen	Host
1	Waxes and Cuticle	*Albugo candida*	Brussels sprout
		Puccinia graminis	Berberries
		Melampsora lini	Linseed
		Colletotrichum cooffeanum	Coffee
2	Structure of natural opening		
	Broad cuticular ridge projection over the stomata	*Xanthomonas campestris pv.citri*	Mandarins citrus
	Mechnical (Sclerenchymatous) tissues	*Entyloma oryzae*	Rice
	Late openining of stomata	*Puccinia graminis tritici*	Wheat
3	Thick and tough outer cell wall (Presence of silicic acid and lignified epidermal cells)	*Pyricularia oryzae*	Rice

b. Pre existing biochemical defense

Plants have specific metabolic mechanisms within their cells that give them defense against infections. The term "biochemical defense" refers to the defense mechanism that plants have adopted. When a plant has biochemical defenses in place before a pathogen attacks. Pre-existing biochemical defense is the term for this.

i. Release of antimicrobial compounds

Typically, plants release organic materials from their roots and above-ground sections. These materials can include sugars, amino acids, organic acids, enzymes, glycosides, and more. Exudates contain a number of chemicals that have inhibitory properties. These inhibitory chemicals either directly impact the microorganism or promote specific groups to take control of their surroundings and act as the pathogen's adversaries. As an example, some linseed cultivars are resistant to *F. oxysporum f. sp. lini* infection due to the presence of HCN in their root exudates. It has also been found that HCN is present in maize and sorghum roots. The presence of HCN in sorghum root exudates may be the reason of the decrease in *Fusarium udum*-induced pigeonpea wilt in plots where sorghum is grown in combination with pigeonpea. Onion red scale (red varieties) contain catechol and protocatechuic acid, which can exude in drops and provide protection against *Colletotrichum circinanas* (onion smudge disease) attacks. Exudates from the roots encourage the establishment of rhizobacteria that promote plant growth (PGPR), which inhibits plant diseases and creates resistance.

ii.The plant cells contain an inhibitor

Certain preexisting inhibitory compounds found in plant cells primarily serve a protective function against the specific pathogen. Certain plants produce two kinds of antimicrobial compounds: phenols and quinones. Additional chemicals produced by plants that are linked to disease resistance include lactones, cyanogenic glucosides, saponins, terpenoids, stilbenes, and tannins. Plant defensins are another class of defense chemicals that disrupt the nutrition of pathogens and impede their growth. Defensins that are secreted can produce an antimicrobial microenvironment that aids in the germination of seeds, and defensins that have accumulated in flowers, leaves, and tubers can offer protection against viruses spread by insects.

iii. Lack of essential factors

(a) **Recognition factors:** If a plant's surface cells don't contain certain recognition factors, the plant may not become infected by a pathogen. The pathogen may not cling to the host surface or develop infection-causing substances like enzymes or structures like appresoria and haustoria if it does not identify the plant as one of its hosts. These recognition molecules include different kinds of glycoproteins, polysaccharides, and oligosaccharides.

(b) **Vital nutrients and growth factors**: The obligatory parasites, or host-specific infections, especially fungi, require certain substances, and the host they infect is the only place where these compounds are present in sufficient amounts. Therefore, the other types are inadequate hosts for such a virus because they lack these particular chemicals. As an illustration, the fungus *Venturia inaequalis* is the cause of apple scab disease. Due to mutation, some races of this recovered pathogen are no longer able to synthesis a growth factor. Such altered races are shown to be non-pathogenic to apples.

(c) **Common antigen in host plant:** The likelihood of disease in the host is determined by the presence of a shared antigen (protein) in both the pathogen and the host. However, the host becomes resistant to that specific infection if the antigen is lacking from the host and only present in the pathogen. For example, rust disease is more likely to affect linseed types that share an antigen with their pathogen, *Melampsora lini*. On the other hand, linseed types that do not contain the antigen but still contain the pathogen are resistant to it. The cotton leaf spot disease caused by *Xanthomonas compestris pv. malvacearum* is comparable.

Table: Pre-infectional biochemical defense mechanisms against plant pathogens

S. No	Defense barrier	Pathogen	Host
1	**Inhibitors released by the plant in the environment**		
	Glucoside	Fusarium spp.	Flax
	Malic acid	*Ascochyta rabiei*	Gram
	Catechol and Protocatechuic acid	*Colletotrichum circinans*	Red scale of onion
	Cutin acid	*Gloesporium limetticola*	Citrus
2	**Phenolic substances**		
	Polyphenols	*Venturia inaequalis*	Apple
		Pyricularia oryzae	Rice
		Venturia pirina	Pear
3	**Inhibitory substances present in the plant cells**		
	Chlorogenic acid	*Streptomyces scabies*	Potato
	Tuliposide	*Fusarium oxysporum f.sp. tulipae*	Tulip
	Avenacin	*Ophiobolus graminis*	Wheat
	Arbutin	*Erwinia amylovora*	Pear
4	**Absence of nutrients required by the pathogen**		
	Histidine	*Erwinia amylovora*	Potato
	Sugar	*Ceratocystis ulmi*	Elm
		Alternaria solani	Potato
		Cochliobolus sativus	Barley
	Choline	*Venturia inaualis*	Apple
	Myoinositol	*Rhizoctonia aurecina*	Plum

2. Post infection or Active or Induced defense

a. Post infection or Induced structural defense

The host, in some way, usually develops specific defense structures inside of them in response to the reaction occurring between host and pathogen once the disease successfully penetrates the pre-existing defense barriers of the host. There are four main sorts of these developed structures:

i. **Induced histological defense**: As a result of tissue differentiation or chemical substance deposition in tissues surrounding or preceding the pathogen, certain defense structures are created inside the host. We refer to these defensive mechanisms as induced histological defense. These could exist in different levels:

   (a) **Cork layer formation**: The production of cork layers is a component of plants' inherent healing mechanism. Cork layers obstruct the passage of water and nutrients from the healthy to the diseased area, starving the pathogen of sustenance and occasionally causing it to weaken or even die. Additionally, it prevents the pathogen's toxin compounds from entering the body and harming the host's healthy tissues beneath the surface. For instance. Two notable examples include Rhizophus rot of sweet potatoes and common scab of potatoes *(Streptomyces scabis*).

   (b) **Tyloses formation**: Tyloses are created when parenchymatous xylem cell walls protrude via pits into xylem vessels. Though they can often restrict the flow of water through the vessels, leading to water stress in the plant, tylose development can also limit the spread of pathogenic propagules in the xylem. *Fusarium oxysporum f. sp. batatas* is one example of a sweet potato.

   (c) **Gum deposition**: Gums that shield wounds from infection are frequently secreted by injured tree trunks. Examples include rice leaf spot (*Heminthosporium oryzae*) and blast disease (*Magnaporthe grisea*).

   (d) **Abscission layers**: These are the spaces between layers of the host cell and the mechanisms that allow older leaves and mature fruits to fall off. Plants may also employ this as a defensive mechanism, dropping off pathogens and infected or invaded plant tissues or components. Fruit trees frequently have shot holes on their leaves.

ii. **Introduced cellular defense:** Only a limited function is played by the cellular defense structures, or modifications to the cell walls. The types listed below are typically observed:

   - Apposition of carbohydrates (secondary wall synthesis and papillae development).
   - The hyphal covering around the haustorium, called calcine deposition, prevents the pathogen, *Phytophthora infestans*, from coming into contact with the host cells.
   - Structural proteins

iii. **Induced cytoplasmic defense:** When a pathogen's hyphae are able to enter a certain cell, the cell's cytoplasmic material organizes itself so that the cytoplasm becomes dense and granular and specific organelle-like structures form in it. Consequently, the penetrating hyphae break down into tiny granular entities, halting the hyphae's continued development. Perhaps the final line of structural defense, cytoplasmic defense structures work well against weak parasites, slow-growing diseases, and certain symbiotic interactions.

Table: Post-infectional morphological and structural defense mechanisms against plant pathogens

S. No	Defense barrier	Pathogen	Host
1	Formation of cork layer	*Rhizoctonia solani*	Potato
		Coccomyces prunophorae	Plum
	Formation of tyloses	*Fusarium oxysporum f.sp. batatas*	Sweet potato
		Verticillium- alboatrum	Tomato
3	Formation of abscission layer	*Xanthomonas arboricola pv. pruni*	Sour cherry
		Clsterosporium carpophilum	Peach
3	Sheathing of hyphae	*Phytophthora infestans*	Potato
4	Gum deposition	*Pyricularia oryzae*	Rice
		Drechslera oryzae	Rice
		Physalopora cydoniae	Apple

iv. **Induced hypersensitive defense:** Hypersensitive cell death is another widespread mechanism used by hosts to prevent the spread of a pathogen. Infected cells and those surrounding them "suicide", preventing further spread, and in some cases, killing the pathogen. Hosts also frequently employ hypersensitive cell death as a defense against the spread of infection. The pathogen is sometimes killed by the "suicide" of infected cells and the surrounding cells, which stops the infection from spreading further. It frequently coincides with the start of additional reactions, like lignification and the production of antimicrobial substances. The nutritional needs of the particular pathogen and the timing, scope, and location of the host response determine whether hypersensitive cell death is effective as a resistance strategy.

b. **Post infection or Induced biochemical defense:** These mechanisms restrict the spread of the pathogen after infection is established and contain the damage to host tissues.

v. **Role of Phytoalexins:** Phytoalexins are low molecular weight antibiotics produced by many (but not all) plants in response to infection. Phytoalexins inhibit the growth of bacteria and fungi *in vitro* and *in vivo*, and production of these antibiotics during an infection can induce resistance to consequent infections by that pathogen. Over 350 phytoalexins are known in over 100 plant species.Pterocarpanes, cryptophenols, isocoumarins, isoflavenoids, sesquiterpenes, and other substances are among them. Any portion of the plant can produce phytoalexins, albeit distinct phytoalexins can accumulate in various organs. Typically, closely related plant species produce multiple structurally related phytoalexins, allowing the plant to defend itself against invasive diseases with a deadly cocktail. In order to create a hazardous microenvironment within the infected cell and, presumably, inhibit the onset of disease, phytoalexins are synthesized in cells around an infection site and transferred to the infected cell wrapped in lipid vesicles. The accumulation of phytoalexins is frequently linked to hypersensitive cell death, despite the fact that phytoalexins are only synthesized by living cells. Additionally, certain plants have the ability to store dormant sugar-conjugates called phytoalexins in vacuoles, which can be swiftly broken down and released in the event that the plant's defense mechanisms fail.

Table: Some important phytoalexins

S. No	Phytoalexins	Producer Plant	Chemical nature
1	Resveratrol	*Vitis viifera*	Stilbene
2	Trifolirhazin	*Trifolium pratens*	Isoflavonoid
3	Isocoumarin	Carrot root	Isocoumarin
4	Orchinol	*Orchis militaris*	Dihydrophenanthrene
5	Isomeamarone	*Ipomea batata*	Sesquiterpene
6	Rishitin	Potato tuber	Sesuiterpene
7	Lubimin	Potato tuber	Sesuiterpene
8	Camelexins	*Aradiopsis thaliana* *Camalenia sativa*	Thiazoyl substituted indole camalexin
9	Brassinins	*Brassica rapa*	Indole ring linked to a dithiocarbamate
10	Pisatin	*Pisum sativum*	Isofalvonoid
11	Maackiain	*Cicer arietinum*	Isofalvonoid
12	Phaseollidin	*Phaseolus vulgaris*	Isofalvonoid

II. Role of phenol compounds:

Commonly found in plants, phenolic substances have an aromatic ring with a hydroxyl substituent. Bacteria, viruses, and fungi are all poisoned by phenols. They inhibits:

- Growth of mycelia
- Spore germination
- Generation of pectic and cellulotic enzymes
- Pectic and cellulotic enzyme activity, as well as
- Fungal production of toxins

The phenolic compounds are main toxic chemicals produced to inhibit pathogen or its activities. Some of these are preformed toxic chemicals while others may be *de novo* synthesized or modified to more toxic forms, viz., conversion of phenols to more toxic quinones. When a pathogen infects the plant, the latter either accelerates the synthesis of these compounds adjacent to infection site or accelerates the flow of pre- existing amount from healthy tissue towards the infected one and plant defend itself.Some of the individual phenolic substances have been identified as the key defense chemicals in plants.

It has been determined that a few specific phenolic compounds are the main defensive compounds found in plants. They are as follows:

Phenolic substances	**Crop**
Chlorogenic acid	Potato
Catechol and Pyrocatechic acid	Onion
Avenacin	Oats
Hordatine	Barley
Tomatine	Tomato
Tuliposides	Tulip
Catechin and Isoquercitin	Cotton
Umbelliferone and Scopoletin	Sweet potato

III. Role of Pathogenesis-related proteins (PRPs)

Plant proteins that are induced in pathogenic and associated circumstances are known as pathogenesis-related proteins. Resistance to pathogens is increased when they are present prior to infection. These unique proteins show that the plant has achieved systemic acquired resistance and are maximum accumulated 7–10 days post-infection. PRPs build up in the vacuole and in the gaps between cells. When faced with a difficult pathogen, intercellular PRPs provide the initial line of defense. If this is ineffective and tissue is damaged, the release of vacuolar PRPs provides the second line of defense by engulfing the pathogen in lytic enzymes that contain antiviral, antifungal, and antibacterial properties.

Table 16: Families of pathogenesis –related proteins (PRPs)

S. No	Protein family	Protein activity	Targeted pathogen sites
1	PRP-1	Unknown	Active against oomycetes
2	PRP-2	Β-1,3 glucanase	Cell wall glucan of fungi
3	PRP-3	Endochitinase	Cell wall chitin of fungi
4	PRP-4	Endochitinase	Cell wall chitin of fungi
5	PRP-5	Taumartin-like	Active against oomycetes
6	PRP-6	Proteinase inhibitor	Active on nematodes and insects
7	PRP-7	Endoroteinase	Microbial cell wall dissolution
8	PRP-8	Endochitinase with lysozyme activity	Cell wall chitin of fungi, and mucopeptide wall of bacteria
9	PRP-9	Peroxidase	
10	PRP-10	Ribonuclease activity	Viral RNA
11	PRP-11	Endochitinase	Cell wall chitin of fungi

S. No	Protein family	Protein activity	Targeted pathogen sites
12	PRP-12	Defensin	Antifungal and antibacterial activity
13	PRP-13	Thionin	Antifungal and antibacterial activity
14	PRP-14	Lipid transfer proteins	Antifungal and antibacterial activity
15	PRP-15	Oxalate-oxidase	Produces H_2O_2 that inhibits microbes and also stimulates host defense
16	PRP-16	Oxalate-oxidase-like with suuperdismutase	Produces H_2O_2 that inhibits microbes and also stimulates host defense

IV Inactivation of enzymes and toxins: It is commonly known that pathogen toxins and enzymes have a role in pathogenesis. Compared to biotrophs and hemibiotrophs, necrotrophs use more of these chemicals to harm host tissue. The defense mechanism of resistant plants involves the inhibition of enzymes by proteins, tannins, and phenols. Phenolics neutralize the enzymes of pathogens, rendering them useless, however they do not have antifungal properties. In immature grape fruits catechol- tannin is known to inhibit enzymes produced by *Botrytis cinerea.*

In resistant varieties of plants, which get diseased as a result of toxins produced by the pathogens, the pathogenic toxins released are detoxified by the ability of metabolic processes. For example, *Pyricularia oryzae*, while causing blast disease of rice, releases picolinic acid and pyricularin toxin inside the host. into other compounds. Thhough susceptible varieties get affected by these toxic substances, the resistant varieties convert 40-60% of picolinic acid into its methylester and N-methylpicolinic acid, pyricularin into other compounds. These resistant compounds are non toxic to rice plants.

Table: Defense through detoxification of pathogen toxin

Detoxifying compound	Pathogen toxin	Disease	Pathogen
Chlorogenic acid	Piricularin	Blast of rice	*Pyricularia oryzae*
Dehydroascorbic acid	AK toxin	Black spot of pear	*Alternaria kikuchiana*
High concentration of cysteine		Wilt of cotton	*Fusarium oxysporum f. sp. vasinfectum*
Fusaric acid		Wilt of tomato	*Fusarium oxysporum* f. sp. *lycopersici*

Systemic acquired resistance

Systemic acquired resistance, also known as induced resistance, is defined as a plant's enhanced resistance to multiple pathogens after infection by a single pathogen. As such, it differs significantly from the particular antigen-antibody mechanism of resistance observed in the mammalian immune response. Systemic acquired resistance lessens the severity of subsequent diseases rather than conferring immunity in and of itself. The emergence of a slowly growing necrotic lesion and other localized responses to infection, the release of a phloem-translocated signal from the infection site, and the subsequent priming of the plant against additional attacks—which enables a faster response in the event of future infections—are typically necessary for the development of systemic acquired resistance. It is still uncertain what exactly sets off systemic acquired resistance, although it most likely involves a complicated signal transduction pathway that is mediated by several stress signals. Through its interactions with salicylic acid-binding proteins, which can increase the production of reactive oxygen species or trigger the expression of certain genes, salicylic acid plays a crucial function. Around necrotic lesions, salicylic acid levels rise and stay high in plants that have developed tolerance. However, the signal that is translocated systemically throughout the plant is not salicylic acid per se.

Induced Systemic Resistance (ISR): ISR operates without the activation of the SA and PR genes. It has been demonstrated that ethylene and jasmonic acid work together to activate genes that code for protective proteins including plant defensins and proteinase inhibitors. Rhizobacteria called *Pseudomonas fluorescens*, which are not pathogenic, activate the JA-dependent ISR pathway. ISR is active against the oomycetous leaf pathogen *Peronospora parasitica*, the fungal root pathogen *Fusarium oxysporum f.sp. raphani,* the bacterial leaf diseases *Xanthomonas campestris pv. campestris* and *Pseudomonas syringae pv. tomato* in Arabidopsis.

Significance

- Induce a set of defense gene expression.
- Dramatically accumulated after infection.
- Both have been recognized as inducers of plant defense for many years.
- Exogenous application evokes the protection against pathogen attack.
- Deficient mutants exhibit reduced defense gene expression and enhance disease management.

Systemic Acquired Resistance (SAR) & Induced Systemic Resistance (ISR)

Features	ISR	SAR
Definition	Resistance activated in response to artificial application of non pathogenic strains of pathogen, PGPR, and abiotic stresses	Resistance activated in response pathogenic infection.
Pathway	Jasmonic acid dependent pathway	Salicyclic acid dependent pathway
Signalling molecule	Jasmonic acid	Salicyclic acid
Defense gene involved	CHIB, HEL,PDF1,2	PR-1 and PR-2
Pathogen associated	Operates against necrotrophic & facultative parasites	Operate against biotrophic pathogens.

Model Practice Question

A.Objective Questions

a. Multiple choice questions

1. Post-infectional structural defense mechanism is
   (a) Histological defense mechanism (b) Cellular defense mechanism
   (c) Cytoplasmic defense mechanism (d) All above
2. Post-infectional cellular defense structure is
   (a) Carbohydrate appositions and callose deposition
   (b) Hyphal sheathing
   (c) Structural proteins
   (d) All the above
3. Lignin and silicic acid of the epidermal cell of rice leaves provide protection against
   (a) Rice blast (b) Brown spot of rice
   (c) Bacterial blight of rice (d) Leaf streak of rice
4. Phenolic compounds produced in response to late blight of potato are
   (a) Orthodiphenol & Scopeletin (b) Caffiec acid & Umbelliferon
   (c) Wyerone & Wyeronic acid (d) None of these
5. The characteristic feature of SAR is
   (a) Expression of resistance against organisms having wide host range
   (b) Accumulation of SA
   (c) Production of PR-proteins
   (d) All above
6. The host-pathogen system in which lignified cell wall provide effective barrier to hyphal penetration is
   (a) Potato-*Phytophthora infestans* (b) Wheat-*Septoria nodorum*
   (c) Radish-*Peronospora parasitica* (d) All above
7. The pathogen credited with the discovery of SAR is
   (a) *Tobacco mosaic* virus (b) *Phytophthora infestans*
   (c) *Ceratocystis fimbriata* (d) *Blumeria graminis*
8. Callose deposition is a device of defense in plants against
   (a) *Blumeria graminis f.sp. hordei* (b) *Alternaria solani*
   (c) *Xanthomonas campestris* (d) *Phytophthora infestans*
9. The enzyme involved in the synthesis of phytoalexin, phenolics and lignin is
   (a) Phenylalanine Ammonia Lyase (b) Peroxidase
   (c) Cutinase (d) Lipase

10. Amputative resistance' is due to the formation of
    (a) Abscission layer (b) Tyloses
    (c) Cork layer (d) Papilla
11. The correlation of hypersensitive sensitivity with disease resistance was put forwarded by
    (a) Ward (b) Stakman
    (c) Kuc (d) Hutchenson
12. The pathogen used by Mullar and Borger in their classical experiment on phytoalexin production during host pathogen interaction is
    (a) *Phytophthora infestans* (b) *Phytophthora parasitica*
    (c) *Melampsora lini* (d) *Tobacco mosaic virus*
13. Resistance activated in response to artificial application of non pathogenic strains of pathogen, PGPR, and abiotic stresses is called
    (a) ISR (b) SAR
    (c) TR (d) AR
14. Resistance activated in response to pathogenic infection is called
    (a) ISR (b) SAR
    (c) TR (d) AR

Answer

Q. No.	Answer	Q. No.	Answer
1	(d) All above	8	(a) *Blumeria graminis f.sp. hordei*
2	(d) All the above	9	(a) Phenylalanine Ammonia Lyase
3	(a) Rice blast	10	(a) Abscission layer
4	(a) Orthodiphenol & Scopeletin	11	(a) Ward (1902)
5	(d) All above	12	(a) *Phytophthora infestans*
6	(d) All above	13	(a) ISR
7	(a) Tobacco mosaic virus	14	(b) SAR

b. True/False

1. Chickpeas' high level of hairiness on their leaves and pods confers resistance against Aschochyta blight.
2. The presence of HCN in sorghum root exudates inhibit pigeonpea wilt in plots where sorghum is grown in combination with pigeonpea.
3. Onion red scale (red variants) contain catechol and protocatechuic acid, provide protection against onion smudge disease.
4. Pathogen enzymes and toxins play a role in pathogenesis.
5. Necrotrophs utilize a greater amount of chemicals to damage host tissue than do biotrophs and hemibiotrophs.
6. ISR operates against biotrophic plant pathogens.
7. SAR operate against necrotrophic and facultative plant pathogens.

8. Hypersensitive cell death is a widespread mechanism used by hosts to prevent the spread of a pathogen.
9. Phytoalexins inhibit the growth of bacteria and fungi *in vitro* and *in vivo.*
10. The signal molecule produced in ISR is salicylic acid.

Answer

Q. No.	Answer	Q. No.	Answer
1	False	6	True
2	True	7	True
3	True	8	True
4	True	9	True
5	True	10	False

B. Descriptive Questions

a. Short Answer

1. What do you mean when you refer to "constitutive" defense mechanisms?
2. The citrus variety known as Szinkum is immune to the pathogen that causes citrus canker. Which defensive tactic facilitates this?
3. Which opening in the lingnified epidermis allows the pathogen that causes blast disease to penetrate rice plants?
4. Why are red-scale onion cultivars resistant to smudge disease but white-scale onion varieties are prone to it?
5. What is meant by a "oxidative burst"?
6. Who gave the basic concept of phytoalexins? Which phytoalexins were discovered initially?
7. What is the term SAR, or systemic acquired resistance? Which pathogen is thought to have contributed to the discovery of SAR?
8. Write on how toxins contribute to pathogenesis.

b. Long Answer

1. Summarize the biochemical defense mechanisms that plants have in place after infection.
2. Make an attempt at these:

a. Hypersensitivity reaction
b. Tyloses and disease causing vascular wilt
c. Plant production of phytoalexins
d. PR Proteins
f. ISR, or Induced systemic resistance

3. Talk about the many structural defense tactics that plants have developed in response to pathogen attacks.

4. Describe the subsequent
a. Implementation of genes
b. Gene pyramiding
c. Describe plant defense and the several defense systems that plants employ.
d. Explain the various stages of the infection process.

5. Write notes on:
a. Physical blockage of xylem vessels.
b. Protein metabolism in diseased plant tissue.
c. Effect on transcription by pathogens.
d. Effect on translation by pathogens.

12

Epidemiology: Factors Affecting Disease Development

Epidemiology

Epidemiology is the science of disease in populations, according to Vanderplank. The focus of epidemiology is on population phenomena. The host population and the pathogen population are the two most significant populations. Furthermore, host-pathogen interactions that result in disease and crop losses are a focus of epidemiology. Environmental factors frequently influence these connections. Therefore, the effects of both biotic and abiotic environments are studied in epidemiology. In conclusion, human activity has had a significant impact on hosts, diseases, and habitats; these effects are becoming more and more significant in the modern era. Therefore, epidemiology also addresses how disease is caused by humans and occasionally how plant diseases affect human health. This whole thing can be represented by the disease tetrahedron.

Elements of an epidemic: a pathogen, a susceptible host, and a favorable environment.All three of these need to be present for disease to manifest.Disease exists where these three things come together. Components are present; an epidemic won't occur unless all three remain present.Occasionally, a fourth factor of time is included since epidemics can also be significantly influenced by the timing of an infection's onset and the duration of the infection's viable conditions. Disease cannot occur if any of the requirements are not satisfied, such as the presence of a susceptible host and a pathogen but an unfavorable environment for the pathogen to infect and cause disease.

Purpose of study of epidemiology

At least two goals are achieved by studying epidemiology:

- Gaining knowledge about how diseases behave throughout time and space
- Applying this knowledge to disease management.

Factors affecting disease developemnt

Pathogen	Host	Environment
Presence of pathogen	Type of crop	Temperature
Mode of dispersal of pathogen	Degree of genetic uniformity of host plants	Rainfall / Dew
High birth rate of pathogen	Genetic make-up	Leaf wetness period
Low death rate of pathogen	Host age	Soil properties
Survival efficiency	Host nutrition	Wind
Reproductive fitness	Host population structure	Fire history
Adaptibility of pathogen	Nature of propagation	

1. Host factors

a. **Type of crop**: Compared to perennial crops like fruit and forest trees, epidemics typically spread far more quickly among annual crops like cereals, pulses, oilseeds, and vegetable crops. Certain fruit and forest tree epidemics, such as pear decline, citrus tristeza, dutch elm disease, and chestnut blight, take a year to manifest.

b. **Degree of genetic uniformity of host plants**: When genetically uniform plants are grown over a vast area, there is a greater chance that a new pathogen race could emerge and attack their genome, perhaps causing a serious plant disease outbreak. This phenomena has been repeatedly reported in many situations, such as Texas male sterile cytoplasm-carrying corn susceptible to southern corn leaf blight. Cross-pollinated crops typically have the lowest rates of epidemic development, self-pollinated crops have intermediate rates, and vegetatively propagated crops have the greatest rates.

c. **Level of genetic resistance or susceptibility of the host**: The emergence and spread of epidemics are significantly impacted by both vertical and horizontal resistance. Plants that harbor horizontal resistance will most likely contract the disease; nevertheless, the pace of disease and epidemic development is contingent upon the resistance level and environmental factors. Host plants with vertical resistance prevent pathogens from settling in them, epidemics are prevented from happening. The optimal substrate for the establishment and growth of new infections is provided by susceptible host plants devoid of genes conferring resistance against the pathogen.

d. **Host age**: Plant change in their susceptibility to disease with age. For instance, Damping off and roots rots, downy mildews, peach leaf curl, rusts, smuts, bacterial blights and viral infections, the host plants are susceptible only during their growth period and becomes resistant during adult stage. In some other diseases such as flower or fruit blights caused by fungal pathogens like Alternaria, Penicillium, Glomerella and Monilinia and in all post harvest infections, plant parts(fruits) are resistant during growth and early adult stage but become susceptible near riepning.

e. **Host nutrition:** Plants must have a balanced diet in order to grow and develop properly. Biotrophic fungi have a relative advantage in favorable nutritional environments, but necrotrophic and perthotrophic fungi have an advantage in unfavorable environments. The type and amount of fertilizers have influence on the development of plant diseases.

f. **Host population structure:** The kind and quantity of plant diseases are largely determined by crop density, rotation, and sequence. A high crop density will affect the microclimate and increase the effectiveness of inoculum dissemination and infection. These two elements will contribute to the onset of diseases.

g. **Propagation nature:** Propagation nature influences the course of disaeases development. Plants propagated vegetatively have higher disease chances.

2. Pathogen factors

There are certain important pathogen factors that affect the course of epidemic. The pathogen's level of virulence is one of the key elements that determine how an epidemic develops.

a. **Level of virulence:** Compared to pathogens with lower virulence, virulent pathogens that may infect the host quickly ensure a faster creation of more inoculums and, consequently, diseases.

b. **Mode of dispersal of pathogen:** Spreading epidemics more frequently and widely may be aided by the pathogen inoculums' quick and simple dissemination from their source to healthy hosts in the field. Diseases such as mildew, rusts, and leaf spots often go viral because the pathogen's spores are discharged into the atmosphere and swiftly spread across long distances—up to several kilometers. On the other hand, most nematodes and soil-borne diseases such as Fusarium and Verticilium often have two to four reproductive cycles per growing season. Since, the dispersal of such pathogens is limited both in space and time, only localized and slower developing epidemics are caused.

c. **High birth rate of pathogen:** A pathogen's rapid reproductive cycle or high birth rate are important factors in the start and progression of epidemics. Certain fungi pathogens can generate vast amounts of spores through asexual means. These spores can cause epidemics in their respective regions if they spread quickly over great distances and are supported by favorable environmental conditions.

d. **Low pathogen death rate:** Epidemics where the causative organism is systemic and shielded by plant tissues are linked to low pathogen death rates. As a result, there are ultimately higher probability of widespread infections in a given area and significantly lower likelihood of high mortality.

e. **Adaptibility of pathogen:** This is another element that is essential to the spread of epidemics. When pathogens are able to adapt to unfavorable environments, epidemics are almost a given. In order for epidemics to spread, the pathogen's propagule units must be distributed by outside organizations.

E. Environmental Factors' Role

If a pathogen is present on a certain plant, it usually won't result in a major disaeases unless the environmental conditions are favorable. This encompasses both the soil (edaphic) environment and the airborne environment.

i) Aerial environment

Temperature, moisture content, and pollution are characteristics of the air that affect the development of disease.

a. **Temperature**: The latent or incubation, period (the time between infection and the appearance of disease symptoms), the generation period (the interval between infection and sporulation), and the infectious period (the interval during which the pathogen continues to produce propagules), are all impacted by temperature. Higher temperatures hasten the disaease cycle and hasten the spread of epidemics. When paired with temperature data, the leaf wetness period can be used to schedule preventative measures like spraying and forecast when certain diseases will erupt (i.e., when an infection period will occur).

b. **Moisture:** For bacteria, fungi, and pathogenic fungus, moisture is essential. Most airborne spores require a time of leaf wetness to germinate, thus rain splash plays a major role in the spread of some fungi and almost all bacteria. Propagules are disseminated at a period when they are most likely to be able to germinate as well when they are dispersed using water. The length of leaf wetness affects the infection's success because germination and infection are time-consuming processes. Temperature affects how long an infection takes to spread. Since germination and infection are typically expedited in warmer temperatures, a longer duration of leaf wetness is typically required to establish an infection at lower temperatures.

c. **Pollution:** Air pollution is one recently discovered element of the aerial environment that can affect plant disease. High pollution concentrations can influence the onset of diseases and, in severe situations, directly harm plants by producing acid rain.

d. **Soil environment** The soil environment has a major role in influencing soil-borne diseases by controlling the moisture content that pathogens require for germination, survival, and motility. The temperature of the soil affects both germination and the effectiveness of infection. Disease development may be influenced by the soil's organic matter level and fertility. A lack of nutrients weakens a plant's defenses, yet other diseases, such rusts and powdery mildews, do well on well-fed plants. Particularly low organic matter soils are better suited for the growth of other diseases.

F. Role of Humans activities:

Plant disease epidemics are influenced by humans and farmers both directly and indirectly. The kind, quantity, timing, and density of plants grown in a given region are all determined by humans. Inaccurate decisions on these matters could result in the outbreak of an epidemic in that region. In a similar vein, persistent monoculture, heavy fertilizer use, excessive irrigation, and inadequate field cleanliness are examples of cultural practices that humans have embraced and which encourage serious epidemics in a particular location. Planting a specific variety or using a specific chemical may cause the selection of virulent strains that might cause epidemics because they are either able to overcome the variety's resistance or are resistant to the chemical.

Types of Epidemics

Monocyclic epidemics: Monocyclic epidemics are caused by pathogens with a low birth rate and death rate meaning they only have one infection cycle per season. They are typical of soil born diseases such as *Fusarium* wilt of pigeonpea. Mathematical equations used to predict epidemic development in time is

$$dx/dt = r\ (x_a\text{-}x)$$

Where, dx/dt= rate of increase of disease; r =infection rate; x =number of infected individuals; t =time; x_a= total population size.

Examples: r for *Verticillium* wilt of cotton= 0.02 units/day, 'r' for Phymatotrichum root rot of cotton= 1.60 units per year.

Polycyclic epidemics: Pathogens with the ability to reproduce several times in a season are the source of polycyclic epidemics. The most common causes of this are airborne disaeaseses like powdery mildew. Mathematical equations used to predict epidemic development in time is

$$dx/dt = xr\ (1\text{-}x)$$

Where, dx/dt= rate of increase of disease; r = exponential rate of disease increase; x = amount of disease on a scale of 0-1 and t =time under consideration during which host and pathogen have interacte.

Examples: r for late blight of potato = 0.3 to 0.5 units per day 'r' for stem rust of wheat = 0.3-0.6 units per day.units/day, 'r' for *Phymatotrichum* root rot of cotton= 1.60 units per year.

Polyetic epidemics: Some disaeaseses take years to manifest symptoms and begin to reproduce as a pathogen. These kinds of outbreaks are known as polyetic epidemics, and they can be brought on by both monocyclic and polycyclic diseases. One example of a polyetic epidemic brought on by a polycyclic pathogen is apple powdery mildew, whereas Dutch elm disease is a polyetic epidemic brought on by a monocyclic pathogen.

Differences between monocyclic and polycyclic plant disease epidemics

Characters	Monocyclic	Polycyclic
Birth rate	Low	High
Death rate	Low	High
Rate of increases	Like simple interest	Like compound interest
Servival	Long lived	Short lived
Reproduction	Reproduces only one generation per growing season	Reproduce new batches of propagules several times per growing season

To accurately convey the potentialities and magnitudes of disease, epidemiologists employ specialized language.

Epidemiologists use specific terminology to accurately communicate magnitudes and potentials for disease.

- **Disease progress curve**: Periodic observations on disease severity, in any field or area, when plotted against time on a graph paper gives curved lines representing progress of disease over time . These cumulative disease curves are known as Disease progress curve DPC.
- **A saturation curve**-monocyclic disease
- **A sigmoid curve**- polycyclic disease
- **A bimodal curve**-polycyclic disease affecting different organs
- **Disease-gradient curve;** A graph that plots disease vs. distance from an inoculum source.
- **Epidemic;** A disease increase in a population; usually a widespread and severe outbreak of disease.
- **Epidemiology;** The study of factors affecting the outbreak and spread of disease.

Course of epidemic

The course of epidemic follows two distinct phases viz.,

i. Progressively destructive phase
ii. The decline phase

i. Progressively destructive phase

While some epidemics spread quickly, others spread more slowly. Slow epidemics, also known as epiphytotics, are typically brought on by systemic pathogens in a population. The pathogen slowly multiplies in accordance with the features of the basic interest disease. They have a shorter incubation and sporulation period and fall into the group of low death rates. Conversely, environmental conditions have a major impact on the quick epiphytotics.

ii. Decline phase

An epidemic's early stages see rapid spread, infecting new hosts with diseases.

Following the emergence of a saturation stage, it exhibits an independent decline. The absence of susceptible crop stages, unfavorable weather, and a decline in disease virulence could all be reasons for no outbreaks. Typically, the disease targets the hosts during a certain developmental phase. After a plant passes through this stage, its susceptibility to pathogens either decreases or disappears entirely. In these circumstances, the epidemic wanes. Unfavorable meteorological conditions for disease development could potentially be the cause of the epidemic's decline. As a result, the epidemic will lessen and the disease's potential to spread in the future will be curbed.

Slow and rapid epiphytotics

The characteristics of the host, pathogen, and weather determine the type of epidemic. Epidemics are known as "tardive" because they can develop slowly. Between these intermediate phases, a "explosive" epidemic—which spreads quickly—may occur.

i. Rapid epiphytotics

There is rapid epiphytotics in annual crops. High birth rate non-systemic infections are the cause of it. In a brief period of time, the pathogen produces several generations. Unlike slow epiphytotics, rapid epiphytotics is primarily controlled by environmental conditions. The disease spreads quickly, reaching a peak in a short amount of time, and then sharply declining when the weather turns bad, the host matures and becomes resistant, or when the pathogen's propagules are constrained in their ability to spread. such as apple scab. The control of this kind of epiphytotic is achieved through chemical dusting or protective spraying.

ii. Slow epiphytotics

Populations of perennials, such as trees, have slow epiphytotics. The infected host lives for a few years before passing away. Slow epiphytotics has most of the characteristics of a simple interest disease. Most of the causal agent is systemic. The infection grows slowly. They travel far more slowly between plants. These pathogens have a low fatality rate. The ideal practice in slow epiphytotics is crop sanitation. For instance, a swollen cocoa stalk. From tree to tree, and even less from one garden to another, this disease spreads very slowly.

Measurement of Plant Disease and Yield Loss

While measuring disease, one is interested in measuring

- **Disease Incidence**; the proportion of a plant community that is diseased.
- **Disease Severity**; the proportion of plant area that is affected.
- **Yield loss**: the proportion of the yield that the grower will not be able to harvest because the disease destroyed it directly or prevented the plants from producing it.

Disease and crop loss assessments are essential for evaluating the economic impact of a disease and the benefit of particular control strategies. There is no point in implementing a management practices if it will cost more that the increased crop yield will return. The growth of the crop, its yield potential, the development of the disease and its impact on yield all have to be measured to predict the impact on yield of particular levels of disease. This information can be combined with predictions of likely disease levels to determine whether preventative treatments

Assessment of the effect of disease on crop yield normally involves five steps:

- Developing a descriptive growth stage key for the particular crop species .
- Developing methods to assess the incidence and severity of disease.
- Developing statistically sound methods of sampling crop populations for assessment of the amount of disease.
- Estimating the negative impact of particular levels of the disease on crop yield and quality, and
- Evaluating the economic benefit from various methods available for reducing the amount of disease.

Model Practice Question

A.Objective Questions

a. Multiple Choice Questions

1. Increase or decrease in disease per unit area is
   (a) Infection rate (b) Disease progress
   (c) Infection period (d) Latent period
2. If a disease is spreading on continental or global scale, it is known as
   (a) Pandemic (b) Sporadic
   (c) Endemic (d) Epiphytotic
3. Pathogen requires many years to complete one life or disease
   (a) Polycyclic (b) Polyetic
   (c) Monocyclic (d) None of these
4. Which of the following is wrong about monocyclic plant disense epidemic?
   (a) Low birth rate (b) High death rate
   (c) Soil borne diseases (d) Long reproductive cycle
5. The effect of one or more environmental factors, which make plant vulnerable to attack by a pathogen is
   (a) Predisposition (b) Disease proneness
   (c) Pre-conditioning (d) All of these
6. Infection of a host plant by inoculum arising from that same plant is
   (a) Alloinfection (b) Autoinfection
   (c) Primary infection (d) Secondary infection
7. A sigmoid curve is characteristic for

(a) Polycyclic disease (b) Monocyclic disease
(c) Polyetic diseases (d) Both a and b

8. Simple interest type of disease growth is shown by
(a) Loose smut of wheat (b) Karnal bunt
(c) Wilt of sugarcane (d) All of these

9. Wart disease of potato and Karnal bunt of wheat are example of
(a) Endemic disease (b) Epidemic disease
(c) Sporadic disease (d) Pandemic disease

10. Rapid epidemics occur in
(a) Annual crops (b) Vegetative crops
(c) Biannual crops (d) Perennial crops

Answer

Q. No.	Answer	Q. No.	Answer
1	(a) Infection rate	6	**(b)** Autoinfection
2	(a) Pandemic	7	**(a)** Polycyclic disease
3	(b) Polyetic	8	(d) All of these
4	(b) High death rate	9	(a) Endemic disease
5	**(d)** All of these	10	(a) Annual crops

b. True/False

1. Epidemiology is the science of disease in populations.
2. Epidemics typically spread quickly among annual crops than perennial crops.
3. Epidemics are avoided when host plants possess vertical resistance.
4. Rapid reproduction of a pathogen is not critical elements in the initiation and spread of epidemics.
5. A saturation curve develops in monocyclic diseases.
6. A bimodal curve develops when a polycyclic disease affects several organs of crop.
7. The disease-gradient curve is a graph that shows the relationship between an inoculum source's distance and the disease.
8. Dutch elm disease is a polyetic epidemic brought on by a monocyclic pathogen.
9. Monocyclic epidemics are caused by pathogens with a high birth rate and death rate .
10. Pathogens with the ability to reproduce several times in a season are the source of polycyclic epidemics

Answer

Q. No.	Answer	Q. No.	Answer
1	True	6	True
2	True	7	True
3	True	8	True
4	False	9	False
5	True	10	True

Descriptive Questions

a. Short Answer

1. List the four diseases that are affected by soiltemperature, along with the pathogen for each.
2. Why does the late potato blight spread more easily in humid, cold weather?
3. List a few diseases that alkaline soil is more conducive to.
4. How can a nitrogen shortage promote the onset of disease?
5. List a few plant diseases that can infect plants when the relative humidity is low.
6. Why does cotton wilt occur in heavy clay soils yet root rot disease develops in sandy or loamy soils?
7. List the four diseases that might arise from alkaline soil.

b. Long Answer

1. What is epidemiology? Plant disease epidemiology is taken into account in the context of agricultural meteorology.
2. What is the cause of plant disease epidemics? Explain the many factors that cause late blight of potato and wheat rusts to spread into epidemics.
3. Slow and rapid epiphytotics: what are they? Identify the environmental variables that led to the establishment of the plant disease epiphytotic.
4. Describe how the environment influences a disease's development.
5. Describe in detail how moisture and temperature affect the development of disease in crop.
6. What do rapid and slow epiphytotics mean? Identify the environmental elements that led to the development of an epiphytotic.
7. Make notes about:
   (a) The impact of nutrition and host on the development of plant disease.
   (b) The impact of light on the onset of plant disease.
   (c) The impact of O_2 and CO_2 concentration on the onset of plant disease.

13

Survey, Surveillance and Vigilance

Definition

Sampling: Disease assessment activity within plots and fields.

Survey: To move over a specific place for some observation.

Surveillance: Repeated or sequential survey of some place/ locality for some observation to ascertain the changes /fluctuations in the objects of study .

Principle of Survey: Principle of survey relates that no management practice should be followed against a pest unless its presence is confirmed at the population level that exceeds the economic threshold to make the management tactics economically desirable.

Objectives of Survey

(1) To know the prevalence of disease in a crop in a particular area.

(2) To study the damages caused by a particular disease and assess the economic importance of the disease.

(3) To correlate the disease incidence with the weather factors, vegetation and soil factors.

(4) To give suitable forecasting to the farmers about the outbreak of file disease.

(5) To device suitable control measure in advance by cultural and chemical means.

Kinds of Survey

1) **Qualitative Survey**: includes identification of different diseases.

2) **Quantitative Survey**: involves the estimation of level of diseases or extent of lossesdue to disease .

Types of Survey:

1) **Random:** Samples are collected from any crop at any place suspected to be infested.

2) **Planned:** It is further of two types

a. **Extensive Survey:** A vast area is taken in to consideration to obtain a preliminary assessment of disease.

b. **Intensive Survey:** A particular area is surveyed to obtain accurate and detailed knowledge about the disease.

Methods of Survey

(i) Fixed plot survey

- Select two villages and fix two fields per village.
- Select five sampling plots per field of one acre and it should be three metres away from the bunds.
- The size of the sample plot should be 1 m^2 Select 20 plants or leaves perplot and observe for the disease intensity using the standard disease score chart having 0-9 grades.
- Work out the percent disease index(PDI) using the standard formula. Survey the field every week and prepare the report.

(ii) Roving survey

- Select four villages and in each village, fix two fields.
- Observe and score 1 00 plants or leaves in each field by walking across, starting from south west corner to north east corner.
- Work out the PDI using the standard formula. Survey the fields every week and prepare the report.

Design of Survey

- An impression of the health situation of an area or country can be obtained only be means of a survey.
- In a survey, fields to be sampled are selected so that together they represent the whole area.
- The design of the survey is not easy because it requires a compromise among various requirements, such as following
  - **Good statistical sampling**: The samples of fields is representative for the whole area.
  - **Transportation of personel**: Since the observers have to visit the sample fields.
  - **Time**: A survey having value only when a large area is covered within a short period.
  - **Money**: As surveys tend to be costly.

Requirement for Successful Survey

The essential requirement for success are

1. Record forms:

- All forms should be readily understood and convenient for use in the field and for office processing.
- The only information that should be sought is that which the respondents are aware of and willing to provide.
- Good questionnaires are brief; more questions degrade the quality of the information provided.

2. Selection of staff, training and supervision

- Quality of staff must be appropriate to the work.
- All surveyors should be trained enough to know exactly what is expected of them and why.
- Adequate supervision is also needed to ensure that difficulties are resolved and a good standard of work maintained.

3. Establishing contact –public relations

- Effort must be made on the collaboration of farmers and local officials to secure their trust and interest.
- This can be done by advance publicity through digital media.

4. Data processing and reporting

- Before a survey is start, a plan should be made for processing the data and ensuring the essential clerical and computing facilities.
- Coding and computing should be carefully checked, because errors at this stage can invalidate much useful work.
- Consideration should be given to reporting of the results back to local officials and participating farmers by regional reports or discussing meetings.

5. Sample

- Sample must be sufficient to provide estimates to required accuracy.
- Survey should provide information from truly representative samples of the region under study, not just from farms convenient to visit or known to the extension services.

6. Method of observation and measurement

- Methods used by the field staff are consistent and reproducible and as objective as possible.
- Methods must be also acceptable to the farming community.

Disease Surveillance Reports

(i) **White Report:** It contains the disease/pest surveillance detail in a particular block/division/district for a week.

(ii) **Yellow Card:** It should be sent when the disease/pest occurrence attained half the level of Economic Threshold Level (ETL).

(iii) **Red Card:**It should be sent when the disease/pest occurrence exceeded the Economic Threshold Level.

The weekly data on disease incidence along with the weather parameters collected from fixed plot and roving surveys are reported to the Joint Director of Agriculture in each district. These data collected from the farmers'fields and research centres in the district are pooled and interpreted jointly by the plant protection scientists of the university and Deputy Director of Agriculture (Plant Protection) every week and the action to be taken is decided. After analysing the weekly survey reports and the data on the weather factors, the message on the severity level of disease and control measures to be adopted are disseminated to the farmers through mass media. So surveillance and monitoring methods must be continuously adopted to trace the diseases, their movement and spread with a view to prevent the epidemics.

Pilot survey and Pre test

- A pilot survey, under similar conditions to those of the main survey, is essential, to check all aspects of the plan and obtain realistic estimates of the time and resources needed.
- It should be done soon enough for necessary modifications and for retraining.
- Pilot survey is practicable for small survey (50 or so).

Aerial Survey

- Aerial survey can be of use when
    - Early infection foci have to be located.
    - Sudden disaster must be assessed immediately.
    - The terrain is inaccessible or the area too large.
- In these cases, ground survey is not satisfactory.
- Aerial photography in black and white is already classical; the same is true for black and white infrared sensitive film.
- Color film sensitivity to near infrared reflected radiation has given excellent results in some cases.
- Advanced techniques such as multispectral scanning (MSS), together with a good deal of computer science have been used.
- Aerial survey can be used for the purpose of disease detection, disease assessment, yield and loss assessment.

Vigilance

Vigilance in plant disease management involves monitoring for plant pathogens and responding effectively when they are detected. Here are some ways to monitor for plant disease:

1. Airborne inoculum monitoring
2. Plant disease sensing
3. Wearable sensors

1. Airborne inoculum monitoring

This involves monitoring for plant pathogens that are introduced from other areas, such as by air, seeds, or transplants. This monitoring is usually done over a long period of time at the same sites.

2. Plant disease sensing

This is the science of using remote or proximal sensing to detect and diagnose plant disease.

3. Wearable sensors

These sensors can detect transpiration from plants without affecting their growth. They can also be used to monitor temperature and humidity levels around plants.

Model Practice Question

A.Objective Questions

a.Multiple Choice Questions

1. Disease assessment activity within plots and fields is called .
   (a) Sampling (b) Survey
   (c) Surveillance (d) All of the above
2. To move over a specific place for some observation is called.
   (a) Sampling (b) Survey
   (c) Surveillance (d) All of the above
3. Repeated or sequential survey of some place/ locality for some observation to ascertain the changes in the objects of study .
   (a) Sampling (b) Survey
   (c) Surveillance (d) All of the above
4. A particular area is surveyed to obtain accurate and detailed knowledge about the disease.
   (a) Random survey (b) Extensive planned survey
   (c) Intensive planned survey (d) Qualitative survey
5. Samples are collected from any crop at any place suspected to be infested.
   (a) Random survey (b) Extensive planned survey
   (c) Intensive planned survey (d) Qualitative survey
6. A vast area is taken in to consideration to obtain a preliminary assessment of disease..
   (a) Random survey (b) Extensive planned survey
   (c) Intensive planned survey (d) Qualitative survey
7. The report contains the disease/pest surveillance detail in a particular block/ division/district for a week.
   (a) White report (b) Yellow card
   (c) Red card (d) Blue report
8. The report should be sent when the disease/pest occurrence attained half the level of Economic Threshold Level
   (a) White Report (b) Yellow Card
   (c) Red Card (d) Blue report

9. Aerial survey can be of use when
   (a) Early infection foci have to be located.
   (b) Sudden disaster must be assessed immediately.
   (c) The terrain is inaccessible or the area too large.
   (d) All of the above
10. The report should be sent when the disease/pest occurrence exceeded the Economic Threshold Level.
   (a) White Report (b) Yellow Card
   (c) Red Card (d) Blue report

Answer

Q. No.	Answer	Q. No.	Answer
1	(a) Sampling	6	(b) Extensive planned survey
2	(b) Survey	7	(a) White report
3	(c) Surveillance	8	(b) Yellow card
4	(c) Intensive planned survey	9	(d) All of the above
5	(a) Random survey	10	(c) Red card

b.True/False

1. Pilot survey is impracticable for small survey.
2. Aerial survey can be used for the purpose of disease detection, disease assessment, yield and loss assessment.
3. Qualitative survey includes identification of different diseases.
4. An impression of the health situation of an area or country can be obtained only be means of a survey.
5. When conducting a survey, fields to be sampled are chosen in a way that collectively they represent the whole area..
6. Quantitative survey involves the estimation of level of diseases or extent of losses due to disease.
7. The purpose of the survey is to establish a correlation between the incidence of the disease and weather, vegetation, and soil condition.
8. An impression of the health situation of an area or country can be obtained only be means of a survey.
9. We choose four villages for the roving survey and fix two fields in each the village.
10. The goals of the survey to ascertain the frequency of disease in a crop in a specific region.

Answer

Q. No.	Answer	Q. No.	Answer
1	False	6	True
2	True	7	True
3	True	8	True
4	True	9	True
5	True	10	True

2. Descriptive Questions

a. Short answers

1. Define survey and surveillance.
2. What are the objectives of survey?
3. What methods are commonly used in plant disease surveys?
4. Write down the principles of plant disease survey.
5. What role do farmers and the public play in plant disease surveillance?

b. Long answers

1. What is the importance of plant disease surveys and how do they contribute to agricultural productivity?
2. Explain the difference between a plant disease survey and a plant disease surveillance program. What are the key components of each?
3. Classifies of Surveys? How would you create the survey›s design?
4. What are the challenges faced during the surveillance of emerging plant diseases, and how can they be mitigated?
5. What is the future of plant disease surveillance, considering the advancements in artificial intelligence, big data, and genomic tools?

14

Crop Loss Assessment and Models

Disease Measurement

Plant diseases and their consequences on humans, plants, and the environment are the main topics of plant pathology. Plant pathologists have long overlooked two critical areas: the assessment of plant diseases and their effects on agricultural productivity. The observation made by Yaman in 1914 appears highly relevant even now, nearly a century later. Since pests and diseases can have a direct or indirect impact on crop economy, estimating yield losses from these sources has taken on particular importance in the current era of commercial agriculture. Crop losses and disease incidence were recognized by Chester (1959) as crucial elements of crop economy. Predicting agricultural yield or food quality has multiple benefits for farms and society as a whole. The overall plan's effort on yield loss estimation and measurement should include the following stages.

1. The workers should be well-versed in the typical growth stages and normal morphology of crop plants.
2. The workers should be aware of the characteristic symptoms of the disease at various stages (syndrome).
3. To make reliable and general conclusions, the disease's progress in the area should be monitored for a few years.
4. It is important to document even the smallest information regarding crop husbandry, the agroclimatic zone, and exact meteorological data throughout the season.

The term "pathometry" was first used by EC. Lagre in 1953 to refer to the assessment or measurement of diseases, and it has since become a sub-discipline of plant pathology. Pathometry include the measuring of diseases, estimation of yield loss, forecasting of crop disease, and prediction of their consequences. Growers, researchers, and administration can obtain crucial information from all these closely related operations. Pathometry is the quantitative study of suffering plants that includes both qualitative and quantitative evaluation of plant disease,

Objective of Disease Measurement

Scientists and growers alike may find the knowledge on disease measurement useful since it offers a rational and trustworthy foundation for making exact decisions in the following areas:

1. Development of epidemics in time and space.
2. Examining the variables that influence the onset of disease.
3. Appropriate techniques for disease assessment can be used to evaluate management strategies, fungicides, and host resistance.
4. Time and money might be allocated for different problems based on the priorities set by administrators and scientists. Through the use of disease information, growers may make wise decisions on crop protection, threshold levels and yield loss.

Disease Parameters

Plant diseases have been measured or assessed using three parameters. Crop yield loss has been observed as an indirect indicator of plant diseases, whereas disease incidence, prevalence, and severity are direct measures of disease. Furthermore, the initial inoculum amount may also indirectly indicate the extent of the disease.

Disease incidence. It is the measure of proportion of plant population diseased at a time and place. Numbers of infected plants (whole plant) or plant parts (leaves, twigs, stems, fruits, flower, roots) that are infected are counted and disease incidence is expressed as percent of infected plant in a sample or population. It is a qualitative measure of disease, counting a plant is diseased or not irrespective of amount of disease. The term disease prevalence tells about occurrence of disease large geographical area, eg. Barley yellow dwarf virus, which was prevalent in cereal crop throughout UK during 1990. Prevalence is synonym for incidence and use by surveyors and scouts. Smuts, wilts, fruit rots are some of the diseases generally expressed as incidence. Disease incidence can be equated to yield loss if "unit" is completely lost/damaged. Disease incidence (%) can be calculated using following formula

$$\text{Disease incidence} = \frac{\text{No. of diseased plants x 100}}{\text{Total no. of plants examined}}$$

Disease severity. It is the measure of sickness of diseased plant. It is a quantitative trait, which measures amount of disease on a plant in terms of intensity of symptoms or damage. Not all plant diseases are simple to measure. Various disease assessment keys have been developed standardized to ensure uniform measurement at different time and place by various workers. Disease severity (DS) can be calculated by using the formula:

$$\text{Disease severity} = \frac{\text{Area of plant tissue affected} \times 100}{\text{Total area}}$$

Disease index. The disease index (DE), which has two components—disease incidence and disease severity—was equated with total disease by Chiarappa in 1971. There have also been literary references to "disease intensity" as a disease measurement criterion. Quantitative estimations' severity was characterized by

Mckinney (1923) as a numerical rating. Other crucial components have emerged, contingent on the nature of the disease (symptoms).

$$DI = \frac{0(X_0) + 1(X_1) + 2(X_2) + \cdots + n(X_n)+}{X_0 + X_1 + X_2 + \cdots + X_n \text{ (maximum grade)}} \times 100$$

X represents no. of entries within a grade/class and 0-n are grades or classes of disease, as per rating scale.

Methods for Disease Assessment

Precise evaluation of the disease enables comparison of the state of the disease in various treatments carried out at various times and places

Descriptive keys. The descriptive keys are created by summarizing the traits of the disease, such as its number or characteristic, and by providing a description of the late blight of potatoes, which was first identified by the British Mycology Society in 1945 and is still present today. The intensity of symptoms is the only indicates the presence of host pathogen interaction, Sometimes it may me mileading as certain levels of disturbance or plant damage may go unnoticed. Relative yield loss from vital diseases is known to be higher than the degree of visualsymptoms observed.

Table: Key for the Assessment of Potato Late Blight caused by *P. infestans*

	Description
0	Not seen on field.
0.1	Only few plants affected here and there upto 1 or 2 spots in 12 yards radius
1	Up to 10 spot per plant, or general light spotting
5	About 50 sports per plant or up to 1 leaflet in 10 attacked
25	Nearly every leaflet with lesions, plant still retaining normal form ; field may smell of blight but looks green although every plant is affected.
50	Every plant affected and about ½ of leaf area destroyed by blight ; field looks green flecked with brown.
75	About 3/4th of leaf area destroyed by blight; field looks either predominantly brown nor green.
95	Only a few leaves left green but stems green.
100	All leaves dead, stem dead or dying.

Standard area diagrams. A set of pictures (diagrams, sketches or photographs) having different amount of disease can help in measurement of disease under field conditions. The pictures of different grades are matched with the plant samples for rating disease. Cobb (1892) developed a scale for measurement of wheat rust, where rust severity is scored from 0 to 100% (actual leaf area covered by rust fungus is only 0 to 37%). The idea of Cobb's scale was used by many workers for developing scales for different diseases. Large (1966) and James (1974) reviewed the information on standard area diagrams. Eye as a sensitive visual device has its limitations and according to Weber-Fechner law visual acuity of eye is proportional to the logarithm of the intensity of stimulus. Capitalizing on the fact Horsfall and Barrat (1945) developed grading system with a ratio of two. This system has grades both ways from 50%. The H-B system has been widely used by workers (James, 1974). The proposed grades are as follows:

Upgrades: 0-3%, 3-6%, 6-12%, 12-25% and 25-50%,

Downgrades: 50-75%, 75-85%, 85-94%, 94-97% and 97-100%

Growth stage keys. Plant growth stages have significant bearing on development of plant disease and resulting crop yield losses. Growth stage keys are required to describe the developmental stages of the crop. These are usually based on detailed phenological studies of plants and are of immense value to mention the growth stage at which disease severity/intensity is recorded.

Inoculum-Disease Relationship

This is an indirect method of assessing plant diseases. It is presumed that surviving primary inoculums will initiate disease if susceptible host crop is planted under suitable environment. The amount of disease will be be proportional to the amount of primary inoculums present. This is being practiced for diseases caused by nematodes and other soil borne fungi.e.g *Rhizoctonia solani, Sclerotium rolfsii,* Pythium spp. Fusarium spp., Modules have been devised for estimation of cereal rusts based on uredospore count in air.Zadoks, (1972) justified use of inoculums as parameter for measuring plant disease.

Crop Yield Loss Disease Relationship

Once the amount of disease has been determined, the subsequently step is to assess, either experimentally or statistically, the effect of different levels of disease on the crop yield. Experimental assessment involves setting up test crops in which the level of disease is controlled. Monitoring crops with different levels of disease allows the comparison of epidemic progress and crop yield under different disease conditions. A relationship between disease parameters and yield can then be formulated, allowing prediction of crop loss for a certain level of disease at a particular point in the crop's growth. This method is dependent upon the assumption that the treatments used to keep disease at a certain level have no effect on crop yield themselves. This might not always be the case. Likewise, assessments using susceptible and resistant cultivars rely on both cultivars having a similar yield to start with. Again, this might not be true, and must be determined first in a disease-free environment. All aspects of experimental design need to be cautiously considered in order to gain an accurate assessment of the effect of disease on crop yield. For instance, harvesting methods under experimental conditions are often more efficient than under field conditions, giving the appearance of a higher crop yield. From crop loss assessment studies, models can be devised. Crop loss models are usually based on one of three types of disease assessment: disease at a critical point in development, disease at multiple points in development, and disease throughout crop development.

Models for Crop Yield Loss Estimation

Various methods have been proposed to forecast yield losses, depending on the type of host-pathogen relationship, the nature of damage induced, and the disease assessment technique used. Some of the models utilized in plant pathology are as follows:

Critical Point Model; To establish a relationship with the crop, the disease severity at a critical point is used. Typically, these models employ linear regressions with disease included as an independent variable. When the key stage for disease rating can be satisfied, these models perform satisfactorily for a certain diseases. For instance, the severity of wheat stem rust and rice blast at development stages 11–12 (first–second leaf unfold, the potato cyst nematode(*Globodera rotochiensis*).

$$Y= a+bx$$

The models' primary shortcoming is their failure to take into consideration epidemiological variables that influence disease progression and yield reduction.

Multiple Point Model; This approach differs from the first model in that crop yield losses are calculated as the product of disease ratings and regression coefficients, and disease is scored at various points during disease progression. Using multiple regression analysis, the dependent variable, Y (% yield loss), is reliant on the independent variables, X, X_2, and Xn, which are periodically measured.

$$Y = b_1 X_1 + b_2 X_2 + b_3 X_3 + ... + b_n X_n$$

Area Under Disease Progress Curve (AUDPC)

During the course of a disease's development, changes in epidemiological parameters can have a significant impact on disease progress curves. These are all taken into account by the AUDPC. Since the severity and duration of the disease determine crop damage and production loss. At key phases, there may be cases with the same AUDPC but different levels of disease. There will be variations in yield losses. This represents one of these models' drawbacks.

Large E.C. (1952) developed a method for predicting tuber yields depending on the course of the disease during his traditional research on late blight of potatoes. Research pertaining to mean bulking curves (MBC) has demonstrated that tuberization ceases at 75% leaf damage. It was noted that this figure ranged from 40 to 50 percent for a few other cultivars. When 75% of the foliage was damaged, according to a location's disease progress curve, can be found out. To predict the predicted yield or yield losses, this can be read from MBC.

Recent approaches in crop yield estimation

1. Simulation models have been devised for estimating crop growth and production under various disease/pest pressure scenarios.
2. The area under the disease progress curve was used to describe the crop yield loss. This has been further adjusted such that yield losses are better

explained by multiplying the relative area under the disease's progress curve by the impact of the disease on bulking rate. After 53 epidemics were used to evaluate the model, 80% of data showed disparities between projected and observed losses of less than 5%.

Monitoring of Plant Diseases

Plant diseases are dynamic systems. Variations in the state of disease throughout time and space are frequently seen. Therefore, it's necessary to keep a watchful eye out to learn about any changes in the disease's status. The most popular approach used by government organizations, the pesticide industry, and operations consultants/scouts is disease monitoring. The monitoring's objectives could be:

1. Scouts reporting disease cases they have encountered.
2. Farmers should inspect and assess a crop before choosing where to plant it.
3. Planners and researchers alike can benefit from the data on disease patterns and progression.
4. Computer databases are used to generate periodic alerts and warnings for diseases that pose a high threat (such as late potato blight).
5. Researching the emergence and spread of epidemics on a continental scale, such as wheat rust and late potato blight, can aid in the development of preventive strategies against potentially catastrophic diseases.
6. Frequent crop monitoring can help identify the following:
   - Appearance of new virulence in pathogen.
   - Development of less sensitive strains.

Model Practice Questions

A. Objective Questions

a. Multiple Choice Questions

1. Who recognized that crop losses and disease incidence are crucial elements of crop economy.
   (a) Ward (b) Stakman
   (c) Chester (d) Hutchenson
2. EPIPRE is a simulation model related to
   (a) Cereal rusts (b) Late blight of potato
   (c) Southern corn leaf blight (d) Apple scab
3. The term "pathometry" was first used by
   (a) EC. Lagre (b) Stakman
   (c) Chester (d) Hutchenson
4. Assessment or measurement of diseases is called
   (a) Pathometry (b) Pathology
   (c) Epidemiology (d) Geography

5. Model based equation used to represent disease development in time is
   (a) Logistic model (b) Exponential model
   (c) Monomolecular model (d) All the models
6. Pathometry include
   (a) Measuring of diseases (b) Estimation of yield loss
   (c) Forecasting of crop disease d) All of the above
7. The method used for the assessment of yield losses due to diseases is
   (a) Critical point model (b) Multiple point model
   (c) Area under disease progress curve (d) None of these
8. The measure of proportion of plant population diseased at a time and place
   (a) Disease incidence (b) Disease severity
   (c) Disease index (d) All of the above\
9. The quantitative trait which measures amount of disease on a plant
   (a) Disease incidence (b) Disease severity
   (c) Disease index (d) All of the above
10. The parameter used for the assessment and measurement of plant disease is
   (a) Disease incidence (b) Disease severity
   (c) Disease index (d) All of the above

Answer

Q. No.	Answer	Q. No.	Answer
1	(c) Chester	6	(a) All of the above
2	(a) Cereal rusts	7	(d) None of these
3	(a) EC. Lagre	8	(a) Disease incidence
4	(a) Pathometry	9	(b) Disease severity
5	(d) All the models	10	(d) All of the above

b.True/False

1. Disease incidence and disease seventy are the two components that make up the disease index.
2. Disease severity is the measure of sickness of diseased plant.
3. The area under the disease progress curve (AUDPC) was used to describe the crop yield loss.
4. Disease incidence is a qualitative measure of disease.
5. Crop yield loss has been observed as an indirect indicator of plant diseases.
6. Pathometry includes both qualitative and quantitative evaluation of plant disease.
7. Critical point model employ linear regressions with disease included as an independent variable.

8. Simulation models have been devised for estimating crop growth and production under various disease pressure scenarios.
9. Disease incidence, prevalence, and severity are direct measures of disease.
10. Disease severity is a quantitative trait.

Answer

Q. No.	Answer	Q. No.	Answer
1	True	6	True
2	True	7	True
3	True	8	True
4	True	9	True
5	True	10	True

2. Descriptive Questions

a. Short answers

1. Describe pathometry.
2. What are the objectives of measuring disease?
3. Describe the incidence of diseases
4. Discuss about the disease's severity.
5. How do you monitor plant diseases?

b. Long answers

1. Discuss in details regarding the disease assessment procedure.
2. Describe in detail the models used to estimate crop yield losses.
3. Explain in details about disease parameters.

Principles and Pre-requisites of Forecasting of Diseases

Definition

Fore casting involves all the activities in ascertaining and notifying the grower of community that conditions are sufficiently favourable for certain diseases, that application of control measures will result in economic gain or on the other hand, and just as important that the amount expected is unlikely to be enough to justify the expenditure of time, energy and money for control. The above statement made explicit distinction between positive forecast and negative forecast and both have value for growers as well as society in general.

Positive forecast: Employs need based chemical sprays, provides adequate protection to crop and reduces damage to environment.

Negative forecast: Avoids unnecessary chemical sprays, no risk to the crop health and no disruption of environment.

Importance of Forecasting

- Forecasting informs the growers whether the conditions are not favourable and the disease is unlikely to be intense enough. This helps growers to save the expenditure in terms of energy, time and money by not applying unnecessary management measures.
- It informs the growers about when the conditions are going to be sufficiently favourable for economically important diseases of an agroclimatic zone. This help growers to apply disease management timely that results in economic gain.
- It helps growers to plan advance preventive measures against likely losses due to the occurrence of economically important severe diseases.
- Government and other organizations initiate necessary steps for the timely stocking of chemicals, equipments, etc. when they are timely warned by the possibility of an epidemic by the forecasting's.

Requirements for disease forecasting

The following requirements are necessary to make useful and successful disease forecasts

- The disease must cause economically significant damage in terms of yield loss or quality.
- Control measures must be available at an economically acceptable cost.
- Disease should not be regular feature.
- Reliable means of communication with farmers.
- Growers must have sufficient manpower and equipments to apply control measures when disease warning is given.
- Farmers should be adaptive and have purchase power.

Information's needed for disease forecasting

The information's required for forecasting are:

1. Host Factors

a. Prevalence of susceptible varieties in the given locality.
b. Response of host at different stages of the growth to the activity of pathogen.
c. Density and distribution of the host in a given locality.

2. Pathogen factors

a. Amount of primary inoculum in the air, soil or planting material
b. Dispersal of inoculums
c. Spore germination
d. Infection
e. Incubation period
f. Sporulation on the infected host

g. Re-dispersal / Dissemination of spores
h. Perennating stages
i. Inoculum potential and density in the seed, soil and air

3. Environmental factors

a. Temperature
b. Humidity
c. Light intensity
d. Wind velocity

Methods of disease forecasting

Disease forecasting requires field observations on the pathogen characters, collection of weather data, variety of the crop and certain investigations and their correlations. Usually the following methods are employed in disease forecasting.

1. Forecasting based on amount of initial inoculum

Amount of primary inoculum present in seed, soil or vector can be estimated and disease prediction can be made without much problem. This criterion is more important for monocyclic diseases where the secondary infection does not take place and the amount of primary inoculums is related to disease severity and damage. It is being used in case of loose smut of wheat and soil borne inoculums of *Rhizoctonia, Sclerotium, Verticillium*, nematodes can be estimated and disease can be forecast. Presence of loose smut of wheat, ergot of pearl millet and viral diseases of potato can be detected in the seed lots at random by different seed testing methods. Seed testing methods can be used to determine potential disease incidence and enable decision to be made on the need for chemical seed treatment.

2. Forecasting based on weather conditions

Prevailing weather conditions *viz*., temperature, relative humidity, rainfall, light, wind velocity etc., have become a major criteria for diseases forecasting for polycyclic diseases where in addition to amount of primary inoculums, the multiplication and dispersal of secondary inoculums are weather dependent. Forecasting of late and early blight of potato, wheat rust, anthracnose, apple scab, powdery mildew of cucurbits and downy mildew of grapes are made considering the prevailing weather conditions which determine the amount of disease and damage of the crop.

3. Forecasting based on correlative information

Weather data of numerous years are collected and correlated with the intensity of them diseases. The data are compared and then the forecasting of the disease is done. Forecasting criteria developed from comparisons of disease observation with standard meteorological data have been provided for diseases like fire blight of apple and powdery mildew of barley.

4. Use of computer for disease forecasting

In some developed countries forecasting of disease is made by the use of computers. The computers are fed with weather data collected on the farm by individual farmers. The computer then processes the data, determines whether an infection period is imminent, likely to occur or can not occur, and a recommendation is made to the farmer as to whether or not to spray and what materials to apply. This system gives the results quickly. One such computer based programmes in the USA is known as 'BLITECAST' for potato late blight.

Examples

a. Late blight of potato

Late blight of potato had been instrumental in development of diverse fundamental concepts in plant pathology and disease forecasting is no exception. The potato growers in Europe were intelligent enough to recognize blight weather (Moderate temperature and lots of moisture: rain, dew or humidity) much earlier than Van Everdingen postulated Dutch rules in 1926. Followings are four conditions correlated with weather that could predict about appearance of late blight in Netherlands:

- Night temperature below dew point at least 4 hours.
- Mainimum temperature of 10^{0}C or slightly above.
- Clouds on the next day.
- Rainfall during next 24 hours of at least 0.1 mm.

Notable contributions in late blight forecasting

Van Everdingen postulated Dutch rules in 1926

Krause, et al. (1970), BLITECAST

Singh, et al. (1980), JHULSACAST for Indo Gangetic plains

Fry and Apple (1983), Integrated Host Resistance and fungicide weathering in BLITECAST

Dommermuth (1988), Phytoprog I. Late blight warning service

Runno and Kopple (2002), NEGFRY

Grunwald (2002), Modified and validated SIMCAST

Negative Forecast

- When action and warning thresholds can be determined, it is also possible to determine that no action needs to be taken, at least within a certain period of time is called negative forecasting.
- An official negative forecasting system is operated in Germany for potato late blight under the name PHYTOFROG. BLITECAST also contains the option of negative forecasts.

- Negative forecasting system will become more popular in the future
- For example, cereal crops, particularly rice and wheat, are treated increasingly against various diseases. However, because of the relatively low margin of profit and the very large acreages involved, negative forecasts can prevent control costs from exceeding the limits of economy as well as forestall long term, unpredictable environmental and health hazards from chemical.

Supervised control

- It is the form of disease management system often, but not always, relying on the application of pesticides under the direction of a specialist.
- It is generally based on the assessment of disease intensity, crop loss and other ecological considerations.
- Since supervised control aims at the most efficient and least hazardous use ofchemicals, it is an essential component of many integrated control systems.
- BLITECASTS is a more advanced form of supervised control.
- Supervised control is more common in fruit growing areas and in large estated where specialized supervisors are available.

Model Practice Questions

A. Objective Questions

a. Multiple Choice Questions

1. Who postulated Dutch rule for forecasting of late blight of potato
   - (a) Van Everdingen (b) Krause, et al
   - (c) Singh, et al (d) Fry and Apple
2. Who developed JHULSACAST for Indi Gangetic plainsfor late blight of potato
   - (a) Van Everdingen (b) Krause, et al
   - (c) Singh, et al (d) Fry and Apple
3. Phytoprog I. late blight warning service developed by
   - (a) Van Everdingen (b) Krause, et al
   - (c) Singh, et al (d) Dommermuth
4. BLITECAST developed by
   - (a) Van Everdingen (b) Krause, et al
   - (c) Singh, et al (d) Fry and Apple
5. Forecast system based on the amount of primary and secondary inoculum
   - (a) Apple scab (b) Cereal rusts
   - (c) Sugarbeet yellows (d) All of these
6. The first computer simulation programme to forecast plant disease
   - (a) EPIDEM (b) EPIVEN
   - (c) EPIDEMIC (d) BLIGHTCAST

7. The system used for the early prediction of late blight of potato is
   (a) Dutch rules
   (b) Beaumont and Stainland's method
   (c) Wallin and Hoyman's method
   (d) All these
8. The method used in disease forecasting is based on
   (a) Weather conditions
   (b) Amount of propagules of pathogens in the air, soil or planting material
   (c) Change in the host physiology
   (d) All of the above
9. Which model is appropriate for most polycyclic diseases
   (a) Monomolecular model
   (b) Exponential model
   (c) Logistic model
   (d) Gompertz model
10. The first computer simulation program was designed to simulate epidemic of
   (a) Apple scab
   (b) Cereal rusts
   (c) Late blight of potato
   (d) Early blight of potato

Answer

Q. No.	Answer	Q. No.	Answer
1	(a) Van Everdingen	6	(a) EPIDEM
2	(c) Singh, et al	7	(a) All these
3	(d) Dommermuth	8	(d) All of the above
4	(b) Krause, et al	9	(a) Logistic model
5	(d) All of these	10	(d) Early blight of potato

b. True/False

1. EPIPRE is an example of a multiple disease/pest forecasting system.
2. Forecasting models are often based on a relationship like simple linear regression where x is used to predict y.
3. Polycyclic epidemics such as potato late blight are usually best modeled by using the logistic model.
4. Monocyclic epidemics may be best modeled using the monomolecular model.
5. Plant disease forecasting is a management system used to predict the occurrence or change in severity of plant diseases.
6. Monomolecular model is appropriate for modeling of monocyclic epidemics .
7. Exponential model is used in early stage of polycyclic disease.
8. Forecasting based on amount of initial inoculums done for loose smut of wheat.
9. BLITECAST also contains the option of negative forecasts.

10. Supervised control is more common in fruit growing areas and in large estate where specialized supervisors are available.

Answer

Q. No.	Answer	Q. No.	Answer
1	True	6	True
2	True	7	True
3	True	8	True
4	True	9	True
5	True	10	True

2. Descriptive Questions

a. Short answers

1. Define Plant disease forecasting.
2. Write down the importance of plant disease forecasting in agriculture.
3. Differentiate between positive and negative forecasting.
4. Discuss about supervised control.
5. What are the main prerequisites for disease forecasting?
6. Explain dutch rule.

b. Long answers

1. Explain the concept of plant disease forecasting and its significance in modern agriculture. How does forecasting help in managing plant diseases more efficiently?
2. Describe the role of weather data in plant disease forecasting. How do specific weather conditions influence the development and spread of plant diseases?
3. What is Plant disease forecasting. Write down the importance, method of plant disease forecasting .
4. Explain the role of computer in plant disease forecasting
5. How can climate change impact the future of plant disease forecasting? Discuss the potential challenges and opportunities associated with a changing climate.
6. What are some examples of successful plant disease forecasting programs or initiatives globally? How have these programs improved agricultural productivity and sustainability?

15

Principles & Methods of Plant Disease Management

The two major pillars of disease management strategies are prevention, or prophylaxis, and cure. The plant is protected from disease as a result of the first principle of prevention, which comprises disease management strategies used prior to infection. Avoidance, exclusion, eradication, resistance, and protection are some of its components. The plant is treated for the disease according to the second principle of curative action, which applies to any measure used after the plant becomes sick.It involves counseling.

Principles of Plant Disease Management

Prophylaxis	Avoidance	Prevent disease by selecting a time of the year or a site where there is no inoculum or where the environment is not favorable for infection
	Exclusion	Prevent the introduction of inoculums
	Eradication	Eliminate, destroy or inactivate the inoculums
	Protection	Prevent infection by means of a toxicant or some other barrier to infection
	Resistance	Utilize cultivars that are resistant to or tolerant of infection
Curative	Therapy	Cure plants that are already infected

1. **Pathogen avoidance:** Preventing disease by planting when or where the inoculum is lacking or ineffective because of unfavorable environmental circumstances.
   - Choice of geographic area
   - Selection of field
   - Choice of time of sowing
   - Disease escaping varieties
   - Selection of seed and planting stock
   - Modification of cultural practices
2. **Exclusion of inoculum:** Keeping the inoculum from entering or growing in a field or other location where it is not present.
   - Seed treatment ·
   - Certification and inspection

- Insect vector eradication
- Quarantine

3. **Pathogen eradication:** This involves lowering, inactivating, getting rid of, or destroying the inoculum at its source, either from a particular area or a single plant where it has already taken hold.
   - Biological control of plant pathogen
   - Crop rotation
   - Removal or destruction of diseased plant organs
   - Rouging
   - Eradication of alternate and collateral hosts
   - Sanitation
   - Heat and chemical treatment of disease plants
   - Soil treatments
4. **Resistant:** Make use of cultivars that can tolerate or resist disease. If resistant varieties are available, you can manage the disease for free.
5. **Protection measures:** These include keeping the plant and the pathogen from becoming infected by creating a chemically toxic barrier.
   - Chemical intervention
   - Chemical management of insect vectors.
   - Environmental modification.
6. **Therapy:** Lessening an infected person's disease severity.

Chemotherapy: Applying chemicals to a diseased or infected plant to stop the disease from spreading is known as chemotherapy.Examples include heating or chemically treating vegetative material, such as woody cuttings, bulbs, and corms, to get rid of any nematodes, viruses, bacteria, or fungi that have been established in the plant material.

Heat therapy: Rice seeds treated with hot water at 51–53°C for 15 minutes after being submerged in cool water for a day to manage *Aphlenchoides besseyi*-caused white tip disease

Tree surgery: After cutting off diseased sections in horticultural crops, apply chaubatia paste.

Although these guidelines still hold true in the context of contemporary ideas for managing plant diseases, they have some serious flaws, such as. The objectives of plant disease control, such as exclude, prevent, and eliminate, are expressed in absolute terms and suggest a zero disease outcome. In this view, controlling plant diseases is not feasible and, for the most part, not even viable. To be sure, we don't need to completely eradicate a disease; all we need to do is slow down its progression and keep it from getting worse than it needs to. We should consider plant disease management rather than plant disease control.

These concepts fail to account for the dynamics of plant disease, which refers to variations in disease incidence and severity over time and geography. We require a way to quantitatively evaluate how different control strategies, both separately and in combination, affect the course of the diseases.

Lastly, these plant disease control approaches frequently place an emphasis on methods without properly integrating them into a comprehensive overall plan.

Does this imply that we should do away with the conventional wisdom? Naturally, no; all we need to do is include them into a well-thought-out overarching plan grounded in epidemiological principles.

The difference between strategies and tactics: Although the two concepts have similar definitions in dictionaries, a strategy is often an overarching plan for achieving a certain goal, while tactics are the concrete ways in which a given strategy is put into practice. A tactic is what is known as a strategy at one level of attention and a strategy at another.

Strategy of disease management

It is commonly acknowledged that eliminating a disease entirely from the ecosystem and the planet is exceedingly challenging and goes against the natural order. Consequently, the goal of disease management should be to keep the level of disease below what is acceptable in order to prevent harm from being done. In light of this, the disease management approach ought to incorporate all currently recognized techniques while accounting for the subsequent elements.

1. **Inhibition of pathogens:**Since it is impossible to completely eradicate a pathogen, greater focus should be placed on creating technologies that keep its population below levels that are harmful.
2 **Tolerable losses:** refers to the degree of disease that can be tolerated without causing financial hardship. For instance, a two percent disease infection in crops grown for seed is acceptable, but a five percent disease infection is acceptable in crops grown for human consumption.
3. **Economical management:** The cost-benefit ratio should be taken into account before implementing the disease management approach. A high enough return must be obtained for each unit of money spent.
4. **Long term management:** In order to cut costs and environmental damage, the disease management plan should aim to apply techniques that have a long-term impact.
5. **Collective strategy:** A single farmer cannot manage disease because pathogens can spread from field to field. Therefore, for successful control, a collective area-wise approach is required. A battle cannot be won based just on the amount of weaponry at hand. Only a well-thought-out plan that makes optimal use of weapons in a coordinated fashion at the appropriate moment and using the right techniques can lead to success. Comparably,

if an integrated strategy is used and all five of the fundamental elements covered above are taken into consideration, the fight against diseases can be won.

Methods of Plant Disease Management

1. Physical management
2. Cultural management
3. Host plant resistance
4. Legislative management
5. Biological management
6. Chemical management

Model Practice Questions

A. Objective Questions

a. Multiple Choice Questions

1. Avoiding disease by planting at times when or where inoculum is absent or ineffective due to unfavorable environment conditions.
   (a) Avoidance of disease (b) Exclusion of inoculum
   (c) Eradication of disease (d) Immunization
2. Preventing the inoculum from entering or establishing in the field or area where it does not exist.
   (a) Avoidance of inoculum (b) Exclusion of inoculum
   (c) Eradication of inoculum (d) Immunization
3. Reducing, inactivating, eliminating or destroying inoculum at the source, either from a region or from an individual plant in which it is already established.
   (a) Avoidance of inoculum (b) Exclusion of inoculum
   (c) Eradication of inoculum (d) Immunization
4. Preventing infection by creating a chemical toxic barrier between the plant and the pathogen is comes under protection measures
   (a) Avoidance of pathogen (b) Exclusion of inoculums
   (c) Eradication of pathogens (d) Protection
5. Disease management tactics applied before infection
   (a) Prevention (b) Cure
   (c) Regulation (d) Eradication
6. Disease management tactics applied after infection
   (a) Prevention (b) Cure
   (c) Regulation (d) Eradication
7. Avoidance of pathogen includes
   (a) Choice of geographic area (b) Selection of field
   (c) Disease escaping varieties (d) All

8. Exclusion of inoculum includes
   (a) Seed treatment (b) Inspection and certification
   (c) Quarantine (d) All
9. Reducing severity of disease in an infected individual
   (a) Chemotherapy (b) Heat therapy
   (c) Tree surgery (d) All
10. Altering the effectiveness of the pathogen by selection or introduction of resistance genes in the plant.
   (a) Avoidance of pathogen (b) Exclusion of inoculums
   (c) Eradication of pathogens (d) Disease resistance

Q. No.	Answer	Q. No.	Answer
1	Avoidance of disease	6	Cure
2	Exclusion of inoculum	7	All
3	Eradication of inoculum	8	All
4	Avoidance of pathogen	9	All
5	Prevention	10	Disease resistance

b. True /False

1. Eradication of pathogens includes biological management of plant pathogens.
2. Plants with viral infections are rouged under the heading "eradication of Pathogens."
3. Disease-escape variants fall under the category of pathogen avoidance.
4. Treatment with seeds is classified as inoculum exclusion.
5. The section on pathogen eradication includes quarantine.
6. The approach of cross protection is typically employed to manage viral diseases.
7. Disease management might be viewed as reactive approach.
8. Correct diagnosis of a disease is not necessary to identify the pathogen, which is the real target of any disease management program.
9. Chemotherapy is the application of chemicals to an infected or diseased plant that stops the infection.
10. Complete eradication of disease is against the law of nature.

Q. No.	Answer	Q. No.	Answer
1	True	6	True
2	True	7	False
3	True	8	False
4	True	9	True
5	True	10	True

B.Descriptive Questions

a. Short answer

1. Write down the difference between control and management.
2. Prevention is better than cureprove with examples.
3. Why collective approach is very effective in management of plant diseases.
4. Explain immunization.
5. Explain disease escape.

b. Long answer

1. Distinguish between i.control and management ii. prevention and cure iii. strategies vs tactics.
2. Why the term management is preferred over the term control?
3. Discuss principles of plant disease management.
4. Write down the major critical shortcoming of principles of plant disease management.
5. What is therapy.Name some fruit trees which can be controlled by tree therapy.
6. Discuss in details the general principles of plant disease management with suitable examples.
7. Elaborate strategies of plant disease management .
8. What is plant disease management? Give the basis of the principles of plant disease management.
9. Distinguish between the following: i. Exclusion and Eradication ii. Protectant and Eradicant iii. Prophylaxis and Immunization iv. Chemotherapy and Thermothearapy

16

Host Plant Resistance

Definition

Those traits that allow a plant to withstand or recover from pathogen infections in situations when other plants in the same species would suffer more harm.

or

Resistance is an organism's innate capacity to survive or resist a pathogen, such as in the case of crop plants.

Classification of Resistance

Numerous pathogen species harm every species of plant. However, resistant plants that have evolved a variety of methods that are categorized differently manage to survive.

1. Based on existence

a. Preformed

Even in the absence of the pathogen, when the plant already possesses resistance. It is referred to as passive, axenic, or performed resistance. An instance of mildew resistance in a barley cultivar Negrete.

b. Induced

It is referred to as active, apergic, or induced resistance when the plant does not naturally possess resistance in the absence of the pathogen but instead develops resistance in response to its contact.

For instance, cotton seedlings are vulnerable to many pests because they lack gossypol; yet, if infected by Verticillium, the seedlings develop resistance to these pests because gossypol is generated through induction.

2. Based on type of host response

a. Immune

Being immune means being 100% free from disease or being immune from infection. No symptoms appear. For example, potatoes can withstand karnal bunt.

b. Resistance

When a certain level of host-pathogen interaction is elicited, resistance is present. As a result, resistance is only partially effective; in contrast to immunity, which is asymptomatic, resistance always manifests some symptoms.

c. Tolerance

The ability of the host, either innate or learned, to withstand disease is known as tolerance. Disease tolerance is the ability to minimize yield losses even in the presence of disease.

2. Based on epidemiological terms

a. Vertical resistance

A plant variety is said to have "vertical" or "perpendicular" resistance when it is more resistant to certain races of pathogen than others.

b. Horizontal resistance

Host resistance is referred to as "horizontal" or "lateral" when it is equally effective against all pathogen races.

Table: Major difference between horizontal & vertical resistance

Horizontal resistance	Vertical resistance
Minor	Major
Polygenic	Monogenic
Race nonspecific	Race specific
Quantitative	Qualitative
Multiple gene	Multiple allele
Durable	Nondurable
Non hypersensitive	Hypersensitive
Low to moderate	High
General resistance	Specific resistance

Disease Escape or Apparent Resistance

It describes how host plant kinds that are susceptible to a disease are free from it because of only environmental conditions. It mostly happens as a result of avoiding touch and preventing infection.

It may be achieved by

- Earlier varieties
- Modify the planting date
- Modification to the planting location
- NPK application that is balanced
- Management of disease carrier

For instance, a short-lived or early-maturing variety may be protected from significant disease attacks (such as potato late blight and wheat rust).

Gene for Gene hypothesis

According to Gene for Gene hypothesis for every resistance gene present in the host, the pathogen has a gene for virulence. Susceptible reaction would result only when the pathogen is able to match all the resistance genes present in the host with appropriate virulence gene. In most of the pathogens, virulence is recessive to avirulence. The hypothesis is proposed by HH Flor in Linseed rust (*Melampsora lini*).

Mechanism

1. Compatible reaction

The alleles for susceptibility (r) in the host and those for virulence (a) in the pathogen produce specific compounds, which interact with each other to produce the susceptible response. A lack of interaction between the products of r and a genes specifiy resistant response of the host. It is found in heterotrophic pathogens.

2. Incompatible reaction:

- This type of gene for gene relationship is found in biotrophic pathogens.The alleles for resistance in the host (R) and those of avirulence in the pathogen (A) produce specific compounds which recognize each other , when these specific compounds interact , they produce resistance response in the host
- The lack of interaction between the products of the genes for resistance and avirulence produces susceptible reaction in the host.

	Virulence Gene	
Resistance Gene	A_1	a_1
R_1	R	S
r_1	S	S

- Hypersensitive reactions on the part of the host are typically linked to incompatible reactions.

Mechanism of Disease Resistance

1. **Mechanical:** Anatomical or mechanical characteristics of the host could stop infection. For example. Wheat and barley's closed blooming habit shields them from ovary-infecting fungal spore infection.
2. **Hypersensitivity**: An immunological response, such as localized cell death, is observed in biotrophic species. For instance, yellow rust.
3. **Nutritional**: Unfavorable physiological factors inside the host typically limit the decrease in development and spore production.

Source of Disease Resistance

- A known variety
- Germplasm collection
- Related species
- Mutations
- Somaclonal variation
- Unrelated organisms

Method of breeding for development of resistance

The techniques used to breed disease-resistant varieties of a character are not all that different from those used to breed other characters. The subsequent techniques are employed:

1 Introduction,
2. Selection
3. Hybridization followed by Selection
4. Mutation
5. Somaclonal variation
6. The use of genetic engineering

Techniques to combat viral infections

- Coat protein virus: the most effective
- Sat-RNA
- A flawed viral genome
- Antinsense construct of critical viral genes.
- Ribozymes

Techniques for combating bacterial and fungal infections

Through the following transgens' expression:

- Genes encoding target enzymes that are insensitive.
- Genes indicating the deactivation of toxins
- Antibacterial peptide expression
- Bacterial lysozyme expression
- The gene governing synthetically induced cell death
- The manifestation of phytoalexins.
- Genes encoding proteins that inhibit ribozyme
- Heterologous thionin expression

Resistance Breeding

In resistance breeding, the biology of the pest and pathogen must be taken into account in addition to the crop plant breeding system and the resistance genes. Finding and utilizing powerful genes that the stabilizing selection acts against is necessary for the application of particular resistance.

Strong genes can be applied in several contexts.

- Singly
- Deployment
- In combination
- Multiline

a. Use of single resistance gene

One single gene at a time is utilized. Because of the crop's uniformity brought about by this technique, the pathogen population is strongly selected against. As a result, using a single R-gene is dangerous and has drawn criticism for not offering long-term resistance.

b. Gene Deployment

It refers to strategic use of major gene, by deploying these over space and time to avoid phenomena of boom and bust.

Advantages

- It aids in the prevention of disease.
- Preserve the diversity of the main genes across various geographic regions.

Drawback

- Difficult to identify distinct genes with comparable efficacy.
- Numerous plant diseases, including bacterial blight of rice, rusts of wheat, oat crown rust, barley leaf rust, bean rust, and others, have been managed by the application of the gene deployment technique.

c. Pyramiding of genes

Pyramiding or stacking of genes occurs when two or more known resistance genes are inserted into a single variety.

Advantages

- Offer resistance to a shared racial range.
- `Increases resistance's breadth rather than its depth to offer additional resistance.
- Offers long-lasting defense against crown rot in oats.

Disadvantages

- Possibility of super race development.
- Tedious.
- Agronomic performance of developed cultivar.

d. Multilines

Multiline varieties are blends of multiple purelines with distinct genes for disease resistance, and that share similar height, flowering and maturation dates, seed color, and agronomic traits. In situations where new races are constantly being produced, multiline varieties seems to be an effective strategy for controlling disease like rusts. Three wheat multiline varieties have been released in India: KSML3, MLKS 11, and KML 7406.

Boom and Burst Cycle

When as new variety carrying vertical resistance is developed it is resistant to the prevalent pathotype of the pathogen in question, therefore it becomes popular rapidly and occupy a large area under the concerned crop this period of boom continues till the pathotype specific to the VR gene employed in the variety evolves and or increase in frequency to cause on epidemic in the fields occupied by the variety. Now susceptible variety loss popularity and area under it declines sharply thus the variety is effectively busted by the virulent pathotype. This has been called as boom and bust cycle by Priestley One way to end the boom-burst cycle is via

- R-gene use that is localized
- Cycles of the R-gene
- Applying polygenic resistance
- Combination of all.
- Multigenic resistance
- Multiline variety

Vertifolia effect

The process of neglecting of losing horizontal resistance in the course of breeding for vertical resistance has been termed as 'vertifolia effect'. There are two explanations for this failure:

- Varieties that use oligogenes for resistance typically have low levels of horizontal resistance.
- The pathogen has the capacity to develop virulent pathotypes.

Pre-emptive breeding

Breeding to develop resistance to potential patho types is known as anticipatory or preemptive breeding for resistance. Its effectiveness rests on the breeder's capacity to foresee the pathogen phenotypes (pathopypes) that are likely to be significant in the future.

Model Practices Questions

A. Objective Questions

a. Multiple Choice Questions

1. The decline in the level of horizontal resistance due to continuous breeding for vertical resistance is
   (a) Vertifolia effect (b) Pasteur effect
   (c) Boom and bust cycle (d) Apparent resistance
2. The ability of a plant to sustain the effects of a disease without suffering serious injury or crop loss is
   (a) Resistance (b) Tolerance
   (c) Susceptibility (d) Innate immunity
3. Hypersensitivity is applicable to the plant diseases caused by
   (a) Fungi (b) Bacteria
   (c) Viruses (d) All these

4. When two or more known resistance genes are introduced into a single variety it is called
   (a) Pyramiding of gene (b) Gene deployment
   (c) Vertifolia effect (d) Boom & Bust cycle
5. Strategic use of major gene, by deploying these over space and time to avoid phenomena of boom and bust is called
   (a) Pyramiding of gene. (b) Gene deployment
   (c) Vertifolia effect (d) Boom & Bust cycle
6. When the resistance is already present in the plant even the absence of the pathogen. it is known as
   (a) Axenic resistance (b) Gene deployment
   (c) Vertifolia effect (d) Induced resistance
7.The boom burst cycle can be broken by
   (a) Localized usage of R- gene (b) Cycling of R- gene
   (c) Use of polygenic resistance (d) Combination of above
8.The mechanism of disease resistance in plant are
   (a) Mechanical (b) Hypersensitivity
   (c) Nutritional (d) All
9. The gene for gene hypothesis is proposed in disease
   (a) Linseed rust (b) Wheat rust
   (c) Pea rust (d) Barley rust
10. Apparent resistance in crop is achieved by
   (a) Early varieties (b) Change date of planting
   (c) Change in the site of planting (d) All
11. The inherent ability of an organism (i.e., the crop plant) to resist or withstand the pathogen is called
   (a) Resistance (b) Tolerance
   (c) Susceptibility (d) Endurance
12. Absence of infestation or injury to host plant due to transitory process like incomplete infestation. This pertains to few individuals of host.
   (a) Host evasion (b) Induced resistance
   (c) Disease Escape (d) Boom & Bust cycle
13. Host may pass through the most susceptible stage quickly or at a time when insects are less or evade injury by early maturing. This pertains to the whole population of host plant.
   (a) Host evasion (b) Induced resistance
   (c) Disease escape (d) Boom & Bust cycle

14. Strategic use of major gene, by deploying these over space and time to avoid phenomena of boom and bust is called
    (a) Pyramiding of gene. (b) Gene deployment
    (c) Vertifolia effect (d) Boom & Bust cycle
15. Major gene, race -specific, seedling, monogenic, differential, specific and pathotype specific resistance is
    (a) Vertical resistance (b) Horizontal resistance
    (c) Apparent resistance (d) Disease Escape
16. Cytoplasmic inheritance is governed by
    (a) Chloroplast DNA (b) Plasmid DNA
    (c) Mitochondrial DNA (d) All of these
17. The plant pathogen regarded as genetic engineer of plant kindgdom is
    (a) *Agrobacterium tumefaciens* (b) *Erwinia amylovora*
    (c) *Neurospora crassa* (d) *Pseudomonas syringae*
18. Non-differential resistance that is uniform, effective against all the races of a pathogen and reduces the rate of disease development is
    (a) Horizontal resistance (b) Vertical resistance
    (c) Apparent resistance (d) Non-host resistance
19. SAR govern by
    (a) Salicylic acid (b) Jasmonic acid
    (c) Ascorbic acid (d) Lactic acid
20. Breeding for resistance to future pathotypes is called
    (a) Introduction (b) Selection
    (c) Hybridization (d) Pre emptive breeding

Q. No.	Answer	Q. No.	Answer
1	(a) Vertifolia effect	11	(a) Resistance
2	(b) Tolerance	12	(c) Disease escape
3	(d) All these	13	(a) Host evasion
4	(a) Pyramiding of gene.	14	(b) Gene deployment
5	(b) Gene deployment	15	(c) Viroid
6	(a) Axenic resistance	16	(a) Vertical resistance
7	(d) Combination of above	17	(a) *Agrobacterium tumefaciens*
8	(d) All	18	(a) Horizontal resistance
9	(a) Linseed rust	19	(b) Jasmonic acid
10	(d) All	20	(d) Pre emptive breeding

b.True /False

1. Multiline variety appears to be useful approach to control diseases like rusts where new races are continuously produced.
2. The concept of boom and bust cycle was given by Priestley.

3. The gene for gene hypothesis is proposed by HH Flor.
4. Immunity is exempt from infection or hundred per cent freedom from disease
5. Resistance is partial, that is,always some symptoms appear.
6. Monogenic resistance is easy to incorporate into plants by breeding.
7. Polygenic resistance is more sensitive to environmental fluctuations.
8. All Tms cytoplasm of maize used in hybrid seed production is susceptible to T race of southern corn blight.
9. Vertical resistance reduces the effective amount of initial inoculums from which the epidemic starts.
10. Horizontal resistance may delay or reduce infection, colonization of the plant and/or production of spores by the pathogen.
11. Sick plots are created for testing resistance of soil borne diseases.
12. Vertical resistance is also called juvenile plant resistance.
13. Horizontal resistance is also called minor gene resistance.
14. Effectiveness of horizontal resistance declines as environment favors pathogen or predisposes host.
15. Vertical resistance exerts strong selection pressure on pathogen population leads to emergence of new races.

Answer

Q. No.	Answer	Q. No.	Answer
1	True	9	True
2	True	10	True
3	True	11	True
4	True	12	True
5	False	13	True
6	True	14	True
7	True	15	True
8	True	16	-

A. Descriptive Questions

a. Short Answer

1. Differetiate between horizontal and vertical resistance.
2. Write down the mechanism of disease resistance.
3. What are the disadvantages of growing genetically uniform crops over large areas.
4. Differentiate between disease escape and tolerance.
5. Name four diseases (with their pathogens) that exemplify vertical resistance.
6. Define pathotype, pathodeme, esodemic, exodemic.

7. What are Hrp proteins?
8. Write in brief about cross protection.
9. What is durable resistance? Give five examples of durable resistance.
10. Name two diseases in which the soil moisture determines disease escape.
11. Discuss about i.Gene pyramiding ii.Gene deployment iii. Apparent resistance.

b. Long Answer

1. What are resistance genes? Classify them on the basis of their structure and function. Discuss their mechanism of action and evoloution.
2. Give a details account of genetic resistance in plants.
3. What is host plant resistance? Classify resistance on the basis of existence, host response and epidemiological competence.
4. Write short notes on a. Boom and Bust cycle b. Vertifolai effect c. Anticipatory breeding d. Multiline.
5. Who first proposed 'gene-for-gene concept'? Name the disease involved with its pathogens.
6. Attempt the following:
   a. Boom and bust cycle
   b. Genetic recombination in viruses
   c. Cytoplasmic resistance
   d. Differences between VR and HR
   e. Physiological races
   f. Avirulence genes.

17

Cultural Management

Introduction

The manipulation or alteration of crop production processes for the purpose of managing plant diseases is known as cultural management. It's possible that cultural customs predate agriculture itself. Early on in the history of agriculture, growers understood from their observations and experiences that cultivating a specific crop species or variety repeatedly on a given plot of land frequently led to crop disease. In fact, cultural practices are now seen as crucial backup strategies for managing plant diseases in modern agriculture. By making appropriate adjustments to crop production techniques, the environment can be changed to make it less conducive to pathogens and their pathogenesis. Accordingly, cultural behaviors that affect disease control are preventative in nature; they work to lower the inoculum's activity and density.It provides the chance to modify the host's condition, the environment, and the behavior of the causative agent in order to accomplish cost-effective disease management. at decreasing the inoculum's activity and density.It provides the chance to modify the host's condition, the environment, and the behavior of the causative agent in order to accomplish cost-effective disease management.

Procedures for disease management through cultural practices

The following three headings outline the steps for managing disease using cultural customs:

A. Adjustment of cultural practices

Getting a good crop stand and planting healthy seeds in healthy soil are the major goals of changing cultural practices. The process for managing disease by modifying cultural customs is outlined under the following headings.

1. **Summer ploughing:** During the summer, deep plowing buryes the soil-borne fungal inocula. Because summer temperatures are greater, fungal propagules, sclerotia, and other spore/conidia on plant tissues perish when exposed to sunlight. Infected self-sown plants, volunteer hosts' plants, weed hosts, regrowths from the plant roots, and further alternate and alternative hosts are also eliminated. This prevents the disease from spreading. The best way to manage groundnut blight (*Corticium rolfsii*) is to till the soil down to a depth of 20 cm.

2. **Adjustment of sowing time:** A decent return can be obtained by modifying the sowing dates appropriately. The susceptible stage of the crop's growth, the soil's properties, and other variables that are conducive to the pathogen's maximum activity should all be taken into account when determining when to sow. Wheat crops planted later tend to become less infected than wheat sown on regular dates. Red rot of sugarcane can be prevented by planting on days that aren't too cold or overcast. Compared to crops sown later in the month, rapeseed planted in mid- to late-August is more susceptible to leaf spot (*Alternaria brassicae*) attacks. Root rot and blight are common in peas and gram that are planted soon after rains, when soil temperature and moisture levels are high. These diseases decrease from november to december as the soil cools and loses moisture. Late sowing aids in crop preservation in locations where these diseases are severe. Wheat stem rust harms late-planted wheat more severely than early-planted wheat. because ear development and illness initiation occur at the same period. It is recommended to sow between January and April or October and December to avoid the finger millet neck blast attack. October sowings of peas and chickpeas typically experience severe root rot and wilt (a combination of Fusarium, Rhizoctonia, and Sclerotium). The infections are less severe or virtually nonexistent when these crops are sown later. The *Aphis craccivora* transmits the groundnut rosette.
3. **Selection of seeds and seed materials:** The main source of infection is seeds and seed materials. The lack of an early inoculum in seeds undoubtedly aids in postponing or reducing the disease's incidence. It's a method of prevention. Seeds can spread diseases such as foot rot of papaya, brown spot of rice, loose smut of wheat, short smut of sorghum, bacterial blight of rice, bacterial blight of cotton, and leaf crinkle of blackgram. Tubers, setts, rhizomes, corns, grafts, and budwoods are the transfer points for virus diseases and black ring of potatoes, foot rot of ginger, foot rot of betelvine, Panama disease of bananas, red rot of sugarcane, cassava mosaic, bunchy top, and virus diseases of fruit trees. A unique technique for obtaining disease-free potato seed is called "tuber indexing." Nurseries and seed merchants selling potato seed tubers frequently engage in this technique. It is advised to use seeds rather than rhizomes or suckers to control the cardamom disease known as "katte."
4. **Provision of drainage facilities:** Proper leveling of the field prior to sowing or planting is particularly important since fungi such as Pythium, Phytophthora, Rhizoctonia etc. are favoured by water stagnation in different regions of the field. In order to prevent rice sheath blight, drainage must be further improved. Before planting, drainage channels must be provided for orchard crops like citrus, jackfruits, mango, etc. in the garden. Planting seedlings in raised beds is one strategy used to reduce damping-off diseases

of vegetables and other crops. Ginger foot rot can also be prevented by using the nursery's raised bed method.

5. **Depth of sowing:** Seed-transmitted diseases are largely influenced by the depth of sowing. Wheat plants are shielded from flag smut caused by *Urocystis tritici* by shallow planting in moist soils. Deep planting could delay the emergence of seedlings, making them more susceptible to damping off before they emerge. Early emergence causes tissues to lignify sooner, making them more resilient to infections carried by the soil.
6. **Spacing:** Increased spacing inevitably modifies the microclimate beneath the crop canopy, potentially creating an ideal setting for the emergence of diseases. In crowded crops, cotton, bol rot is a common occurrence. Because the crop canopy is more humid, closer spacing encourages the spread of numerous airborne diseases. Blister blight of tea and early and late groundnut blight are more common in dense canopies. Black rot of cabbage spreads early and is more closely spaced. Certain systemic disorders may be decreased by crowded stands. If there is less fungal inoculum in the soil, cotton wilt caused by *Verticillium alboatrum* will occur in crops that are closely planted. Similar to this, rice planted farther apart inhibits rice tungro virus transmission, especially in cases where the vector population is lower. Wider spacing and avoiding shadow both lower the likelihood of tobacco powdery mildew. In crops that are closer together, late blight on potatoes and downy mildew on grapevines spread quickly. Tomato leaf curl virus, which is spread by *Bemisia tabaci,* is less severe in densely planted areas than it is in widely separated areas.
7. **Altering the soil pH:** The environment's changed pH creates a barrier that keeps the infection out. Less than 5.2 pH is extremely low, which is detrimental to the common scab bacterium on potatoes.However, when the pH of the soil is 7.0 or higher, the club root disease that affects cabbage cannot survive and spread.
8. **Crop rotation:** The earliest and least expensive technique used in agriculture to remove specific diseases from contaminated soil is crop rotation. Growing the same crop in the same field year after year allows harmful organisms to persist in the soil through a process known as monoculturing. The crop's soil-borne diseases readily proliferate in the soil and grow in number. The soil eventually becomes so severely infected that the specific crop cannot be grown there. If a crop is grown continually in a field, it has been observed that virus infections of crop plants and their vectors become more prevalent after each harvest. However, it is anticipated that the pathogen will starve to death in the absence of sustenance when immune, resistant, or non-host crops are cultivated for a set amount of time following a susceptible crop in the field, leading to a decrease in the population of such pathogens. For instance, crop rotation using paddy or sugarcane is useful in preventing

Panama wilt (*Fusarium oxysporum* f.sp. *cubense*) in bananas, while crop rotation using paddy or green manures is useful in preventing red rot in sugarcane (*Colletotrichum falcatum*). It is advised to rotate cereal crops such as finger millet, fox-tail millet, or pearl millet to manage Macrophomina root rot in pulse crops. Crop rotation mostly affects the next crop and is essentially a preventive approach.

9. **Mixed cropping:** Growing two or more crops concurrently on the same plot of land is known as mixed cropping. Materially, mixed cropping aids in the prevention of several diseases. It is possible to alter the soil environment to one that is unfavorable to the pathogen by carefully choosing the crops for the mixture. One way to prevent cotton root rot is to cultivate cotton with moths.Moreover, a mixed crop's fewer host plants mean a smaller susceptibility region for airborne foliar pathogens. As a result, there are fewer primary infections and fewer secondary inoculums produced to aid in the disease's transmission. This slows the rate at which diseases are controlled. The Phyllosticta blight of pulse crops has been effectively combated by cultivating pulses in combination with cereals such as pearl millet and sorghum.
10. **Intercropping:** Another tool for the management of some soil-borne illnesses is intercropping. It is important to choose intercrops carefully so that they don't share any pathogens.For instance, sorghum intercropping decreased the prevalence of wilt (*F. udum*) in pigeonpea fields. The viruses' ability to travel through soil may be hindered by the roots of non-host plants. Toxic compounds that could inhibit the growth of diseases affecting the main crop may be released from their roots. The pigeonpea wilt fungus, *F. udum*, is poisonous to the hydrocyanic acid (HCN) found in the root exudates of sorghum.
11. **Barrier cropping:** To keep virus vectors away from a crop that is shorter in height, taller crops might be planted. Yellow mosaic incidence and vector population can be effectively decreased by barrier cropping, which involves planting three rows of maize, sorghum, or pearl millet around the primary crop, black or green gram. Growing three rows of kale or barley in cauliflower seed beds as barrier crops and undersown beet steckling against beet yellows and cauliflower mosaic disease, respectively, is another excellent example. It is believed that the approaching aphids land on the kale or barley and investigate for a short while, which causes the non-persistently transmitted virus to be shed from the plants. The other barrier crops that are being examined for these crops are maize or sunflower.
12. **Decoy crop and trap crop:** Non-host crops called decoy or hostile crops are cultivated with the intention of wasting the infection potential of soilborne diseases. This is accomplished by reawakening dormant fungal propagules, parasitic plant seeds, etc. when the host is not present. Trap crops are the

pathogen's host crops; they are planted to draw infections but are meant to be harvested or eliminated before the viruses have a chance to finish their life cycle. To lessen sorghum's downy mildew, fodder sorghum can be used as a trap crop.

13. **Mulching:** Plant diseases are frequently reduced by mulching or coating topsoil with organic wastes. In the field, mulches with non-host origin should be utilized. These mulches are known to release chemicals that limit the development of parasites and nematode predators in the underlying soil. It has been discovered that applying mulch to the soil surrounding a crop plant, or reflective surfaces, effectively inhibits the activity of aphid vectors. Mulch made of aluminum strips or grey or white plastic sheets has been proven to be effective in protecting crops against watermelon mosaic virus in the Imperial Valley of California, as well as against CMV and PVY in red peppers and summer squash in Israel. In Israel, tomato crops affected by the white fly-transmitted tomato yellow leaf curl virus have been effectively managed through the application of straw mulches. It's thought that the straw's color draws in whiteflies, who are then destroyed by the heat that reflects off of it. The drawback of straw mulches is that their yellow color ultimately fades; however, if straw is substituted with yellow polythene sheets, longer-lasting control might be achieved.

14. **Irrigation water management:** The goal of irrigation is to saturate the soil around a crop so that roots may readily absorb nutrients and water. If too much water is given to the soil, it could directly impact pathogen activity or indirectly impact disease incidence by affecting the host. Irrigating the field lowers the temperature of the soil, eliminates stress, and suppresses disease. Excessive irrigation prolongs the juvenile stage of plants, making them more vulnerable to fungal diseases like Pythium. Therefore, it is advised to provide irrigation water on a frequent basis but in small amounts to minimize the likelihood of damping off in nurseries.

    When there is an overabundance of water present, the roots' ability to breathe is hindered, and a large number of soluble salts build up in hazardous concentrations around the roots and base of the stem. This makes the roots more susceptible to illness. Guttation is increased by irrigation. Many diseases, like Xanthomonas campestris on Brassica spp. and Helminthosporium spp. on cereals, use guttation droplets on leaves as a medium for growth and penetration. The pathogens that require moist soil to (i) activate their resting structures and (ii) facilitate the propagulation of these propagules are the ones that directly benefit from an abundance of water. Thus, bacterial cells and Pythiaceous fungal zoospores spread readily when there is an abundance of free water present. Irrigation should therefore be avoided throughout the plant stage when these pathogens can harm the crop. Drainage and irrigation water transport soil-found chlamydospores, oospores, smut spores, and mycelium from one field to another. The primary means of transmission for diseases such stem rot, sheath blight, bacterial blight, damping off of vegetables, and Macrophomina root rots of numerous crops are irrigation and drainage water. Therefore, care should be taken to ensure that irrigation water from a diseased field is not used to irrigate a healthy field.

15. **Management of plant nutrients:** When given in excess, plant nutrients generally have the potential to either boost or decrease a plant's resistance to disease. Numerous diseases are more common when nitrogenous fertilizers are applied more frequently. Heavy nitrogenous fertilizer applications cause crops to grow robustly with succulent tissue and foliage, but they also make crops very vulnerable to diseases including tobacco mosaic, blast, rust powdery mildew, and several bacterial infections. The ammoniacal form of nitrogen is more conducive to some disorders than the nitrate form. Ammoniacal nitrogen generally favors wilts (Fusarium sp.) and root rots (Rhizoctonia spp.), whereas nitrate nitrogen favors verticillium wilts and root rots caused by Pythium spp. Phosphatic fertilizers applied frequently can postpone the beginning of barley's take-all disease and reduce its severity (Gaeiimannomyces graminis). Applying potassium helps lower the frequency of several crop diseases, most likely by boosting plant production of phenolics. Groundnuts become resistant to Macrophomina phaseolina-caused root rot when potash is applied. Applying calcium to bean roots reduces the lesions caused by R. solani. It results from the production of calcium pectate, which the enzyme polygalacturanase (PG) may act on less readily than pectic acid. Ferric chloride manages rice brown spot, manganese lessens potato late blight, and silicon treatment lessens rice blast.
16. **Fallowing:** By killing the host, letting the pathogen to starve aids in the inoculum's decrease. This approach controls diseases on various crop plants, such as Macrophomina root rot. Wet fallowing encourages the germination of pathogenic propagules in or on the soil, making them vulnerable to saprophyte attack. Verticillium dahliae and Sclerotium rolfsii are two examples. These fungi's sclerotia or microsclerotia become active when the host's root exudates are absent. When the soil is alternately wet and dry, they germinate swiftly. Wet fallowing works well to lower Pythium myriotylum populations when they are not very high. Alternaria solani's saprophytic survival on crop debris is decreased by wet fallowing.

B. Field and plant sanitation

Sanitation of fields and plants is a crucial component of disease prevention through cultural practices. Field plants carry an inoculum that can proliferate on the plant or in the soil over time, potentially negating or lessening the impact of control measures. When the crop is once again cultivated in the field, many viruses that have overwintered or oversummered on plant detritus during the off-seasons come to life. Therefore, it is best to remove plants that are harboring pathogens or plant detritus that is delivering inoculum into the soil as soon as feasible.It has been observed that in the majority of soil-borne illnesses, such as wilt and root rot, the fungus will continue to proliferate as long as there are dead roots, other roots, and other affected areas in the soil. The number of pathogens in the soil quickly decreases after such damaged plant materials are removed. In this way, the following diseases are decreased: bacterial blight of cotton, white rust of crucifers, black spot of rose, powdery mildew of pea and cereals, root rot of beans, downy mildew of pearl millet, sorghum, maize, and peas, Verticillium wilt of cotton, pigeonpea and banana, and Verticillium wilt of cotton. The rice blast and brown spot fungi, the linseed rust fungus (Melampsora lini), and the fungus that causes early potato blight can persist in some regions through dormant stages in infected crop debris. Burning agricultural debris as soon as it is harvested helps to minimize the number of inocula that survive through the debris. One of the foundational principles of cultural control is to prevent the spread of infection from one field to another by people, equipment, or water. Anything that conveys dirt is questionable when it comes to soil-borne diseases; this includes boots, wheels, and water that comes from nearby fields or that travels through drainage ditches from farther fields. When it comes to sap-borne viruses, workers' hands and wheels need to be thoroughly cleaned. while they move from one field to the next. Any garments can potentially harbor this type of infection. It is important to schedule the work so that workers do not switch from older to younger fields on the same day.

C. Eradication

Eradication is the process of getting rid of a pathogen once it has taken hold in the area where the host is developing. The following are crucial techniques used to stop the disease from spreading:

i. Elimination of alternative hosts
ii. Collateral and self-sown overwintering hosts eliminated
iii. Removing the impacted trees or plants
iv. Removal of pathogens from contaminated plant portions through surgery and
v. Elimination of removed plant materials, trash, etc., by various cultural customs,
vi. Eradication of alternate hosts Eliminating alternative hosts contributes to the prevention and control of the black rust pathogen-induced illness in wheat. *Puccinia graminis tritici,* the pathogen that causes black stem rust on wheat, has an alternative host in barberry bushes, where it persists during the winter. Eliminating barberries has two advantages: it prevented the establishment of new physiologic races of the pathogens and eliminated the early spring main inoculum.

vii. Collateral and self-sown overwintering hosts eliminated Many cultivated plant species that are considered weeds or wild counterparts serve as volunteer plants or collateral hosts for economic crops, serving as reservoirs for annual crop diseases. The pathogen uses reservoir hosts to carry on the infection chain. These hosts produce the primary inoculum, which is then transferred to the hosts of cultivated crops. Destroying these pathogen's wild or unprofitable host plants eliminates the principal inoculum sources and lowers the likelihood that the illness will spread to crop hosts. The infection chain and the pathogen's life cycle are broken when these hosts are destroyed.

viii. Removing the impacted trees or plants For some serious plant diseases, it is imperative to remove both the pathogen and the host from the affected area. A successful eradication program is exemplified by the citrus canker (*Xanthomonas axonopodis* pv. *citri*). In 1913, this disease was initially discovered in citrus plants in Florida. In 1915, an eradication campaign was launched. The diseased trees were chopped down and burned, and all citrus nurseries and orchards were examined. Up to 1927, the eradication effort was carried out, and the region was free of citrus canker. Diseased trees were removed and destroyed in order to control peach yellows and peach rosette.

ix. Eradication of pathogens from infected plant parts by surgery: In some circumstances, removal of the afflicted plant sections (tree surgery) is also carried out, which lessens the supply of the primary inoculum. During the winter, the fire blight bacterium (Erwinia amylovora) that causes lesions on apple and pear trees is eliminated. This lessens the quantity of inoculum that can spread to neighboring branches and trees in addition to stopping further spread in the afflicted trees. In addition, tree surgery is performed on coconut trees afflicted with *Dendrophthoe* spp. on citrus, *Phytophthora palmivora* on palm bud rot, *Ceratocystis paradoxa* on stem bleeding disease, and *P. arecae* on koleroga of arecanut.

Model Practice Questions

A. Objective Questions

a. Multiple choice Questions

1. Manipulation /or adjustment of crop production techniques for management of plant disease

   (a) Cultural management (b) Physical management
   (c) Biological management (d) Chemical management

2. The disease controls affected by cultural practices are mostly

   (a) Preventive (b) Curative
   (c) Systemic (d) All

3. Cultural practices that help to reduce disease incidence by avoiding the contact between the pathogen inoculum and the plant is

(a) Date of sowing	(b) Deep ploughing during summer
(c) Depth of sowing	(d) All these

4. Once the pathogen has established in an area, steps taken to destroy or remove is known as

(a) Exclusion	(b) Eradication
(c) Protection	(d) Therapy

5. Roughing is employed to control

(a) Loose smut of wheat	(b) Loose and covered smut of barley
(c) Wilt of arhar	(d) ll above

6.Tuber indexing‘ is a special method to obtain disease free seed materials in potato

(a) Potato	(b) Tomato
(c) Brinjal	(d) Bhindi

7. pH unfavorable to common scab pathogen of potato

(a) <5.2	(b) 6.0
(c) 7.0	(d) >8.0

8. pH unfavorable to club root pathogen of cabbage

(a) 5.5	(b) 6.0
(c) 7.0	(d) 4.0

9. The oldest and cheapest method adopted in agriculture for eradication of certain types of pathogens from infested soil

(a) Crop rotation	(b) Mixed cropping
(c) Companion cropping	(d) All

10. Intercropping of sorghum in pigeonpea field reduced the incidence of

(a) Wilt	(b) Phytophthora blight
(c) Sterility mosaic	(d) Collar rot

11. Root exudates of sorghum is toxic to *F. udum*, the pigeonpea wilt fungus due to secretions of chemical

(a) HCN	(b) HCl
(c) NH_4	(d) CO_2

12. The non-host crops sown with the purpose of making soilborne pathogens waste their infection potential

(a) Decoy crop	(b) Trap crop
(c) Companion crop	(d) Barrier crop

13. Silicon application reduces which disease of rice.
    (a) Blast (b) Bacterial blight
    (c) Sheath blight (d) Sheath rot
14. Removal of alternate host barberry helps to prevent and check the spread of wheat disease
    (a) Black rust (b) Brown rust
    (c) Yellow rust (d) Flag smut
15. Silicon application reduces which disease of rice.
    (a) Blast (b) Bacterial blight
    (c) Sheath blight (d) Sheath rot

Q. No.	Answer	Q. No.	Answer
1	Cultural management	9	(a) Crop rotation
2	Preventive	10	(a) Wilt
3	(d) All these	11	(a) HCN
4	(b) Eradication	12	(a) Decoy crop
5	(a) Loose smut of wheat	13	(a) Blast
6	(a) Potato	14	(a) Black rust
7	(a) <5.2	15	(a) Blast
8	(c) 7.0	16	-

b.True/False

1) Cultural management offer an opportunity to alter the environment,the condition of the host, behavior of the causal agent to achieve economic management of disease.
2) The main purpose of adjusting of cultural practices is to *seed healthy seeds in healthy soil and to obtain a healthy crop stand.*
3) Groundnut blight is controlled by ploughing the soil to a depth of 20 cm.
4) Rapeseed sown in mid to late August is more liable to attack by leaf spot (*Alternaria brassicae*) than late-sown crops.
5) Peas and chickpea sown in October usually suffer heavily from root rot and wilt.
6) Foot rot of ginger is also controlled by following the raised bed system of nursery.
7) Deep planting may cause delay in the emergence of seedlings, which may be vulnerable to pre-emergence damping off.
8) Early and late blight of groundnut are more in dense canopy.
9) Crop rotation is essentially a preventive measure and has its effect mainly on the succeeding crop.
10) Crop rotation with sugarcane or paddy is effective in the control of Panama wilt' of banana.

11) Fodder sorghum can be raised as a trap crop to reduce downy mildew of sorghum.
12) Aluminium strips are used as mulch to protect red peppers against CMV and PVY.
13) Potassium application reduces the disease incidence in many crop diseases probably by increasing phenolics synthesis in plants.
14) Wet fallowing reduces saprophytic survival of *Alternaria solani* on crop debris.
15) Field and plant sanitation is an important method of disease control through cultural practices.

Answer

Q. No.	Answer	Q. No.	Answer
1	True	9	True
2	True	10	True
3	True	11	True
4	True	12	True
5	False	13	True
6	True	14	True
7	True	15	True
8	True	16	-

B.Descriptive Questions

a.Short Answers

1. Soil Solarization.
2. Write down the basic principles of cultural practices for disease control.
3. Explain the role of wet fallowing in disease management with suitable examples.
4. Write down the role of mixed cropping ind disease management with suitable eaxaples.
5. Write down the role of field and plant sanitation in disease management.

b.Long Answers

1. Give a detail account on use of cultural practices for plant disease management.
2. Describes the role of field and plant sanitation in plant disease management.
3. Explain the following: i. Field and plant sanitation ii. Organic amendment of soil
4. How plant diseases are influenced by macro and micro nutrients.
5. Name five plant diseases which can be controlled by field and plant sanitation.

6. Discuss in details the plant disease management by modification of agricultural practices.
7. Which type of disease can effectively managmed by crop rotation ?
8. What is the advantage of deep summer ploughing in plant disease management.
9. Descrive the role of mixed cropping and fallowing in plant disease management.
10. What is cultural management? Write down the procedures for disease management through cultural practices.

18

Physical Control

Definition

- Physical control is the use of physical measures to manage plant diseases.
- Also known as thermotherapy or physical therapy
- Its main purpose is curative.

Principles

The pathogens present in seed, soil, or planting material are the fundamental ideas behind physical control and thermotherapy.

- Pathogens are removed or rendered inactive at temperatures that do not pose a threat to the host's tissues.
- Heat does not immobilize the pathogen; rather, it causes inactivation. Temperature affects the rate of pathogen inactivation; the greater the temperature, the faster the inactivation.

Methods

To lessen or completely eradicate primary inoculums that could be found in seed, soil, or planting material, physical techniques are used.

i. Hot water treatment (HWT)
ii. Hot air treatment
iii. Aerated steam treatment (AST)
iv. Moist hot air treatment
v. Sun heat therapy (SHT)

1. Hot water treatment (HWT)

- A common method for reducing seed-borne pathogens, particularly bacteria and viruses, is hot water treatment.
- For example.

Table: Control of seed borne pathogens through Hot –water treatment

Crop	Disease	Causal organism	Seed Treatment
Rice	White tip	*Aphlenchoides besseyi*	51-53°C for 15 min after dipping for 1 d in cool water.
Pearl millet	Downy mildew	*Sclerospora graminicola*	55°C for 10 min
Cauliflower & Cabbage	Black rot	*X. campestris pv. campestris*	50°C for 20 min

- The primary disadvantage of the hot water treatment is that if the treatment period is longer than intended, the seeds may be destroyed or lose their ability to germinate.

ii. Hot air treatment (HAT)

- Although hot air treatment is easier to use and less harmful to seeds than hot water treatment, it is less effective.
- It is used to treat soft, succulent canes and has been utilized to combat a number of sugarcane diseases.
- An 8-hour hot air treatment at 54°C effectively eradicates the RSD pathogen without compromising bud germination.
- In a similar vein, heated air at 54°C for eight hours has also been used to control the grassy shoot disease of sugarcane.

iii. Steam and aerated steam therapy (AST)

- When it comes to preventing seed-borne diseases, aerated steam is safer and more effective than hot air.
- Since water vapor has a heat capacity that is around half that of water and 2.5 times that of air, the amount of time and temperature needed for air may differ from that for hot water.
- This method's benefits include less seed loss during germination, simpler temperature control, and no harm to the legumes' seed coat.
- To get rid of the mosaic virus, sugarcane setts are additionally subjected to aerated steam for three hours at 50°C.

iv. Moist hot air treatment (MHAT)

- The grassy shoot disease in sugarcane can be successfully eradicated with this technique.
- The setts undergo three different exposures: eight hours of hot air at 54°C, one hour of aerated steam at 50 °C, and two hours of moist hot air at 54°C.

v. Solar heat treatment (SHT)

- Soil-borne and seed-borne diseases can both be effectively controlled by solar heat treatment.

- It was developed in India to get rid of the pathogen that causes loose smut in wheat.
- In India, Luthra developed a technique in 1953 to get rid of the long-lasting Ustilago nuda infection. The process is commonly referred to as solar energy treatment or solar heat.
- Using this approach, the seeds are spread out and dried in the hot sun for four hours in the afternoon after being soaked in cold water for four hours on a sunny summer day. Following another treatment with 2g/kg of carboxin or carbendazin, the seeds are kept.
- When handling a lot of seed lots, this procedure is quite helpful.

Soil Solarization

- Described as a hydrothermal method of treating soil to control soil-borne diseases.
- This management strategy preserves solar energy for around 40 days by using a transparent polythene sheet to raise soil temperature (by 10–15 °C above room temperature, which is sufficient to kill the majority of soil-borne diseases.

Results of solarization

- An elevation in soil temperature.
- Better physical and chemical properties of the soil.
- Pest management.
- Promotion of advantageous soil organisms.
- Enhanced growth of plants.
- Soil solarization is a successful way to control fungi diseases such damping off, root and stem rots, wilt and blight diseases, and certain nematodes.

Refrigeration

- The low temperatures, which are at or just beyond the freezing point, inhibit the growth and activities of all these pathogens, which are responsible for a range of post-harvest diseases in fruits and vegetables.
- In order to prevent post-harvest diseases, the majority of perishable fruits and vegetables should be shipped and kept in refrigerated cars and establishments.

Model Practice Questions

A. Objective Questions

a. Multiple choice Questions

1. Management of plant diseases using physical means is called

(a) Cultural management (b) Physical management

(c) Biological management (d) Chemical management

2. The disease controls affected by physical practices are mostly
   (a) Preventive (b) Curative
   (c) Systemic (d) All
3. The principles involved in physical control or thermotherapy is that the pathogens present in seed, soil, or planting material.
   (a) Are inactivated or eliminated at temperatures nonlethal for the host tissues.
   (b) Heat causes inactivation and not immobilization of the pathogen by heat.
   (c) The rate at which the pathogen is inactivated is determined by temperature, the higher the temperature, the faster is the inactivation.
   (d) All
4. Physical methods are employed for reduction or elimination of primary inoculums that may be present in
   (a) Seed (b) Soil
   (c) Planting material (d) All
5. Hot air treatment at which temperature effectively eliminates RSD pathogen without impairing the germination of buds.
   (a) 54°C for 8h (b) 50°C for 8h
   (c) 45°C for 8h (d) 60°C for 8h
6. Grassy shoot disease of sugarcane has also been controlled by hot air at .
   (a) 54°C for 8h (b) 50°C for 8h
   (c) 45°C for 8h (d) 60°C for 8h
7. Downy mildew disease of pearl millet has also been controlled by seed treatment with hot water at.
   (a) 55°C for 10 min. (b) 50°C for 10 min.
   (c) 45°C for 10 min. (d) 60°C for 10 min.
8. Black rot disease of cabbage has also been controlled by seed treatment with hot air at .
   (a) 54°C for 20min. (b) 50°C for 20 min.
   (c) 45°C for 20min. (d) 60°C for 20min
9. Sugarcane setts are also exposed to aerated steam at what temperature to eliminate mosaic virus.
   (a) 54°C for 3h (b) 50°C for 3h
   (c) 45°C for 3h (d) 60°C for 3h
10. Solar heat treatment devised in India to eliminate the pathogen of which disease of wheat
   (a) Black rust (b) Brown rust
   (c) Yellow rust (d) Loose smut

11. Who devised a method in India to eliminate the deep seated infection of *Ustilago nuda*
    (a) Blakeslee in 1953 (b) Luthra in 1953
    (c) Butler in 1955 (d) Vander plank 1955
12. Soil solarization results in
    (a) Increased soil temperature (b) Improved soil physical and chemical features
    (c) Control of pests (d) All

Q. No.	Answer	Q. No.	Answer
1	Physical management	7	(a) 55°C for 10min
2	Curative	8	(b) 50°C for 20 min
3	(d) All	9	(b) 50°C for 3h
4	(d) All	10	(d) Loose smut
5	(a) 54°C for 8h	11	(b) Luthra in 1953
6	(a) 54°C for 8h	12	(d) All

b. True/False

1. Hot water treatment is widely used for the control of seed borne pathogens, especially bacteria and viruses.
2. The main drawback in the hot water treatment is that the seeds may be killed or loose its germ inability, if the period of treatment exceeds the specified time.
3. Hot air treatment is less effective than hot water treatment but less injurious to seed and easy to operate.
4. Hot air treatment has been used against several diseases of sugarcane and is employed for treating canes which are soft and succulent.
5. The use of aerated steam is less effective than hot air and safer than hot water in controlling seed borne infections.
6. Solar heat treatment is effective in controlling both seed borne and soil borne diseases.
7. Soil solarization is defined as hydrothermal treatment of soil for management of soil borne diseases.
8. Fungal diseases such as damping off, root and stem rots, wilt and blights diseases and some nematodes have been successfully managed by soil solarization
9. Hot air treatment is more effective than hot water treatment for disease managment but less injurious to seed and easy to operate.
10. Physical control is also called physical therapy or thermotherapy

Answer

Q. No.	Answer	Q. No.	Answer
1	True	6	True
2	True	7	True
3	True	8	True
4	True	9	False
5	False	10	True

2. Descriptive Questions

a. Short answer type questions

1. What is solar heat treatment
2. Why hot air treatment is less effective than hot water treatment.
3. What is soil solarization
4. Write down an advantages & disadvantages of soil solarization.
5. Write down the role of refrigeration in management of post harvest diseases of fruits and vegetables.

b. Descriptive questions

1. Describe the role of regrigeration and radiation in plant disease management.
2. What is physical control ? Write down the different physical methods used plant disease control.
3. How can solar energy be used for the treatment of seeds? In which disease has it been used? Is it more coenveinet than chemical treatment.
4. Distinguish between hot water treatment and moist heat air treatment.Give atleast four examples of plant disease controlled by hot water treatment.
5. What is thermotheary ? What is the scientific principle involved in thermotherapy.
6. What are the objectives of the physical methods of plant disease management? How are they helpful in minimizing disease incidence?
7. Who developed solar heat treatment device? Name the disease for which the device was developed.
8. Write short notes on i.Principle of physical plant disease control ii.Hot water treatment iii. Hot air treatment iv.Steam and aerated steam therapy
9. Explain about moist hot air treatment and solar heat treatment.
10. Why aerated steam is more effective than hot air and safer than hot water in controlling seed borne infections?
11. What is soil solarization? Describe the method, advantage and disadvantage of soil solarization.

19

Legislative Control

1. Biosecurity

A comprehensive and integrated strategy, biosecurity includes frameworks for policy analysis and risk management for the environment, plant and animal health, and food safety.

Fundamental Components

- Introduction of foreign diseases and pests
- Emergence of aggressive biotypes, races, or strains of economically significant native diseases or pests
- Intentional mishandling of diseases or pests
- Biosafety concerns associated with LMOs

2. Plant Quarantine

The Latin word quarantum, which means forty, is where the word "quarantine" originates. It alludes to the 40-day detention period that ships from Middle Ages nations suffering from cholera and the bubonic plague endured. A similar quarantine was initially put in place in Venice in 1374. To prevent agriculture and the environment from suffering avoidable harm from dangerous organisms, implement plant quarantine, limit the entry of plants, plant products, soil, cultures of living organisms, packing materials, and commodities, as well as their containers and modes of transportation.

3. Basic Principles of Plant Quarantine

The main goals of plant quarantine are to prevent the introduction and spread of insects and plant diseases that could be harmful to imported germplasm.

4. Goal of Plant Quarantine

Plant quarantine's objectives are

i. To stop the infection from entering through artificial channels.
ii. If entry is successful, to stop infection.
iii. In the case that infection is successful, to stop pathogens from spreading through natural and man-made channels.
iv. If established, to reduce or postpone the spread.

Pests Introduced in India

Pests	Sicentific Name	Year of Intorudction
Lantana weed	*Lantana camara*	1809
Coffee rust	*Hemilia vastatrix*	1879
Late blight of potato	*Phytophthora infestans*	1883
Wooly aphid of apple	*Eriosoma lanigerum*	1889
Grape downy mildew	*Plasmopara viticola*	1910
San Jose Scale	*Quadrapidiotus perniciosus*	1911
Water hyacinth	*Eichhornia crassies*	1914
Potato tuber moth	*Phthorimaea operculella*	1937
Banana bunchy top virus		1940
Wart of potato	*Synchytrium endobioticum*	1953
Congress grass	*Parthenium hysterophorus*	1956
Golden nematodes	*Globodera rostochiensis & G pallida*	1961
Apple scab	*Venturia inaequalis*	1978
Downy mildew of sunflower	*Plasmopara halstedii*	1984
Peanut stripe virus		1987
Sunflower necrosis	Sunflower necrosis virus	1997
Cotton mealy bug	*Phenococcus solenopsis*	2006
Papaya mealy bug	*Paracoccus marginatus*	2009
Tomato leaf minor	*Tuta abosoluta Meyrick*	2014
Root knot nematode	*Meloidogyne enterolobii*	2016
White rust of chrysanthemum	*Puccinia horiana*	2016
Fall armyworm in maize	*Spodoptera frugiperda*	2018

5. Prerequisites of Plant Quarantine Regulations

Rules pertaining to plant quarantines contain specific requirements, such as

- The only diseases and pests that need to be taken into account are those that endanger important crops or forests.
- It should be created with pest control in mind, not to obstruct trade or the achievement of other goals.
- It should only be used in accordance with the law.
- Quarantine personnel ought to possess the necessary training and expertise.
- International cooperation between the public and professionals is required for quarantine rules to be implemented effectively.

6. Component of Plant Quarantine

Programs for plant protection and quarantine (PPQ) typically consist of three parts.

- Excluding pests and diseases that require quarantine and have a significant economic impact.

- Suppression and complete elimination of newly imported exotic diseases and pests via both natural and artificial routes.
- Support for plant product exporters to comply with import nations' quarantine regulations.

7. Pest Risk Analysis

The likelihood that a pathogen or pest of quarantine significance will infiltrate through a man-made pathway is known as pest risk. Risk is frequently described as low, medium, or high.

- Low risk refers to the minimal likelihood of a pathogens or pest entering.
- High risk denotes a high likelihood of disease or pest entry.
- An acceptable risk level is one in which the advantages of taking a risk are sufficient to make it worthwhile to take the risk in the first place, provided that safety precautions are taken.

8. Plant quarantine methods

The of plant quarantine activities are:

i. Total Embargo

- To totally ban some plants and plant products from entering a nation where extremely harmful pests or diseases are present.
- In actuality, though, it is challenging to accomplish due to the increasing flow of varied genetic material between nations.

ii. Partial Embargo

- Be applicable when a disease or pest that requires quarantine in an importing country is known to exclusively exist in a certain region of the exporting country.
- There is an internal plant quarantine service that runs efficiently and can keep the pest or disease contained in this area.

ii. Examination, handling, and certification at the place of origin

- It entails treating and inspecting a certain commodity when it comes from that nation.
- And certification in line with the importing nation's quarantine regulations.

iii. Examination at the entrance

- As soon as plant material arrives at the designated port of entry, it must be inspected.
- And, if required, be treated prior to the same-related procedure.

iv. Post-entry plant quarantine facilities are used

- It entails cultivating propagating material from invasive plants in limited or isolated environments.

9. Plant Quarantine in India

Sl .No	National legislations
1	Destructive Insects and Pests (DIP) Act of 1914
2	Plant Quarantine (Regulations of Import into India) Order 2003
3	Livestock Importation Act1898
4	The Environmental (Protection) Act 1986
5	Biological diversity Act 2002

- The Central Government's Destructive Insects and Pests (DIP) Act of 1914, which controls the entry of foreign diseses and pests into the nation and their migration between states and union territories. The Act also gives state governments the authority to create appropriate regulations and publish notifications for the interstate transportation of plants and plant materials. Plant quarantine regulations are those guidelines.
- State-specific Agricultural Pests and Diseases Acts, which work to control or stop the spread of pests and diseases inside of a state or union territory.

10. Pesticide Management Bill, 2020

On February 12, 2020, the Union Cabinet passed the Pesticides Management Bill, 2020. It takes the place of the Insecticide Act of 1968, which the Minister of Information and Broadcasting, Prakash Javadekar, stated was "age-old and needs immediate rewriting." The Pesticides Management Bill, 2017 seeks to primarily

- Regulate the following areas: import, manufacture, export, storage, sale, transit, distribution, and pesticide quality and use.
- Ensure that high-quality pesticides are available.
- Only permit its use after determining its safety and effectiveness.
- Reduce the amount of pesticide residues that contaminate agricultural products.
- Raise user understanding of the prudent and safe usage of pesticides.

To take the required steps to prevent pesticide use from endangering people, animals, or the environment by continuing, restricting, or outright banning its use.

According to the bill, before any person, organization, or agency produces, synthesizes, imports, exports, uses them for commercial purposes, stores, distributes, or transports them, or packs them, they must register the pesticides at the Plant Quarantine and Pesticides Management Center. The center will only do this after the necessary requirements have been satisfied. A person, organization, or agency's certificate may be suspended or revoked at any moment if they violate the terms and conditions stated in it or if any registered pesticide seems to have a negative impact on the environment, human health, or both.

Highlights of the Bill

- The draft measure offers state governments more authority to address the problem and take enforcement action against the sale of counterfeit, inferior, and mislabeled pesticides. It also suggests raising the penalties for such sales.

- The proposal includes provisions allowing for the two-year temporary registration of new pesticides in India in the event of "national exigency."
- The maximum penalty for a violation (such selling fake or illegal pesticides) is INR 50 lakh in addition to up to five years in jail. Prior to this, the penalty could only be up to three years in jail and INR 2,000.
- The draft stipulates that anyone who "uses" a pesticide in violation of this Act's restrictions will face consequences.
- The proposed legislation calls for compensating impacted farmers or users in accordance with the 1986 Consumer Protection Act.
- The measure includes comprehensive provisions for novel molecular registration.
- The requirements for novel molecule registration and licensing have also been tightened.
- It has provisions for compensating the farmers as well as a wider category of offenses.
- The plan mandates that the state government notify the center of every poisoning case on a quarterly basis.
- Unlike the present law, which permits bans for up to two months and forbids the use of pesticides that are not registered with the center, states may impose six-month bans on chemical pesticides.

11. Agencies involved in plant quarantine

The quarantine guidelines established by the DIP Act are implemented by the Plant Protection Network of the Directorate of Plant Protection, Quarantine & Storage (DPPQ & S) of India. For commercial purposes, this agency manages the bulk import and export of seeds and planting materials.

(a) The Ministry of Agriculture & Farmers Welfare, Government of India, New Delhi is home to the National Plant Protection Organization (NPPO) of India.

(b) The Directorate of Plant Protection, Quarantine & Storage (DPPQ&S), with its headquarters located in Faridabad.

(c) Regional Plant Quarantine Station - 07

(d) Plant Quarantine stations : 59

(e) Central IPM Centers: 12

(f) Central Insecticide lab-01

(g) Regional Pesticide Testing Laboratories 02

(h) The Seaports—34

(i) Air ports: 12

(j) Land custom stations-14

(k) Foreign post office-11

12. Domestic Quarantine

The solitary object of enforcing domestic quarantine is to prevent the spread of these diseases from infected to non-infected areas. Presently, domestic plant quarantine exists in following diseases and pests

Pest/Diseases	Host plant material	Restricted States	
		From	To
Wart (*Synchytrium endobioticum*)	Potato	Darjelling West Bengal	Any other state or place in India
Bunchy top (virus)of banana	Banana planting material	Assam, Kerala, Orissa, TamilNadu &West Bengal	Any other state or place in India &Uninon Territory
Mosaic (virus) of banana	Banana plants & planting material	Maharashtra &Gujarat	Any other state or place in India &Uninon Territory
Apple scab (*Venturia inaequalis*)	Apple planting materials	J&K and Himachal Pradesh	Any other states of India
Potato Cyst Nematodes (*Globodera rostochiensis* &G pallida	Potato	Tamil Nadu	Any other state or place in India &Uninon Territory
Codling moth (*Carpocapsa pomenella*	Apple and Walnut plants including fruits	Ladakh districts	Any other area of Jammu & Kashmir
Fluted Scale (*Icerya purchase*)	Many host plant	Mysore (Karnataka), Madras (Tamil Nadu) & Kerala	Any other state or place in India
Coffee Berry Borer (*Hypothenamus hampei*)	Coffee seeds/plants/ powder	Nilgiri (TN), Kodagu (Karnataka) &Wynad (Kerala)	Any other states of India
San Jose Scale (*Aspidiotus perniciosus*)	Many host plant species	Punjab, UP, Madras (Tamil Nadu) &Kerala,WB,Assam, Orissa, HP, J&K	All other part of India

Model Practice Questions

A. Objective Questions

a. Multiple Choice Questions

1. Management of plant diseases using legal restriction is called

(a) Cultural management (b) Physical management

(c) Plant quarantine (d) Chemical management

2. The term 'Quarantine' is derived from the Latin word *quarantum,* meaning
   (a) 40 (b) 50
   (c) 60 (d) 70
3. Plant quarantine, restrict to protect agriculture and the environment from avoidable damage by hazardous organisms.
   (a) Entry of plants & plant products
   (b) Soil, cultures of living organisms
   (c) Packing materials, commodities,containers and means of conveyance
   (d) All
4. The first such quarantine was imposed in.
   (a) Venice in 1374 (b) Australia in 1374
   (c) Austria in 1374 (d) Purtagal in 1374
5. Strategic and integrated approach encompasses policy frameworks to analyze and manage risks for food safety, animal and health plant life and health including environment
   (a) Plant biosecurity (b) Plant qurantine
   (c) Agroterrorism (d) Biotererorism
6. The basic elements of plant biosecurity are
   (a) Introduction of exotic pests/diseases.
   (b) Emergence of virulent strains/races/ biotypes of economically important indigenous pests/diseases.
   (c) Biosafety issues arising from LMOs.
   (d) All
7. Destructive Insects and Pests (DIP) Act was passed in
   (a) 1914 (b) 1912
   (c) 1910 (d) 1920
8. Introduction of exotic diseases and pests into the country or their spread from one state or union territory to another under
   (a) Destructive Insects and Pests (DIP) Act of 1914
   (b) Plant Quarantine (Regulations of Import into Indi(a) Order 2003
   (c) Livestock Importation Act1898
   (d) The Environmental (Protection) Act 1986
9. Which among the following replaces the existing Insecticide Act, 1968,
   (a) Destructive Insects and Pests (DIP) Act of 1914
   (b) Pesticide Management Bill, 2020
   (c) Livestock Importation Act1898
   (d) The Environmental (Protection) Act 1986

10. Which among the following pests introduced earliest in India
    (a) Lantana weed (b) Coffee rust
    (c) Late blight of potato (d) Wooly aphid of apple
11. The benefits derived after taking risk are high enough to justify taking the risk, with safe guards, in the first place.
    (a) Low risk pest (b) High risk pest
    (c) Acceptable risk pest (d) Pest risk analysis
12. Which among the following pests introduced in 2018 in India
    (a) Fall armyworm in maize (b) White rust of chrysanthemum
    (c) Late blight of potato (d) Wooly aphid of apple
13. Domestic quarantine exist in India for
    (a) Potato scab (b) Potato virus
    (c) Potato late blight (d) Potato wart
14. Once the pathogen has established in an area, steps taken to destroy or remove is known as
    (a) Exclusion (b) Eradication
    (c) Protection (d) Therapy
15. PRA stands for
    (a) Pest restricted area (b) Pest risk analysis
    (c) Pest reserved area (d) Pesticide risk analysis

Answer

Q. No.	Answer	Q. No.	Answer
1	(c) Physical management	7	(a) 1914
2	(a) 40	8	(a) Destructive Insects and Pests (DIP) Act of 1914
3	(d) All	9	(b) Pesticide Management Bill, 2020
4	(a) Venice in 1374	10	(a) Lantana weed
5	(a) Plant Biosecurity	11	(c) Acceptable risk pest
6	(d) All	12	(d) All
13	(d) Potato wart	14	(b) Eradication
15	(b) Pest risk analysis	16	-

b.True/False

1) The basic principles of plant quarantine are to check the entry and spread of potentially dangerous plant pathogens and insects imported along with the germplasm.
2) Low risk pest means that there is little chance that the pathogen or pest will enter.
3) Quarantine regulations are enforced by a country only.

4) Certification is the method of management of pathogen avoiding the contact between the pathogen and the host.
5) Exclusion is the most effective approach to keep pathogens away from the area.
6) The quarantine regulations anticipate that the disease or pathogen is not present in the country.
7) The most used methods for exclusion of the pathogens are regulatory measures.
8) Complete embargo involves absolute prohibition or exclusion of specified plants and plant products from a country infected or infested with highly destructive pests or diseases.
9) Plant quarantine laws were first enacted in India in 1906 in response to concerns about the spread of the Mexican boll weevil.
10) It is not permitted to transport potato tubers from West Bengal to any other state or territory in India because of the possibility of wart spreading.

Answer

Q. No.	Answer	Q. No.	Answer
1	True	6	True
2	True	7	True
3	False	8	True
4	True	9	False
5	False	10	True

B. Descriptive Questions

b. Short Answer

1. Why need of plant quarantine.
2. Explain about Pest Risk Analysis..
3. What is the primary goal of plant quarantine?
4. Why are quarantine stations important in plant quarantine?
5. Give an example of a plant disease that has caused significant damage due to lack of quarantine measures.

b. Long Answer

1. Discuss the economic impact of plant pests and diseases, and explain how effective plant quarantine measures can mitigate these impacts.
2. Explain the concept of pest risk analysis and its importance in plant quarantine.
3. Discuss the methods of plant quarantine.
4. Write short notes on i. Plant Biosecurity ii.Domestic Quarantine iii.Agency involved in Plant Quarantine iv. DIP act.

5. Discuss in details about Plant Quarantine organizations in India.
6. Discuss the role of modern technology in strengthening plant quarantine systems.
7. What is Plant Quarantine? Describe domestic quarantine measures in India and how they are helpful in crop disease management.
8. Explain the importance of public awareness and education in supporting plant quarantine efforts.
9. In which way are plant quarantine measures helpful in the management of plant diseases? Do you justify quarantine measures?
10. Write down the principles and goals of plant quarantine.Give the factors affecting the efficacy of plant quarantines.

20

Biological Control

Definition

Garret (1970) defined biological control as any condition under which or practice whereby survival or activity of a pathogen is reduced through the agency of any other living organism (except man) with the result that there is a reduction in the incidence of the disease caused by that pathogen.

According to Baker and Cook (1974), biological control is the process by which one or more organisms reduce the inoculum density or disease-producing activities of a pathogen or parasite in either their active or dormant stage. This can be done naturally, by manipulating the environment, host, or antagonists, or by introducing one or more antagonists in large quantities.

A basic definition of biocontrol is the use of living organisms to manage diseases

Types of biological control

1. **Partial Biological Control:** If control is achieved at somewhat higher average population densities than it is economically satisfactory, this is sometimes alluded to as partial Biological control.
2. **Substantial Biological Control:** If biological control can be substantially successful economically for a period of time, the only partially so far a subsequent period. Here a degree of Biological control is evaluated on economic yard stick..
3. **Total Biological Control:** If biological agent is successful economically for a very long period of time. So far as such success has not yet been achieved.

Scope of biological control

1. Encourage basic research.
2. Determine the bioagent's potential for use against crop pests.
3. Assist in lowering the risks associated with pollution.
4. The need to expand biological control's horizons.
5. Biological control techniques being adopted by agricultural industries.

Mechanism of Bio control

Major mechanism of biocontrol are

1. **Competition:** There are many different ways that organisms might interact during competition, with one being favored over the other. However, competition has been described as a more or less active demand on the part of two or more organisms in excess of the immediate supply of material or condition if we keep antibiosis or even exploitation limited to their specific mode of action and effects.
2. **Antibiosis:** The state known as antibiosis occurs when one or more metabolites that an organism excretes negatively impact one or more other organisms. In a relationship like this, species A doesn't directly benefit from the chemical component that species B produces; instead, it harms or is hostile to species B. On the other hand, species A might have an indirect advantage over species B in substrate colonization due to their superior ability to compete.
3. **Mycoparasitism:** The process by which one fungus parasitizes another is known as mycoparasitism. Examples of mycoparasitism are common and can be found in all classes of fungus, ranging from upper basidiomycetes to chytrids. Mycoparasitism involves several forms of contact, such as hyphae coiling, penetration, haustoria formation, and hyphae lysis.
4. **Antagonism:** In any interaction of two or more species, antagonism indicates that one or more of the other species' actions negatively impact at least one of the interacting species.
5. **Exploitation**: This refers to the direct harm that species A causes to species B in order to further its own interests (parasitism and predation). The concepts of parasitism and predation both refer to the same process of destroying the host or prey. In parasitism, the host is not quickly destroyed and an etiological relationship of some kind is formed between the parasite and the host. By directly preying on its target, a predator physically eliminates it without developing an etiological attachment.
6. **Hypovirulence:** A pathogen strain infected with ds RNA exhibits a phenotypic of decreased virulence compared to a normal strain.
7. **Induction of host resistance:** Microbes linked with soil and plants provide a range of chemical stimuli that plants respond to. These stimuli have the ability to either initiate or condition plant host defenses through biochemical alterations that strengthen resistance to various pathogen infections in the future.

Some examples of mechanism of biocontrol

S.No	Mechanism	Examples
1	Hyperparasitism / Mycoparasitism	•*Trichoderma harzianum parasitize* mycelia of *Rhizoctonia and Sclerotium.*
2	Antibiosis	• HCN produced by *Pseudomonas fluorescens.* • Bacilysin produced by *Bacillus subtilis.* • Trichodermin produced by *Trichoderma.*
3	Competition	• Exudates/leachates consumption • Siderophore scavenging • Physical niche occupation
4	Induction of host resistance	• ISR mediated resistance in tobacco, Aridiopsis • Rhizobacteria applied to seeds or roots induce systemic resistance response expressed against pathogens infecting aerial tissues.
5	Hypovirulence	• Hypovirulence of fungal pathogens by mycovirus.

Role of Biocontrol

1. Disease management.
2. Growth promoter for plants.
3. Inducing resistance and defensive mechanisms in plants.
4. Genetically modified plants.
5. The use of bioremediation.

Advantages of Biocontrol

- Carry out preventive measures.
- Limited tenacity and aftereffect.
- Mostly self-replicating and biodegradable.
- Effect of delayed knockdown.
- Prone to less resistance.
- Less reappearance of pests.
- Less damaging to advantageous microbes.
- Mostly focused on a certain target.
- Nearly zero waiting time.
- More expensive but fewer applications needed.

Limitations

- Inconsistent field performance and a lack of faith.
- Low quality and short shelf life.
- Chemical pesticide competition.
- High production costs and less agribusiness.
- Expensive and time-consuming registration process.
- Lack of end-user knowledge, confidence, and awareness.

- Propensity to use chemical pesticides.
- Lack of accessible popular literature.
- Limited methods for producing biotic agents in large quantities for both local and foreign applications.
- Lab testing of biotic agents and their effectiveness in the field.
- Greater focus is needed while transferring technologies.

Difference between Biopesticide and Chemical pesticide

Biopesticides	**Chemical pesticides**
These do not harm non target species.	Nontarget species are also harmed.
They do not pollute the environment.	Cause pollution; sometimes serious.
No harmful residues remain in food, fodder and fibers.	Harmful residues may often remain in food, fodder and fibers.
Relatively cheaper.	Relatively costlier.
Insects are expected not to develop resistance to biopesticides.	Insects may become resistant, e.g., Heliothis has become resistant to most insecticides.
Since they are highly specific, correct identification of the pest is essential.	It is often not critical.
High specificity may often make the use of two or more biopesticides necessary.	Often not required.
Performance may be variable due to the influence of biotic and abiotic factors of the environment.	This is not often the case.

List some biocontrol agents for the management of plant diseases

Name of bioagent	**Trade name**	**Target pathogen**
Trichoderma	Ecofit, Bioderma	Soil borne diseases causing pathogens *viz.*, *Sclerotinia, Phytopthora, Rhizoctonia, Pythium, Fusarium, Verticillium*
Paecilomyces lilacinus	Paecil or Bioact	Egg parasite of *Meloidogyne incognita*
Pythium oligandrum	Polygandron	*Pythium altimum*
Ampelomyces quisqualis		Powdery mildew disease
Pseudomonas florescence		*Sclerotium rolfsii, Xanthomonas axonopodis, Fusarium moniliformae*
Pseudomonas syringae	Bio save 100	*Botritis cinerea, Penicillium* spp
Bacillus subtilis	Companion	*Ralstonia, Rhizoctonia, Pythium, Fusarium* and *Phytopthora*
Agrobacterium radio bactor strain	Galltrol-A	*Agrobacterium tumefaciens*
Pasteuria penetrans		Parasite on juvenile of root knot nematode

Most commonly used biocontrol agents

The most commonly used biocontrol agents for control of plant pathogens are

1. Trichoderma.
2. Plant growth promoting bacteria (PGPR).

What is *Trichoderma*?

The most popular BCA used in the control of plant diseases is Trichoderma. *Trichoderma viride*, *Trichoderma virens*, and *Trichoderma harzianum* are the most commonly utilized Trichoderma. Certain plant diseases, including as rhizome rot, wilt, collar rot, damping off, and sheath blight, may be managed by Trichoderma.

Secret of success of *Trichoderma* as BCAs

The success of *Trichoderma* strains as BCAs is due to their

- A high procreative potential.
- Capacity to endure extremely adverse circumstances.
- The ability to use nutrients efficiently.
- The ability to alter the rhizosphere.
- A strong defense against fungi that cause phytopathology.
- As well as effectiveness in fostering defensive mechanisms and plant growth.

Plant Growth Promoting Rhizobacteria (PGPR)

PGPR are the bacteria that encourage plant growth. Pseudomonas, Bacillus, Arthrobactor, Achromobacter, Citrobactor, Enterobacter, and Flavobacterium are the genera under which PGPR are classified. In addition to increasing growth and yields, PGPR have the potential to act as biocontrol agents.

Ways that PGPR promote plant growth

- Boosting other helpful bacteria or fungi.
- Increasing the availability of other nutrients, such as phosphorus, sulfur, iron, and copper.
- Generating plant hormones.
- Managing bacterial and fungal infections.

Method of Application of *Trichodema*

Biocontrol agents are applied through

- Seed treatment
- Soil application

Seed treatment

- Depending on the product quality, treat seeds with a formulation of Trichoderma or Pseudomonas at a rate of 5–10 gm/kg of seed.

Soil application

- Trichoderma or Pseudomonas formulations are applied to soil by mixing them with thoroughly broken-down compost before doing so.
- The needed amount varies between 125 and 250 kg/ha. It is combined with FYM, spread out across the ground, and subsequently harrowed into the soil.
- It is used to treat root rots caused by Macrophomiina root rots, Fusarium wilt, and Sclerotium foot rots.
- Applied at least two weeks before to crop sowing.

Commercially available BCAs

Table: Commercially available BCAs, their trade name (s), and the target

Product/Trade name	Species/strain of ***Trichoderma***	Agency/Company
Ecofit,Trichogourd, Defense SF, Tricho-X, Biogourd	*Trichoderma viride*	*Fusarium, Rhizoctonia, Pythium, Phytophthora,Nectria*
Top shield,Root shield, F-Stop, Trichodex	*Trichoderma harzianum*	*Fusarium, Rhizoctonia, Pythium, Rhizoctonia, Colletotrichum, Monilinia, Plasmopara, Rhizopus, Sclerotinia*
Bioderma	*Trichoderma viride+ Trichoderma harzianum*	*Fusarium, Botryosphaeria*
Ecoderma	*Trichoderma viride+ Trichoderma harzianum*	*Rhizoctonia, Pythium, Phytophthora*
Binap- T&W	*Trichoderma harzianum+ Trichoderma polysporum*	Wood decay fungi
Gilogard and Soil guard	*Trichoderma virens*	*Rhizoctonia, Pythium*
AQ 10 biofungicide	*Ampelomyces quisqualis*	Powdery mildew
Biotrox C	*Fusarium oxysporum* (non pathogenic)	*Fusarium oxysporum*
Galltrol	*Agrobacterium radiobactor* strain 84	*Agrobacterium tumefaciens*
Companion	*Bacillus subtilis strain GB03*	*Pythium, Phytophthora,Fusarium, Rhizoctonia*
Biosave10LP, 110	*Pseudomonas syringae*	*Botrytis, Mucor, Penicillium*
Dagger G	*Pseudomonas fluorescens*	*Rhizoctonia, Pythium*

Model Practice Questions

A. Objective Questions

a. Multiple Choice Questions

1. All kinds of interplay between organisms in which one is favored at the expense of the other
   (a) Competition (b) Hyperparasitism
   (c) Antibiosis (d) Antagonism
2. The condition in which one or more metabolites excreted by an organism have harmful effects on one or more other organism.
   (a) Hyperparasitism (b) Antibiosis
   (c) Antagonism (d) Exploitation
3. An act where one fungus parasitizes the other one
   (a) Hyperparasitism (b) Antibiosis
   (c) Antagonism (d) Exploitation
4. Any association of two or more species, at least one of the interacting species is harmed due to the activities of one or more of the rest.
   (a) Competition (b) Hyperparasitism
   (c) Antibiosis (d) Antagonism
5. When species A inflicts harm by the direct use of species B for its own benefit
   (a) Hyperparasitism (b) Antibiosis
   (c) Antagonism (d) Exploitation
6. Phenomenon of reduced virulence of a pathogen strain than normal one developed as a result of its infection by ds RNA.
   (a) Antagonism (b) Exploitation
   (c) Induction of host resistance (d) Hypovirulence
7. Role of biocontrol
   (a) Disease control (b) Plant growth promoter
   (c) Bioremediation (d) All
8. The success of *Trichoderma* strains as BCAs is due to their
   (a) High reproductive capacity.
   (b) Efficiency in the utilization of nutrient.
   (c) Strong aggressiveness against phyto pathogenic fungi
   (d) All
9. Treating of seeds with biocontrol agents and then incubating under warm and moist conditions until just prior to emergence of radical is referred as
   (a) Seed disinfectants (b) Seed disinfestants
   (c) Seed protect ants (d) Seed biopriming

10. Use by of one species of organism to eliminate or control another species of organism---------------------
    (a) IDM (b) Biological control
    (c) Cultural control (d) Physical control
11. Suppressive soils are known to have high population of
    (a) Antagonistic microorganisms
    (b) Weeds
    (c) Propagules of pathogens
    (d) Nematodes
12. The project directorate of biological control (PDPC) is located at
    (a) Hyderabad (b) Mumbai
    (c) New Delhi (d) Bengaluru
13. Most widely used biocontrol agent is
    (a) *Pseudomonas florescens* (b) *Pseudomonas putida*
    (c) *Bacillus subtilis* (d) *Clostridium*
14. Bioproduct containing *Ampelomyces quisqualis* AQ10 is effective against which of the following diseases.
    (a) Powdery mildew (b) Downey mildew
    (c) Black mildew (d) Late blight
15. An approach to soilborne pest and pathogen management that involves the use of plants primarily from the Brassicaceae family in rotation with cash crops is called
    (a) Biofumigation (b) Greening
    (c) Soil composting (d) Biosuppressing

Q. No.	Answer	Q. No.	Answer
1	(a) Competition	9	(d) Seed biopriming
2	(b) Antibiosis	10	(b) Biological control
3	(d) Hyperparasitism	11	(a) Antagonistic microorganisms
4	(d) Antagonism	12	(d) Bengaluru
5	(d) Exploitation	13	(a) *Pseudomonas florescens*
6	(d) Hypovirulence	14	(a) Powdery mildew
7	(d) All	15	(a) Biofumigation

b. True/False

1. The PGPR besides enhancing growth and yields, are also potential biocontrol agents.
2. Trichoderma is the most common BCAs used in plant disease management.
3. A predator physically eliminates its prey by direct feeding on it without establishing any etiological relationship.

4. In parasitism, some sort of etiological relationship between the parasite and the host is established and the host is not rapidly eliminated.
5. The application of living organism to control diseases is called biocontrol.
6. Performance of biocontrol may be variable in the field due to the influence of biotic and abiotic factors of the environment.
7. Rhizobacteria applied to seeds or roots induce systemic resistance response expressed against pathogens infecting aerial tissues.
8. The product bioderma is combination of *Trichoderma viride*+ *Trichoderma harzianum.*
9. The biocontrol agent *Ampelomyces quisqualis* is effective against powdery mildew disease.
10. In augumentative application the biocontrol agent is applies at one place at lower population, which later multiplies and spreads to other plant parts and provide protection.
11. Control of potato scab achieved by green manuring is actually biological control.
12. Most fungi and bacteria produce siderophores that sequester the scarceiron.
13. The PGPR inhibits the growth of deleterious rhizobacteria by preventing their iron uptake.
14. The antibiotic phenazine -1-carboxylic acid (PC(a) produced by *P. fluorescens* is responsible for soil suppressiveness.
15. Coating seeds with bacteria before sowing is called seed bacterization.

Q. No.	Answer	Q. No.	Answer
1	True	9	True
2	True	10	True
3	True	11	True
4	True	12	True
5	True	13	True
6	True	14	True
7	True	15	True
8	True	16	True

B. Descriptive Questions

a. Short answer

1. Describe biopesticides.
2. Describe the Biopesticide Classification System.
3. What distinguishes chemical pesticides from biological pesticides?
4. PGPR's strategies for fostering plant development
5. Describe Trichoderma.

b. Long answer

1. Define biocontrol. Discuss its scope and limitations to control plant pathogens.
2. Write an essay on biological control of plant diseases.
3. What is biological control of plant pathogens? Give some suitable examples where it has been successfully used to control plant diseases.
4. Write down advantage and diasdvantage of biological control.
5. What is biological control?. Discuss various mode of action of biological control agents.
6. Give the characteristics of an effective biological control agent? How BCAs are screened and developed?
7. Attempts the following:
   a. Suppressive spoils. b. Siderophores
   c. Antibiosis d. Hypovirulence
   e. Future of biological control
8. How do you understand by biopesticides? Give name, dose, and disease managmed by three most commonly used biopesticides.'
9. Discuss in details about suppressive soil.
10. What are siderophores? How siderophores production by biological control agnets affects plant pathogen.
11. What do you understand by mycoparasitism? Give some suitable examples of mycoparasitism.

21

Ecological Management of Crop Environment

Three things must be present for a plant to get diseased: a virulent pathogen, a conducive environment, and a host that is susceptible. All the strategies we use to manage plant diseases work to remove or limit one of these factors, thus breaking the plant disease triangle. The foundation of managing plant diseases sustainably is creating an agro-ecological system that is adverse to pathogen evolution and epidemic development based on interactions between plants, pathogens, vectors, and environment, and beneficial to plant growth and development at the population level.The core of ecological plant disease management is to manipulate the environments of host-pathogen interactions in the favor of hosts through the balanced application of RAER strategy.

1. Resistance
2. Avoidance
3. Elimination
4. Remedy

A. Host Resistance

- The most practical and efficient method for managing plant diseases is host resistance.
- It can be quantitative or qualitative, systematic or local, induced or constitutive.
- Due to the accelerated development of plant diseases under modern agricultural practices, qualitative (or major gene) resistance is quite successful; nevertheless, its effectiveness only lasts a few years after commercial release, especially when utilized in large-scale monocultures.
- Quantitative resistance is more persistent but less successful at avoiding infection; instead, it lessens disease epidemics because it places less selection pressure on pathogens.
- One of the most successful ecological strategies for managing plant diseases has been shown to be increasing host heterogeneity through intercropping or combining crop varieties with various genetic and physiological traits, such as kind of resistance (quantitative versus quantitative).

- In the short term, this strategy not only lowers disease epidemics and boosts production, yield stability, and nutrient efficiency; it also enhances soil fertility and inhibits the evolution of pathogens.
- Potato late and early blights have also been effectively managed with varietal mixtures.
- Plant disease ecological management can also benefit from the application of alternative resistance gene deployment strategies, such as R gene rotation and pyramiding, in addition to mixing or intercropping technologies.

B. Avoidance & Elimination

a. Manage the Pathogen

1. Exclusion

- Keep the pathogens out!
- Make sure that your seeds and transplants are free of pathogens.
- The epidemic of late blight in Irish in 1845 was partially the result of the widespread distribution of the pathogen through infected seeds.
- Consider growing your own transplants/seeds; otherwise, inspect them carefully before you bring them to the farm.
- Saving seed can easily carry over pathogens from the previous year. Only save seed from healthy plants to reduce this risk.
- Although pathogens may occasionally be introduced from a commercial seed source, generally they are the most reliable source .
- Keep your equipment and stakes clean and sanitized. You don't want to be the grower who has bacterial spot in your peppers every year because you reuse your stakes without sanitizing them.
- Sodium hypochlorite at 0.5 percent is effective and must be followed by rinsing and proper disposal of solution. Prior to sanitizing, remove visible organic debris from the stakes and/or equipment.
- Any materials to be used on certified farms must be cleared with your certifier before use to prevent mishaps that could result in losing certification.

2. Eradicate or Reduce the Inoculum

- Crop rotation between plant families can help keep the levels of disease down.
- Rotating to remediate a disease problem can be challenging, especially if the pathogen is long-lived in the soil and/or has a wide host range.
- Rotating between unrelated crops such as beans to sweet corn, lettuce to cucurbits, and cucurbits to crucifers can help avoid the buildup of soilborne pathogens .

- A good rule of thumb is that no crop family should return to the same field or bed for a minimum of three years to avoid soilborne disease buildup.
- Beware, some pathogens create special structures that allow them to survive in the field for much longer.
- Many pathogens can survive on debris over the winter. Tilling in plant residue at the end of the season allows soil microorganisms to break the material down, leaving potential pathogens without a host.

3. Antagonistic Plants

- Certain plants, such as mustards and sudangrass, can kill plant pathogens that live in the soil.
- They contain a chemical and an enzyme in their plant tissue, specifically their cell wall.
- When we mow the plant and crush the tissue, the enzyme reacts with the chemical to create a toxic gas—the same as a fumigant. If you quickly incorporate this crushed plant material into the soil after mowing, the volatilized chemical can kill plant pathogens, nematodes, and weed seeds .
- Farmers in Northampton County tried this method to control plant-parasitic nematodes.
- However, the process can be tricky. We need to make sure we have the right varieties, enough moisture, adequate fertility, and good timing to get the result we want.

4. Hot Water Treatments

- If seed is affected by plant pathogens, use a hot water bath that will both surface disinfect as well as kill pathogens within the seed.
- For example, hot water treatment for eggplant submerses seeds in 122°F water for 25 minutes.
- Be careful to find out the correct temperature and length of time for the treatment. Too cold will not work and too hot will kill the seed .

5. Crop Rotation affects Pathogen Persistence

- Clubroot, caused by *Plasmodiophora brassica*, can be a significant problem in brassica crops. The pathogen can survive in the soil for over seven years, even in the absence of mustard family crops or weeds.
- But clubroot tends to decline more quickly when tomato, cucumber, snap bean, and buckwheat are grown.
- Clubroot was effectively controlled by growing aromatic herbs including peppermint, garden thyme, and summer savory for two to three consecutive years.
- For some brassica crops, resistant varieties are also an option.

b. Create an Unfavorable Environment

1. Keep Leaves Dry

- Most fungi and bacteria that kill plants require wet environments from dew, rain, or irrigation to infect and cause disease.
- If we want to keep them from reproducing, don't give them the environment they like.
- Good air circulation and drip irrigation help keep the leaves dry and the diseases out.
- For example, gray mold in tomatoes is generally not a problem in the field. But, when you pack tomatoes in a high tunnel with little air circulation, it becomes common, especially within the lower portion of the plant.

2. Maintain High-Quality Soil

- Balanced fertility, good drainage, and good soil tilth will all help promote a diverse range of soil microorganisms.
- Diverse microbial communities generally compete with plant-pathogenic organisms in the soil and help keep your plants healthier.
- Plants that are not stressed are less susceptible to disease.

3. Manage Weeds

- Many weeds are also hosts for diseases.
- When your crop is surrounded by weeds, the atmosphere tends to be moist, favoring infection.

C. Remedy

- Spraying pesticides to kill pathogens and/or their insect vectors is an inseparable part of plant disease management when other approaches cannot achieve the required level of pathogen population density reduction and epidemic amelioration.
- However, use of pesticides in an integrated disease management system is not to eradicate the disease completely but to control it to the most appropriate extent as guided by ecological and economical thresholds.
- During pesticide application, factors such as action modes and pathogen resistance should be considered.
- To increase their efficiency of application and reduce negative impacts on the environment, pesticides should be used in combination with disease forecasts and knowledge of the pathogen population genetic structure to determine the best time and frequency of application and to choose the type and utilization dosage of the pesticides.
- Remedy successes could also be achieved by other approaches than synthetic fungicides, such as naturally occurring plant compounds with biological control activity.

- Combining pesticides with other biotic and abiotic approaches such as biological agents, soil pH adjustment and UV irradiation has proved to be very effective in long-term control of tomato and lettuce root rot.

Ecological Engineering for Disease Management in Natural Farming

Ecological engineering for disease management has recently emerged as a paradigm for considering disease management approaches that rely on the use of cultural techniques to effect habitat manipulation and to enhance biological control. The cultural practices are informed by ecological knowledge rather than on high technology approaches such as synthetic pesticides and genetically engineered crops .

Ecological engineering for disease management–Above ground:

- Raise the flowering plants / compatible cash crops along the field border by arranging shorter plants towards main crop and taller plants towards the border to attract natural enemies as well as to avoid immigrating pathogen population.
- Grow flowering plants on the internal bunds inside the field.
- Not to uproot weed plants those are growing naturally like *Tridax procumbens,* Ageratum sp, Alternanthera sp etc which act as nectar source for natural enemies.
- Not to apply broad spectrum chemical pesticides, when the P: D ratio is favourable. The plant compensation ability should also be considered before applying chemical pesticides.

Ecological engineering for disease management – Below ground

- Crop rotations with leguminous plants which enhance nitrogen content.
- Keep soils covered year-round with living vegetation and/or crop residue.
- Add organic matter in the form of farm yard manure (FYM), vermicompost, crop residue which enhance below ground biodiversity.
- Reduce tillage intensity so that hibernating natural enemies can be saved.
- Apply balanced dose of nutrients using biofertilizers.
- Apply Mycorrhiza and plant growth promoting rhizobacteria (PGPR)
- Apply *Trichoderma viride/harzianum* and *Pseudomonas fluorescens* as seed/seedling/planting material, nursery treatment and soil application (if commercial products are used, check for label claim). However, biopesticides produced by farmers for own consumption in their fields, registration is not required.

Natural farming preserves a healthy ecosystem and promotes strong crop development that builds resilience to disease and insect attacks. Among the particular ecologically healing techniques used in natural farming are:

Intercropping: Intercropping is the practice of growing two or more crops concurrently while adhering to the idea of "Resource Concentration." It is founded on the idea of agroecosystem variety for pest and disease management. One of the more common intercropping methods for pest management is strip cropping, which involves growing two or more crops in alternating strips.

Mixed crops (Push-Pull System): In this type of intercropping, a particular "push" component uses semiochemicals to repel insects, while a "pull" component, like a trap crop, draws pest species away from the main crop.

Trap crops: Compared to the primary crop, trap crops are more alluring to insect pests.

Cover crops: Rapidly growing crops, such as living mulches, catch crops, and green manure crops, cover a large area of ground in a short amount of time.v

Flower strips: By planting annual flowering plants in a flower strip, you can continuously supply a significant amount of extra-floral and floral nutrients, which are available to important natural enemies.

Plant corridors: Plant corridors serve the dual purposes of protecting natural enemies and preserving biodiversity.

Model Practice Question

A. Objective Questions

a. Multiple Choice Questions

1. The strategy used to exclude the pathogen is

(a)	Quarantine	(b)	Certification
(c)	Inspection	(d)	All of these

2. Cultural practices that help to reduce disease incidence by avoiding the contact between the pathogen inoculum and the plant is

(a)	Date of sowing	(b)	Deep ploughing during summer
(c)	Depth of sowing	(d)	All of these

3. A natural parasite of powdery mildew fungi is

(a)	*Verticillium laccani*	(b)	*Ampelomyces quisqualis*
(c)	*Darluca filum*	(d)	*Hirsutella thompsonii*

4. The most important and fundamental principle of plant disease control is

(a)	Selection of disease free area	(b)	Host resistance
(c)	Field sanitation	(d)	Use of disease free seeds or planting material

5. A sustainable approach to manage disease by biological, cultural, physical and chemical tools in a way that minimizes economic, health and environmental risks is
   (a) Integrated disease management (b) Pest risk analysis
   (c) Sustainable disease management (d) Biological control
6. Domestic quarantine exist in India for
   (a) Potato scab (b) Potato virus
   (c) Potato late blight (d) Potato wart
7. Once the pathogen has established in an area, steps taken to destroy or remove is known as
   (a) Exclusion (b) Eradication
   (c) Protection (d) Therapy
8. The way to break boom and bust cycle is
   (a) Gene pyramiding (b) Gene deployment
   (c) Use of multilines (d) All of these
9. Roughing is employed to control
   (a) Loose smut of wheat (b) Late blight of potato
   (c) Early blight of tomato (d) Black rust of wheat
10. PRA stands for
   (a) Pest restricted area (b) Pest risk analysis
   (c) Pest reserved area (d) Pesticide risk analysis

Answer

Q. No.	Answer	Q. No.	Answer
1	(d) All of these	6	(d) Potato wart
2	(d) All of these	7	(b) Eradication
3	(b) *Ampelomyces quisqualis*	8	(d) All of these
4	(b) Host resistance	9	(a) Loose smut of wheat
5	(a) Integrated disease management	10	(d) Pesticide risk analysis

b. True/False

1. Natural farming preserves a healthy ecosystem and promotes strong crop development that builds resilience to disease and insect attacks.
2. The most practical and efficient method for managing plant diseases is host resistance.
3. Qualitative resistance works quite well, but its benefits wear off a few years after its commercial release.
4. Ecological disease management strategy lowers disease epidemics, boosts production, yield stability, and nutrient efficiency.
5. Antagonistic plants are those that have the ability to eradicate soil-dwelling plant pathogens.

6. Ecological engineering for disease management rely on the use of cultural techniques to effect habitat manipulation and to enhance biological control.
7. Trap crops are more alluring to insect pests.
8. Plant corridors serve the dual purposes of protecting natural enemies and preserving biodiversity.
9. Crop rotation between plant families can help keep the levels of disease down.
10. The term exclusion means keep the pathogens out.

Answer

Q. No.	Answer	Q. No.	Answer
1	True	6	True
2	True	7	True
3	True	8	True
4	True	9	True
5	True	10	True

2. Descriptive Questions

a. Short answers

1. What is RAER strategy.
2. Write down the importance of resistance in ecological disease management.
3. Differentiate between exclusion and eradication.
4. Discuss about antagonistic plants.
5. How crop rotation affects pathogen persistence.

b. Long answers

1. Discuss in details regarding ecological engineering for disease management in natural farming.
2. How farmers produce an environment that is unfavorable to the development of disease.
3. Discuss in detail about ecologically healing techniques used for disease management in natural farming.

22

Role of Soil Microbiome in Disease Management

Microbiome

A microbiome is the community of microorganisms that can usually be found living together in any given habitat. It was defined more precisely in 1988 by Whipps *et al.* as "a characteristic microbial community occupying a reasonably well-defined habitat which has distinct physio-chemical properties.

Plant Microbiome

The phytomicrobiome, sometimes referred to as the plant microbiome, has drawn a lot of attention lately since it is important to plant productivity and health. A characteristic microbial community occupying a reasonably well-defined habitat which has distinct physio-chemical properties" is called microbiome. Thus, the phrase incorporates not only the microorganisms involved but also their active domain. Plants coexist with a variety of microbial consortiums. The microorganisms known as the plant's microbiota inhabit the endosphere (inside) and episphere(outside)of plant tissues, respectively, and are crucial to the ecology and physiology of plants. The core plant microbiome is thought to comprise keystone microbial taxa that are important for plant fitness and established through evolutionary mechanisms of selection and enrichment of microbial taxa containing essential functions genes for the fitness of the plant holobiont. Plant microbiomes are influenced by environmental factors including climate, land use, and management as well as plant-specific characteristics like genotype, organ, species, and health condition. Certain research has suggested that a plant's microbiome may reflect or be related to its overall health.

Soil Microbiome

A microbiome is a community of microorganisms that live in a specific environment with a microbial structures and produce substances. Humans, animals, plants, soils, oceans, and other habitats all contain these kinds of microbiomes.

Every ecosystem has a different mix of microorganisms, but soils are thought to have the planet's most complex and varied microbiome. One gram of soil contains more than 50,000 species, making the soil microbiome the most genetically diverse community on Earth. Less than 1% of the soil microbiome has been examined to far, despite the fact that its diversity is thought to be essential to the wide range of services and functions that support life. The range of habitats that soils provide for microbial communities is what gives rise to the complexity and diversity of the soil microbiome . Researchers use the plural "soil microbiomes" instead of the single "soil microbiome" to highlight the diversity of environments. Globally, soils are highly heterogeneous, meaning that different sites have different qualities due to a mix of factors including climate, parent material, organisms, and time. Additionally, the environmental conditions inside and between the soil horizons (the layers) can vary greatly within a single soil profile, providing a vast array of homes for microorganisms.

In addition to microbial structures, genetic components, and relic DNA, the soil microbiome also includes soil bacteria. In general, microbes can be categorized into four major groups. These are protists, fungi, bacteria, and archaea, arranged from more to less abundant. Viruses are categorized as microorganisms in certain systems, and as "floating genetic material" in others. The majority of soil taxa, or microorganisms grouped together based on shared traits, are not listed in reference databases and have not yet been characterized. More than 80% of the bacteria and archaea species that are found in soil fit this description. Furthermore, most soil microorganisms (>95%) are dormant during specific periods. There is still much to learn about microorganisms in many systems because microbial research has historically been skewed toward the study of bacteria and fungus.

Factors Affecting the Soil Microbiome

The composition and function of the soil microbiome are influenced by various factors, including:

1. Soil type: Different soil types have different physical and chemical properties, which can affect the types of microbes that can thrive.
2. Organic matter: Organic matter provides food and energy for soil microbes, promoting their growth and diversity.
3. Crop rotation: Rotating crops can help to break disease cycles and promote a healthy soil microbial community.
4. Tillage practices: Excessive tillage can disrupt the soil microbiome and make plants more susceptible to disease.
5. Pesticide use: Some pesticides can have negative impacts on beneficial soil microbes, reducing their ability to suppress disease.

Mechanisms of Disease Suppression by the Microbiome

The soil microbiome contributes to plant disease management through several key mechanisms:

1. Competitive Exclusion: A healthy and diverse microbiome can outcompete pathogens for resources like nutrients, space, and infection sites on plant roots. This makes it harder for pathogens to establish themselves and cause disease.

2. Antagonism: Many members of the soil microbiome produce antimicrobial compounds, such as antibiotics, enzymes, and volatile organic compounds, that can inhibit or kill plant pathogens.
3. Induced Systemic Resistance (ISR): Certain beneficial microbes within the microbiome can trigger systemic defense responses in plants, making them more resistant to a broad range of pathogens. This is like vaccinating the plant against future attacks.
4. Mycorrhizal Networks: Mycorrhizal fungi form symbiotic relationships with plant roots, extending their reach into the soil and improving nutrient uptake. These networks can also protect plants from soilborne pathogens by acting as a physical barrier and by enhancing plant defenses.
5. Suppressive Soils: Some soils naturally suppress certain plant diseases due to the presence of a diverse and active microbial community that is antagonistic to the pathogen. These soils are a treasure trove of beneficial microbes that can be harnessed for disease management.

Applications in Plant Disease Management

1. Biocontrol Agents: Beneficial microbes from the soil microbiome can be isolated, cultured, and formulated as biocontrol agents to suppress plant diseases.
2. Soil Amendments: Organic amendments like compost and manure can enrich the soil with beneficial microbes and improve soil health and disease suppressiveness.
3. Microbiome Engineering: Advances in genomics and other technologies are allowing us to better understand the soil microbiome and manipulate it to enhance plant health and disease resistance. This could involve adding specific beneficial microbes to the soil or modifying the environment to favor beneficial microbes.

Soil Microbiome and Soil Functions

Over the past few decades, the importance of the soil microbiome in healthy soils has come to light more and more. Many soil activities and processes that are vital to the sustainability of agroecosystems are supported by microbes. Carbon dynamics (decomposition and synthesis of organic matter), nutrient cycling (decomposition, transformation, nitrogen fixation, and plant nutrient uptake), soil structure and maintenance (particle aggregation and transport), and biological population regulation (control of pests and diseases) are the primary soil functions supported by microbial intervention that support agricultural production. Microbes also provide essential ecosystem services like the water cycle and climate management through their activities.Because soil microbes alter these soil activities by breaking down organic matter, changing carbon and minerals, and degrading and immobilizing pollutants, they are primarily referred to as "chemical engineers." By doing this, they control the availability and cycling of nutrients (nitrogen, phosphorus, iron, and sulfur), the dynamics of organic matter, and the consumption and emission of trace gases (methane, nitrous oxide, and carbon dioxide). However, microbes can also function as "ecosystem engineers," contributing (although in a limited fashion) to soil structure and formation, as well as "biological regulators," guarding against pathogens, feeding on one another, and controlling pests and disease.

Soil Microbiome and Soil Health

The FAO has recently described the soil microbiome as "a game changer for food and agriculture." Determining how soil microbiome might enhance soil health is one of the primary objectives of soil microbiome research. The microbial communities in the soil are affected by agricultural management practices. Certain actions have the potential to adversely impact the soil microbiome or specific taxa, whilst other methods may enhance the diversity, makeup, and quantity of taxa that are advantageous for the well-being of the soil. These methods usually improve the living circumstances for these microorganisms or introduce particular microbes into soils directly. Six headings have been assigned to the practices: Physical disturbance of the soil, crop diversity, synthetic inputs, biocontrol, organic fertilizers, and direct inoculation.

Soil Microbe and Biocontrol

Utilizing organisms and/or natural resources, biocontrol techniques aim to stop and lessen disease and damage brought on by undesirable species including weeds, diseases, and animal pests. Biocontrol is embedded in a holistic manner interacting between plants, bioagents, and agricultural methods. The application of biocontrol can lower the concentration of hazardous chemicals in the ground and foster a favorable environment for soil microbes through various interactions with biodiversity and the soil. Biodiversity and soil health benefit from these measures' long-term vision and enduring effects. The "volatilization and sequestration of certain inorganic nutrients" as well as the soil microflora can both be promoted by the application of biocontrol.

Biological fertilizers and other invertebrates interact with microbial communities and their introduction may have "an impact on different groups of soil microbes and at different distances from the root." Extensively employed biocontrol agents, like the fungi Trichoderma spp., have drawn interest from researchers due to their positive effects on plant protection as well as their ability to strengthen the soil's defenses against abiotic stresses like drought and nutrient deficiency and combat soil-borne diseases. In addition to microbiomes in particular, the application of biocontrol decreased the possibility of contamination and the likelihood that soils would be exposed to chemical residues.

Soil Microbiome and Disease Suppressive Soil

"Soils in which the pathogen does not establish or persist, establishes but causes little or no damage, or establishes and causes disease for a while but thereafter the disease is less important, although the pathogen may persist in the soil" is called disease suppressive. The best illustrations of how microbiomes protect plants from root infections by soil-borne pathogens are found in disease-suppressive soils. Numerous soil-borne pathogens, such as fungus, bacteria, oomycetes, and nematodes, have been reported to be suppressed by such disease-suppressive soils. Although a number of microbial taxa have been identified as important contributors to the disease-suppressive capacity of soils, the intricacy of the microbiome, along with the underlying processes and microbial characteristics, are still unknown. Members of the Chitinophagaceae and Flavobacteriaceae families become enriched in the plant endosphere during pathogen invasion. In order to accelerate and enhance the plant defensive response, this bacterial population shift induced enzymatic activity linked to fungal cell-wall breakdown and secondary metabolite biosynthesis. Crop rotation and agri-management techniques are more frequently responsible for a soil's ability to slow the spread of disease, even if some soils' physico-chemical characteristics can also play a role in this regard. Soil becomes disease suppressive over time following wheat monoculturing. More recently, separate dominant taxa in these two soil types were clearly demonstrated by a comparative metatranscriptome analysis of the wheat rhizosphere microbiome grown in fields suppressive and non-suppressive to the plant pathogen *R. solani* AG8. Furthermore, polyketide cyclase, terpenoid biosynthesis, and cold shock proteins were all expressed more in suppressive samples . The application of probiotics that include naturally occurring bacterial antagonists and competitors that suppress pathogens has recently emerged as a promising strategy for disease suppression in soil, even though the development of probiotics for the human gut microbiome has already been an established field of research.

The bacterial plant pathogen *Ralstonia solanacearum* was found to be suppressed in the tomato rhizosphere microbiome in a study on the application of probiotic consortia consisting of predetermined Pseudomonas species . Metarhizium, an insect-pathogenic fungus commonly used as biological control agents against crop pests, was added to the rhizosphere of common beans (*Phaseolus vulgaris*) in a different study. This significantly increased the relative abundance of plant growth-promoting taxa like Bradyrhizobium, Flavobacterium, Chaetomium, and Trichoderma while suppressing the symptoms of *Fusarium solani's* root rot disease. The majority of soil suppressive qualities come from the biological processes that occur in soils. Thus, the development of efficient, reliable, and long-lasting disease management strategies will be aided by the understanding of microbial functions in suppressive soils through the use of next-generation sequencing techniques.

Soil Microbiome and One Health

The idea of "One Health" links the health of the environment and people by examining how ecosystems, flora, and animals interact. It believes that enhancing life quality requires multidisciplinary approaches to environmental, animal, and human health problems. The domains of human and veterinary medicine—which are associated with antibiotic resistance and infectious diseases—are where the One Health approach first emerged. When One Health is applied to agrifood systems—as the FAO promotes—improvements in soil health are seen as having positive effects on human health.

It is believed that microbes are essential to One Health. Microbiomes can be found in a wide range of environments. The soil microbiome has a direct and indirect impact on environmental health, which in turn effects human health and well-being. While it has indirect effects on soil-borne diseases, animal social behavior, dust and exposure, geophagia, plant growth and promotion, plant pathogens, soil health, and climate change mitigation, it has direct effects on all of these areas. As was seen in the earlier sections, having healthy microbiomes can improve both human health and the environment by lowering the demand for nutrient inputs and plant protection products in agricultural areas.

In addition, soil microbes control the cycling of vital elements and help plants absorb them. Examples of these elements are zinc and selenium, which are supplied by mycorrhizal fungus.. Antibiotics, immunosuppressants, anti-inflammatory and anti-cancer medications, and vitamins and amino acids that humans cannot synthesize can all be produced more effectively by particular microbes. The impact of regenerative agriculture strategies on enhancing the nutritional quality of crops like elevated zinc concentrations in rice and increases in vitamin C in tomatoes well established.

Model Practice Question

A. Objective Questions

a. Multiple choice questions

1. A characteristic microbial community occupying a reasonably well-defined habitat which has distinct physio-chemical properties is called .
   (a) Microbiome (b) Rhizosphere
   (c) Phyllosphere (d) Rhizoplane
2. Microbes can function as "ecosystem engineers," contributing to
   (a) Formation of soil structure (b) Supplying plants with nourishment
   (c) Controlling pests and disease (d) All of these
3. Soil microbes are referred as
   (a) Chemical engineers (b) Ecosystem engineers
   (c) Biological regulators (d) All of above
4. Microbes support a wide range of soil activities and processes that are essential to the sustainability of agroecosystems.
   (a) Carbon dynamics (b) Nutrient cycling
   (c) Biological population regulation (d) All of the above

5. “Soils in which the pathogen does not establish or persist or establishes and causes disease for a while but thereafter the disease is less important, although the pathogen may persist in the soil” is called
   (a) Suppressive soil (b) Conducive soil
   (c) Antagnositic soil (d) All of these
6. Which pathogen was suppressed in the tomato rhizosphere microbiome with application of probiotic consortia consisting of predetermined Pseudomonas species
   (a) *Ralstonia solanacearum* (b) *Xanthomonas solanacearum*
   (c) *Erwinia carotovora* (d) *Streptomyces scabies*
7. The zone of soil surrounding a plant root where the biology and chemistry of the soil are influenced by the root.
   (a) Microbiome (b) Rhizosphere
   (c) Phyllosphere (d) Rhizoplane
8. Which microorganism is called floating genetic material
   (a) Fungi (b) Bacteria
   (c) Virus (d) Mycoplasma
9. Extensively employed biocontrol agents
   (a) Trichoderma (b) Pseudomonas
   (c) Bacilus (d) Pacilomyces
10. The above-ground parts of a plant, such as the leaves, stems, flowers, and fruit, that microorganisms can inhabit.
   (a) Microbiome (b) Rhizosphere
   (c) Phyllosphere (d) Rhizoplane

Answer

Q. No.	Answer	Q. No.	Answer
1	(a) Microbiome	6	(a) *Ralstonia solanacearum*
2	(d) All of these	7	(b) Rhizosphere
3	(d) All of above	8	(c) Virus
4	(d) All of the above	9	(a) Trichoderma
5	(a) Suppressive soil	10	(c) Phyllosphere

b.True/False

1. Plant’s microbiome may reflect or be related to its overall health.
2. The plant’s microbiota inhabit only the endosphere of plant tissues.
3. Soil microbiome is the most genetically diverse community on earth.
4. Soil becomes disease suppressive over time following wheat monoculturing.
5. The application of probiotics and competitors that suppress pathogens has recently emerged as a promising strategy for disease suppression in soil.

6. Microbes are essential to One Health.
7. The soil microbiome has a direct and indirect impact on environmental health, which in turn effects human health and well-being.
8. Regenerative agricultural techniques have a well-established effect on improving the nutritional content of crops, such as increased zinc concentrations in rice and vitamin C levels in tomatoes.
9. The volatilization and sequestration of certain inorganic nutrients as well as the soil microflora can both be promoted by the application of biocontrol.
10. Soil microbes are called biological regulators," guarding against pathogens, feeding on one another, and controlling pests and disease.

Answer

Q. No.	Answer	Q. No.	Answer
1	True	6	True
2	True	7	True
3	True	8	True
4	True	9	True
5	True	10	True

2. Descriptive Questions

a. Short answers

1. Define microbiome.
2. Discuss about soil microbiome.
3. Explain plant microbiome.
4. Why microbe are called chemical engineers.
5. Discuss about factors affecting the soil microbiome.
6. Why microbe are called ecosystem engineers.
7. Why microbe are called biological regulators.
8. Why FAO described the soil microbiome as "a game changer for food and agriculture.

b. Long answers

1. Describe the function of the soil microbiome in the management of disease, using appropriate examples.
2. Describe the various mechanisms by which beneficial soil microorganisms can suppress soil-borne plant diseases.
3. Explain in detail how soil microbes contribute to biological control using relevant examples.
4. What are some of the key factors that affect the composition and function of the soil microbiome, and how can these factors be managed to promote plant health?
5. How can we harness the power of soil microbiomes for developing novel disease management strategies in agriculture?

23

Suppressive Soils: Concepts & Potentialities for Managing Soil Borne Pathogens

Suppressive Soils

According to Cook and Baker (1983), disease suppressive soils are those in which the pathogen either does not establish or persists, establishes but does not cause damage, or causes some disease damage but the severity of the disease gradually decreases despite the pathogen's persistence in the soil. Predation, competition, parasitism, and antibiosis are the mechanisms that suppress soil. There is ample evidence supporting the idea that both biotic and abiotic components of the soil have a role in disease suppression, even if others contend that the term "disease suppressive" should only be applied in circumstances when there is a definite biological component. Through their effects on soil microbial activity, chemical and physical characteristics of the soil, such as pH, organic matter, and clay content, can either directly or indirectly contribute to the suppression of plant diseases. While these abiotic features of soil can help suppress disease, the activity of soil microorganisms or microbial metabolites frequently determines how suppressive the soil is, either directly or indirectly. The physical characteristics of the soil, the degree of fertility, the variety and populations of soil organisms, and soil management all affect how suppressive the soil is . In addition to directly influencing crop growth by providing water and nutrients, the soil environment also indirectly affects the growth of weeds, pests, and diseases.

Concept of Disease Suppressive Soil

Both broad and specific suppression have been used to define the idea of disease-suppressive soil.

1. **General Suppression**: The overall level of microbial activity in the soil or plant during a critical point in the pathogen's life cycle is directly correlated with the general suppression of the pathogen. The general suppression involves the actions of several resident soil organisms and is non-specific, working against the majority of pathogens, if not all of them.
2. **Specific Suppression**: Specific suppression targets just specific types of pathogens. There have been reports of specific suppressiveness in

Rhizoctonia solani, Thielaviopsis basicola, Phytophthora spp., *Pythium* spp., and *Gaeumannomyces graminis* var. *tritici*. All of the time, a given pathogen produces noticeably less disease in suppressive soils than in other soil types (conducive soils); the effect is eliminated when biocides are applied to the soil, suggesting the presence of microorganisms. While disease suppressive soils do contain general suppression, the focus is typically on the phenomena known as specific suppression and its possible exploitation.

The phenomenon of disease suppressive soils that has been documented for many plant-pathogen systems worldwide. The soils that suppress Fusarium wilts induced by *Fusarium oxysporum* or take-all of wheat caused by *Fusarium graminis* var. *tritici* have been investigated the most among plant/fungal patho-systems across a diversit y of geographic regions.

Advantages of Suppressive Soils

1. Should achieve control on par with the best.
2. Minimizes the public, legal, and environmental problems.
3. The dangers of chemicals for safety.
4. Presently accessible via alternative means.
5. Suitable for usage in reduced or organic pesticide.
6. Systems, thereby enhancing the product's worth .

Disadvantages of Suppressive Soils

1. Disease control does not happen immediately.
2. Requires more meticulous planning and management.
3. May limit flexibility in the early years.
4. Needs novel skills.

Possible mechanisms of disease-suppressive soils

1. Parasitism against pathogens by a group of beneficial microbes/microbial communities.
2. Production of metabolites, toxins, antibiotics and competition for nutrients/ resources/substrates.
3. Activation of disease-resistant genes in the host plant by the beneficial microbes.
4. Improved plant nutrition and soil health.

How long does it take for a soil to become suppressive?

Suppression takes time to manifest. The length of time will depend on the circumstances and the organic residues' return. Additionally, we can categorize pathogen-suppressive soils into two main categories based on the suppression of diseases: induced and natural. Natural suppressiveness is generally unaffected by crop history and is often linked to the physical characteristics of soils. On the other hand, agricultural techniques are the only factor that can cause induced suppressiveness. By introducing these antagonists to formerly favorable soils or substrates, the intriguing prospect of managing plant diseases is made possible by the isolation, identification, and cultivation of the antagonistic microorganism or microorganisms responsible for suppressiveness in soils.

It takes more than just a pathogenic microoganism present in a substrate to seriously harm a host that is susceptible. Additionally, the host's physiological condition and the substrate's characteristics must favor the development of the disease. Recent research has demonstrated that the substrate's composition significantly affects the severity of the disease. Agricultural soils or substrates where a particular soil-borne disease is either missing or only mildly present when the pathogen is present due to natural or artificial introduction are referred to as "disease suppressiveness."

Role of Abiotic Factors in Soil Suppressiveness

1. **Crop Rotation:** Rotations that include a break crop such as grain legume or canola greatly reduce root disease in cereals because these crops do not host the cereal root disease fungi. Canola also provides a second advantage in that it releases chemicals into the soil that destroy other soil organisms and fungus that cause root disease. For the purpose of controlling root diseases, rotations will remain crucial. However, in the long-term trials conducted in South Australia it was discovered that once the degree of soil suppression had increased, rotation's impact on the control of root fungal disease was significantly diminished.
2. **Tillage:** The majority of soil organisms are located in the top 10 cm of the soil. Reduced soil mixing brought on by minimal tillage keeps biota concentrations on the top rather than diluting them at lower depths. The more tillage passes performed, the greater the chance of soil erosion, which results in the loss of topsoil—the home of the soil biota. Soil organisms disappear from a paddock when dirt is gone. A major factor influencing the quantity of carbon (organic matter) that returns to the soil is stubble retention. The main source of energy for a large number of soil biota is plant remains, and in the presence of readily available carbon energy sources, the soil population will grow rapidly.
3. **Micronutrients in suppressiveness:** When a plant is attacked with a fungus, it produces more phenolic compounds and flavonoids that inhibit the fungus at the infection site and in other sections of the plant as a means of bolstering its defenses against the infection. The plant's nutrition has a major role in regulating the synthesis and movement of these substances. Consequently, the quantity of the plants' natural antifungal chemicals near the infection site is decreased when essential nutrients (K, Mn, Cu, Zn, and (B) are deficient in the soil and subsequently in the plants. Numerous

micronutrients—including Fe and Mn—are involved in the metabolism of phenols, from the regulation of the transport of carbohydrates into synthetic pathways (boron) to the ultimate polymerization of lignin. Zinc and boron are two of the micronutrients that are crucial for preserving the structural integrity and regulating the permeability of cell membranes. When these functions are significantly compromised, the permeability of the membranes becomes unstable and leaky. Additionally, zinc and boron have a preventive function against the harmful effects of extremely hazardous oxygen free radicals. Deficits in zinc and boron clearly increase the release of many chemical molecules from both root and leaf cells. Numerous instances from the literature demonstrate how deficiencies in zinc and boron increase a plant's the susceptibility to various diseases, including Fusarium, Rhizoctonia solani, and Phytophthora.

4. **Soil texture and Structure:** The development of cavity spot (Pythium spp.) disease in carrots was linked to poor soil aeration caused by poor soil structure, soil type, or water logging. The pea root rot complex (Fusarium spp.) is known to be affected by compaction, temperature, and moisture of the soils. An increase in soil bulk density due to compaction significantly increased root rot incidence and disease severity and drastically reduced the fresh weight of pea plants due to the disease. Thus, soil texture and structure may have an impact on plant diseases. In several diseases, the relationship between tillage and soil texture is crucial. Tillage had no effect on the severity of early season damping-off (*Phytophthora sojae*), although it did interact with texture . Compared to conventional tillage, it was found more frequently in conservation tillage.
5. **Soil moisture &Temperature;** The degree of soil moisture determines the severity of soil-borne diseases, which are highest close to the saturation threshold. One such instance is Pythium, which results in seed degradation and damping off of seedlings. The pathogen, which multiplies and spreads most readily in moist soils, appears to be the main target of the increased moisture . By lowering the temperature and reducing the amount of oxygen available in moist soil, increased moisture might also weaken the host's ability to protect itself. For example, *Pythium sulcatum* and *Pythium violae* cause the carrot disease known as cavity spot. *Rhizoctonia bataticola* incidence in sunflowers is found to decrease with rising soil moisture and lowering soil temperature.

Role of Biotic Factors in Soil Suppressiveness

The biota of soil in both non-suppressive and suppressive soil: Naturally, the ability to control disease exists in all soils. However, the first step toward suppression is microbial activity, which is dependent on temperature, moisture content, and the carbon to nitrogen ratio in the soil. Conditions that alter an organism's biological activity or its interactions with other organisms might inhibit an organism. Higher microbial activity and comparatively higher suppression are found in warm, wet soils with high carbon to nitrogen ratios. The balance between organisms that cause disease and those that suppress these pathogens will also affect how much suppression occurs. In a year when a significant number of disease-causing organisms have survived, a soil that can suppress disease at moderate levels of disease inoculum may not be able to do so the next year.

Suppressive Soils Management Includes

1. **Incorporating root-colonizing rhizosphere microorganisms**: By making the plant "stronger," these organisms can enhance phytostimulatory and biofertilizing impacts on plant health. Numerous bacteria found in the rhizosphere have the ability to cause a plant's systemic reaction, which activates the plant's defense mechanisms.
2. **Improved agronomic techniques**: It has been suggested that cultural methods be modified in order to reduce the potential for soil inoculum or raise the degree of disease suppression. In fact, a variety of techniques, including crop rotation, biofumigation, intercropping, residue destruction, organic supplements, tillage management techniques, and combinations of those techniques, have been used to achieve disease suppressiveness.

a. **Biofumigation:** This method, which is more suited to cooler regions worldwide, entails fermenting organic waste in plastic, which turns soil into anaerobic conditions and produces harmful metabolites. Pathogenic fungi are rendered inactive or destroyed by both of these processes . Glucosinolates are a class of organic compounds produced by several species of Brassicaceae (Cruciferae) that may be used to allelopathically suppress a variety of soilborne plant diseases. For instance, the nematode *Pratylenchus penetrans* and other Rhizoctonia species were inhibited from infecting roots when soil was amended with *Brassica napus* seed meal. Comparably, the breakdown of leek, garlic, and onion tissues releases sulfurous volatiles like zwiebelanes and thiosulfinates, which are then transformed into disulfides with biocidal properties against nematodes, fungi, and arthropods (Arnault).

b. **Soil solarization:** A technique known as "solarization," or "solar heating," uses solar radiation to raise the temperature of the soil to a point where a large number of plant pathogens will be killed or severely weakened, resulting in a noticeable control of the diseases.

c. **Residue management:** By encouraging general microbial activity, plant residues left on or near the soil surface may help to promote disease suppressiveness. Occasionally, though, the debris not only encourages microbial activity but also aids in pathogen preservation, halting the inoculum density from declining. This is the case for Fusarium sp. causing root and crown rot in maize, *R. solani* causing crown and root rot in sugar beet, and *Macrophomina phaseolina* causing charcoal rot in soybeans.

d. **Organic amendments:** Enhanced edaphic microorganisms, parasitism, or antibiosis by antagonists that also proliferate with organic amendments lead to increased competition for available nutrients and appropriate ecological niches, which is the primary mechanism by which organic amendments mitigate disease. This method of disease control is crucial to sustainable agriculture since it prevents the use of pesticides while reusing natural resources. Furthermore, for applications to broad fields, mass cultivation of biocontrol agents is not necessary. The degree of organic matter decomposition and the physical properties of the soil interact, and in some crops, these interactions can influence the occurrence and severity of disease. Organic debris that has partially decomposed helps the soil's structure. This change leads to enhanced drainage during times of heavy precipitation and better water retention during dry weather. In turn, the better soil conditions cause some degree of natural root rot suppression in wet soils and some degree of wilt disease suppression in dry soils. Examples include the Potato early die complex (Verticillium wilt) in dry soils, which is caused by a complex including worms and Verticillium, and Phytophthora root rot, which is common in wet soils.

Potentialities for Managing Soil & Air Borne Pathogens

Many soilborne pathogens, including Pythium species (which causes damping-off), *Fusarium oxysporum* (which causes vascular wilts), *Gaeumannomyces graminis* (which causes take-all of wheat), *Phytophthora cinnamomi* (which causes root rots of many fruit and forest trees), and *Heterodera avenae* (the oat cyst nematode), develop well and cause severe diseases in some soils, known as conducive soils. Conversely, in other soil types, sometimes referred to as suppressive soils, they grow far less and produce much lesser diseases.

It is not always evident how soils inhibit certain pathogens, although the method may depend on the pathogen and may involve biotic and/or abiotic factors. However, it appears that they largely function in the majority of cases because such soils include one or more microorganisms that are antagonistic to the pathogen. Such antagonists prevent the pathogen from reaching large enough populations to cause serious disease by the antibiotics they produce, lytic enzymes, food competition, or direct parasitization of the pathogen. Numerous types of antagonistic microorganisms have been observed to multiply in suppressive soils; most commonly, however, pathogen and disease suppression has been shown to be caused by fungi, *Trichoderma viride, T. asperellum, T.harzianum Gliocladium catenulatum, Coniothyrium minitans, Talaromyces favus, Saccharomyces cerevisae* or by bacteria, *Bacillus amyloliquefaciens, Bacillus subtilis, Pseudomonas chlororaphis, Streptomyces griseovirides, Streptomyces natalensis* .

By adding microorganisms antagonistic to the pathogen, suppressive soil can minimise the quantity of disease when added to conducive soil. For instance, potato tubers notably free of potato scab were produced when soil was supplemented with soil harbouring a strain of a Streptomyces species antagonistic to *Streptomyces scabies*, the source of potato scab. By growing papaya seedlings in suppressive soil that was inserted into holes in orchard soil that was contaminated with the root rot oomycete *Phytophthora palmivora*, for instance, papaya root rot has been controlled. However, in certain diseases, after a few years of severe disease, continued cultivation (monoculture) of the same crop in a favourable environment eventually results in a decrease in disease through increasing populations of microorganisms antagonistic to the pathogen. For instance, continued cultivation of wheat or cucumber results in lower rates of wheat take-all and Rhizoctonia damping-off, respectively. Similar to prolonged cropping of the watermelon variety "Crimson Sweet," antagonistic species of Fusarium similar to those producing Fusarium wilt of watermelon can develop up, leading to a reduction rather than an increase in Fusarium wilt. Such soils inhibit the spread of disease in the future. It is possible to demonstrate that suppressiveness is caused by antagonistic microorganisms by pasteurising the soil for 30 minutes at 60°C, which fully removes the suppressiveness. After the right crops are ploughed under as soil amendments, a kind of "soil suppressiveness" emerges. These crops, which are often crucifers, supply the time and raw materials needed for the biological elimination of pathogen inoculum by soil-based antagonists. When broccoli plants are mixed into the soil, for instance, lettuce drop, which is brought on by the fungus *Sclerotinia sclerotiorum*, is significantly reduced compared to areas that do not receive this treatment.

Soilborne Pathogens

Numerous fungi that are typically not harmful to plants invade and parasitize (mycoparasitism) or lyse (mycolysis) the mycelium and dormant spores (oospores) or sclerotia of a number of phytopathogenic soil oomycetes and fungi, including Pythium, Phytophthora, Rhizoctonia, Sclerotinia, and Sclerotium. The resting spores of various plant pathogenic fungi are infected by a number of nonplant pathogenic oomycetes and fungi, including certain chytridiomycetes and hyphomycetes, as well as several pseudomonad and actinomycetous bacteria. The mycoparasitic fungus Trichoderma sp., particularly *T. harzianum*, are among the most prevalent. It has been demonstrated that the latter fungus parasitizes the mycelia of Rhizoctonia and Sclerotium, inhibits the growth of numerous oomycetes, including Pythium, Phytophthora, and Fusarium and Heterobasidion (Fomes), and *Coniothyrium minitants*, all of which are destructive parasites and antagonists of *Sclerotinia sclerotiorum* and which effectively control several Sclerotinia diseases; and *Talaromyces flavus*, which parasitizes Verticillium and controls Verticillium wilt of eggplant. In addition to fungi, bacteria of the genera Bacillus, Enterobacter, Pseudomonas, and Pantoea have been shown to parasitize and/or inhibit the pathogenic oomycetes Phytophthora sp., Pythium sp, and the fungiFusarium, *Sclerotium ceptivorum*, and *Gaeumannomyces tritici*; the mycophagous nematode *Aphelenchus avenae* parasitizes Rhizoctonia and Fusarium; and the amoeba Vampyrella parasitizes the pathogenic fungi *Cochliobolus sativus* and *Gaeumannomyces graminis*. Other microbes parasitize plant harmful nematodes. For instance, the bacteria *Pasteuria* (Bacillus) *penetrans* parasitizes the nematodes *Meloidogyne javanica* and Pratylenchus sp. *Verticillium lecanii* parasitizes the cysts of the soybean cyst nematode *Heterodera glycines*, whereas Dactylella, Arthrobotrys, Paecilomyces, and Hirsutella sp. parasitize the cysts of the dagger nematode Xiphenema and the cyst nematodes Heterodera and Globodera are parasitized by nematophagous fungi *Catenaria auxiliaris, Nematophthora gynophila, Verticillium chlamydosporium,* and Hirsutella.

Aerial Pathogens

There are various fungal diseases of aerial plant parts that many other fungi have been demonstrated to oppose and inhibit. For instance, *Venturia inaequalis* ascospore and conidia formation is inhibited by *Chaetomium* sp. and *Athelia bombacina* in fallen and growing leaves, respectively. The cucumber powdery mildew fungus *Spaerotheca fuligena* is parasitized by Tilletiopsis sp., while *Nectria inventa* and Gonatobotrys simplex parasitize two pathogenic species of Alternaria.

Model Practice Questions

A.Objective Questions

a.Multiple Choice Questions

1. The soil in which disease impact is low despite the presence of a susceptible host, pathogen and a suitable environment is called
   (a) Suppressive soil (b) Conducive soil
   (c) Conserve soil (d) Salt affected soil
2. The mechanisms that suppress soil are
   (a) Predation (b) Competition
   (c) Antibiosis (d) All
3. The main benefit of suppressive soil is
   (a) Should achieve control on par with the best.
   (b) Minimizes the public, legal, and environmental problems.
   (c) Presently accessible via alternative means.
   (d) All of the above
4. The main disadvantage of suppressive soil is
   (a) Disease control does not happen immediately.
   (b) Requires more meticulous planning and management.
   (c) May limit flexibility in the early years.
   (d) All of the above

5. The majority of soil organisms are located in the.
   (a) Top 10 cm of the soil (b) Top 20 cm of the soil
   (c) Top 30 cm of the soil (d) Top 40 cm of the soil
6. A class of organic compounds produced by several species of brassicaceae that may be used to allelopathically suppress a variety of soilborne plant diseases.
   (a) Glucosinolates (b) Zymase
   (c) Catalase (d) Invertase
7. The breakdown of leek, garlic, and onion tissues releases sulfurous volatiles like zwiebelanes and thiosulfinates, which are then transformed into disulfides with biocidal properties against
   (a) Nematodes (b) Fungi
   (c) Arthropods (d) All of the above
8. Utilize sun energy to heat the soil to a degree where a significant percentage of plant infections will perish or become significantly weaker, giving rise to a discernible control over the diseases.
   (a) Glucosinolates (b) Biofumigation
   (c) Residue management (d) Soil solarization
9. The practice of altering the nutrient content of soil to induce nutritional stress in the soil microbial population and to inhibit the pathogen's spore germination and proliferation is known as
   (a) Microbiostasis (b) Pathometry
   (c) Microbiome (d) Nectrotroph
10. The significance of crop rotation in developing disease-suppressive soil is colloquially termed as "
   (a) Rotation effect (b) Microbiostasis
   (c) Biochemical effect (d) Physical effect

Answer

Q. No.	Answer	Q. No.	Answer
1	(a) Suppressive soil	6	(a) Glucosinolates
2	(d) All	7	(d All of the above
3	(d) All of the above	8	(d) Soil solarization
4	(d) All of the above	9	(a) Microbiostasis
5	(a) Top 10 cm of the soil	10	(a) Rotation effect

b. True/False

1. Soil conduciveness is the tendency of a soil to inhibit soil-borne disease.
2. Suppressive soils provide the best examples of natural microbe-based plant defense.
3. Streptomyces can produce butanone and dimethyl disulphide that inhibit the spore germination of *Cladosporium cladosporioides*

4. Natural Suppressiveness is generally unaffected by crop history and is often linked to the physical characteristics of soils.
5. Agricultural techniques are the only factor that can cause induced suppressiveness.
6. Soil Suppressiveness is the tendency of a soil to enhance soil-borne disease.
7. Microbiome is a community of microbes in a particular environment.
8. Take-all decline disease is one of the most well-documented instances of induced-specific suppression brought on by wheat monoculture.
9. Abiotic factors in the soil have no effect on disease suppressiveness.
10. Pasteurizing the soil for 30 minutes at 60°C eliminates the suppressiveness, which is generated by antagonistic microbes.

Answer

Q. No.	Answer	Q. No.	Answer
1	False	6	False
2	True	7	True
3	True	8	True
4	True	9	False
5	True	10	True

B. Descriptive questions

a. Short answer questions

1. What is soil suppressiveness?
2. Write down the advantages of soil suppressivenss.
3. Explain the concept of disease suppressive soil.
4. Explain the possible mechanisms of disease-suppressive soils.
5. What is biofumigation?
6. What is soil solarization?

b. Long answer questions

1. What is soil suppressiveness ?Write down the mechanism,advantages and disadvantages of soil suppressivenss.
2. What is biofumigation ?explain the role of biofumigation in development of suppressive soil with examples.
3. Explain in details about the role of abiotic factors in soil suppressiveness with suitable examples.
4. Explain in details about the role of biotic factors in soil suppressiveness with suitable examples.
5. Explain the potentialities of soil suppressiveness for managing soil & air borne pathogens.

24

Plant Growth Promoting Rhizobacteria (PGPR) & their Use in Plant Protection

Microbial activity in the soil is concentrated in the plant rhizosphere, which is the small area of soil that encircles a growing plant's root system. Prokaryotes (bacteria, viruses, and archaea and eukaryotes (fungi, oomycetes, nematodes, protozoa, algae, and arthropods) are among the many microbial taxa that inhabit the rhizosphere. Of these, bacteria and fungi are the most prevalent groups that perform essential ecological roles. According to Kloepper and Schroth's 1978 phrase, "plant growth promoting rhizobacteria" (PGPR) refers to free-living soil bacteria that flourish in the rhizosphere, aggressively colonize plant roots, and promote plant development.

This diverse group of bacteria, which is an essential part of the soil microbiome, is known to manufacture and secrete a variety of regulatory compounds in the vicinity of plant roots, which help to promote plant growth. By aiding in improved nutrient uptake by host plants, guarding against phytopathogenic microorganisms, and fostering resilience to a range of abiotic stresses, PGPRs have an impact on the general health of plants. Various PGPR strains have the ability to boost crop yields, demonstrate biocontrol, improve resistance to foliar diseases, encourage nodulation in legumes, and accelerate seedling emergence.

Reported PGPRs include members of the genera Acinetobacter, Aeromonas, Agrobacterium, Allorhizobium, Arthrobacter, Azoarcus, Azorhizobium, Azospirillum, Azotobacter, Bacillus, Bradyrhizobium, Burkholderia, Caulobacter, Chromobacterium, Delftia, Enterobacter, Flavobacterium, Frankia, Gluconacetobacter, Klebsiella, Mesorhizobium, Micrococcus, Paenibacillus, Pantoea, Pseudomonas, Rhizobium, Serratia, Streptomyces, Thiobacillus, and others.

Characteristic/Qualities of the Perfect PGPR: If a rhizobacterial strain has particular characteristics that promote plant growth and has the ability to increase plant growth after inoculation, it is regarded as a putative PGPR. The following requirements should be met by the perfect PGPR strain

1. It need to be eco-friendly and rhizosphere-competent.
2. After being inoculated, it should significantly multiply and colonize the plant roots.
3. It need to have the capacity to encourage plant growth.
4. Its range of action should be extensive.
5. It ought to get along with other rhizosphere bacteria.
6. It needs to be resistant to physicochemical conditions such as heat, desiccation, radiation, and oxygen.
7. It ought to exhibit enhanced competitive abilities in comparison to the current rhizobacterial communities.

Benefits of PGPR Inoculation to Plants

1. Tolerance to drought stress
2. Tolerance to salinity stress
3. Tolerance to biotic stress (biocontrol)
4. Increased nutrient absorption
5. Seed germination enhancement
6. Biostimulation by phytohormone(s) production
7. Soil fertility enhancement
8. Bioremediation of heavy metals and pollutants
9. Modulation of plant secondary metabolites

Mechanisms of PGPR Action

PGPRs, the major microbial population in the rhizosphere, can promote plant growth either actively or passively. They can serve as biofertilizers, assisting in the tolerance of biotic and abiotic stress and bolstering the nutrition of host plants, to encourage the growth and development of plants. These advantageous bacterial groups provide protection to plants and aid in the promotion of plant growth through a variety of mechanisms of action, such as root colonization, beneficial effects on plant physiology and growth, biofertilization, induced systemic resistance, biocontrol of phytopathogens, etc. A thorough examination has been conducted of the specific mechanisms by which PGPRs work and how they contribute to the encouragement of plant development. Traditionally, PGPRs have been thought to work through two different modes of action: internal and external mechanisms that take place inside the plant, respectively.

Through the provision of phytonutrients like fixed nitrogen or solubilized minerals from the soil (like P, K, Zn, Fe, and other essential mineral nutrients), PGPR can directly improve plant nutrition. Alternatively, it can promote plant growth and development by controlling the levels of phytohormones (like auxins, cytokinins, gibberellins, abscisic acid, and ethylene).

The production of antagonistic substances (such as hydrogen cyanide, siderophores, antibiotics, and antimicrobial metabolites) and lytic enzymes (such as chitinases, glucanases, and proteases) as well as the induction of systemic resistance in plants against a wide range of root and foliar pathogens are some of the indirect effects of PGPRs on plant health . These direct and indirect effects that PGPRs have on host plants make them excellent candidates for formulation and commercialization as phytoprotective microbial products and bioinoculants. Nonetheless, the type of host plant affects the PGPR action's mode and mechanism. Furthermore, abiotic factors such as soil type, composition, history of soil management, and prevailing environmental conditions, as well as biotic factors such as plant genotype, developmental stages, defense mechanisms, and the presence of other members of the microbial community, also affect PGPR action.

PGPR: How does it work

a. Direct plant growth promotion

- Nitrogen fixation
- Phosphate solubilization
- Potassium solubilization
- Siderophore production
- Phytohormone production

b. Indirect effects through control of pathogens

- Competition
- Antibiosis (Antibiotic production)
- Hydrolytic enzyme production
- Induced systemic resistance
- Exo polysaccharide production

c. Direct

1. Biofertilizers

- Solubilization of insoluble potassium
- Phoshate solubilization
- Production of siderophores
- Biological nitrogen fixation

2. Producton of Phytohormones

- Production of zeatin (cytokinins)
- Production of ethylene
- Production of abscisic acid (ABAa)
- Production of GA_3 (gibberellins)
- Production of auxin (indole-3-acetic acid).

d. Indirect

1. Stress tolerance

- ACC deaminase activity (low ethylene levels)
- Biosynthesis of proline and quaternary amines
- Production of antioxidant enzymes (SOD, CAT and POX)
- Productionof antifreeze proteins

2. Biological control

- Production of volatile organice compounds(VOCs)
- Production of antibiotics and antifungal metabolities
- Competitive exclusion and predation
- Production of lytic enzymes (celullase, chitinase and others)
- Acquired and induced systemic resistance (ASR & ISR)

Main overview of interactions between PGPR, Plants, and Pathogens

PGPRs directly promote plant growth by improving nutrient acquisition by the plant and growth augmentation via regulating phytohormone levels. The indirect effects of PGPRs include suppression of phytopathogens and inducing systemic resistance in plants against a wide range of pathogenic microbes.

Roadmap to the Commercialization of PGPR-Based Biofertilizers

By providing environmentally benign substitutes for synthetic agrochemicals like chemical pesticides and fertilizers, the use of PGPRs as biofertilizers to increase crop production and plant growth, improve soil fertility, and biocontrol phytopathogens encourages sustainable agriculture. PGPR-based biofertilizer development and commercialization typically adhere to the following roadmap.

1. Rhizosphere
2. Isolation of Rhizobacteria
3. Labortory and green house screening
4. Product refinement and toxicity analysis
5. Combination of strains and synergistic effect on the crops.
6. Field formulation
7. Product formulation
8. Regulatory approval and certification
9. Marketing of the products

Table: Some commercially available biocontrol products/biopesticdes to control plant diseases

Product/Trade name	**Species/strain of PGPR**	**Target Pathogens/Diseases**
Galltrol	*Agrobacterium radiobactor* strain 84	*Agrobacterium tumefaciens*
Companion	*Bacillus subtilis strain GB03*	*Pythium, Phytophthora,Fusarium, Rhizoctonia*

Histick N/T	*B subtilis Strain* MB 1600	*Fusarium, Rhizoctonia, and Aspergillus*
Kodiak	*Bacillus subtilis* strain GB03	*Fusarium, Rhizoctonia, and Alternaria*
Deny	*Burkholderia cepacia*	*Fusarium, Rhizoctonia, and several nematode*
Intercept	Cepacia	*Fusarium, Pythium, Rhizoctonia*
Biosave10LP, 110	*Pseudomonas syringae*	*Botrytis, Mucor, Penicillium*
Dagger G	*Pseudomonas fluorescens*	*Rhizoctonia, Pythium*
Actino-Iron	*Streptomyces lydicus* WYEC 108 plus iron	Soilborne fungal species of *Pythium, Rhizoctonia, Phytophthora, Verticillium, and Fusarium spp.*
Actinovate	*Streptomyces lydicus* WYEC 108 plus iron	Soilborne fungal species of *Pythium, Rhizoctonia, Phytophthora, Verticillium*, and *Fusarium* spp.
Bio-Save 10 LP Bio-Save 110 Bio-Save 1000	*Pseudomonas syringae* strain ESC-10	Postharvest decay: blue mold, gray mold, mucor rot, and potato dry rot and silver scurf

A Practical Control System Using PGPR

The selection of beneficial bacterial strains is crucial for the biocontrol of plant diseases. Finding beneficial strains of bacteria may be boosted by isolating them from pathogen-suppressive soils. When the severity or incidence of disease is less than expected in relation to the surrounding soil, the suppressive soil is evident .The bacteria should be isolated from the same environment in which they will be employed in order to produce strains that work . Other criteria required to identify the efficient strains are resistance to antibiotics and the capacity to colonize roots .

To ascertain whether a biocontrol agent is compatible with the microflora of a field soil, screening of the agents using a seedling bioassay chamber is necessary. A greenhouse assay can also help in the selection of field-effective strains. The inoculum potential of the pathogen, the ambient factors, and the bacterial dose are the key factors in the creation of the tests in the greenhouse. Numerous elements, including soil temperature, soil moisture content, and soil texture, affect the development and survival of bacteria.When implementing biocontrol, formulation and application techniques are frequently crucial .

PGPR offer a lot of potential for biocontrolling plant diseases, however farmers aren't currently using these rhizobacteria in the field. The most evident causes of the restricted use are the small number of PGPR formulations that are now accessible and their uneven efficacy. Blends of distinct strains are necessary to address variations in the biocontrol efficacy. These rhizobacterium mixes can be applied to seeds as a seed treatment, which could help lower the amount of bacterial inoculum needed. This will promote the bacterial inoculum's systemic dissemination along the surface of the growing root system and their antagonistic action on the root surface during the early stages of the pathogens' root infection. In big containers, rhizobacteria formulations or suspensions can also be used with organic manures.

Model Practice Questions

A.Objective Questions

a. Multiple Choice Questions

1. The parts of a plant above the ground, usually surface of leaves, regarded as a habitat for microorganisms
   (a) Phyllosphere (b) Rhizosphere
   (c) Exosphere (d) Endo sphere
2. The small area of soil that encircles a growing plant's root system
   (a) Phylosphere (b) Rhizosphere
   (c) Exosphere (d) Endo sphere
3. The free-living soil bacteria that flourish in the rhizosphere, aggressively colonize plant roots, and promote plant development are called
   (a) PGPR (b) PGAR
   (c) PAGR (d) RAGE
4. The way in which PGPR indirectly affects pathogen control
   (a) ISR (b) Competition
   (c) Antibiosis (d) All of the above
5. Low molecular weight compounds which possess a high affinity for iron and aid transport into cells is called
   (a) Siderophore (b) Rhizoplanre
   (c) Phylloplane (d) SAR
6. The bacterial, archaeal and fungal communities as well as their genetic material closely surrounding plant root systems
   (a) Phyllosphere microbiome (b) Rhizosphere microbiome
   (c) Rhizoplane microbiome (d) Phylloplane microbiome
7. Which among those shown to achieve success in root colonization as well as prevention of the pathogen *Phytophthora capsici* in cucumber
   (a) *Bacillus subtilis* (b) *Bacillus amyloliquefaciens*
   (c) *Pseudomonas stutzeri* (d) All of the above
8. PGPR can synthesize the phytohormones like
   (a) IAA (b) Gibberellic acid
   (c) Cytokinins (d) All of the above

9. The rhizospheric bacteria that can colonize the plant roots have been termed as rhizobacteria by
   (a) EJ Butler (b) Vander Plank
   (c) Kloepper & Schroth (d) A.D.Bery
10. Which among the following identified as the supreme operative approaches for potential biocontrol prior to *in vivo* applications
    (a) Bacteriocins (b) Siderophores
    (c) Antibiotics (d) All of the above

Answer

Q. No.	Answer	Q. No.	Answer
1	(a) Phyllosphere	6	Rhizosphere microbiome
2	(b) Rhizosphere	7	(d) All of the above
3	(a) PGPR	8	(d) All of the above
4	All of the above	9	(c) Kloepper & Schroth
5	Siderophore	10	(All of the above

b.True/False

1. PGPR play key roles in plant growth nutrient regulation and biocontrol activities.
2. Plant-associated microorganisms is called phytomicrobiome.
3. The talc-based formulation of *Paecilomyces lilacinus*, was found active in reducing the population of *Meloidogyne incognita* in soils cultivated with tomato plants.
4. PGPR stability is influenced by the method, formulations, transportation, and storage conditions.
5. PGPR-based biocontrol provides effective and long-lasting disease management.
6. PGPR can also produce volatile organic compounds (VOCs) that play a significant role in plant growth and induced systemic resistance (ISR) to pathogens .
7. Siderophores are specialized chelating agents of ferric iron that inhibit phytopathogens from gaining access to iron.
8. PGPR can indirectly increase crop stress tolerance.
9. Biocontrol agent growth is often easier to monitor under controlled conditions.
10. The term rhizobacteria has been used to accentuate their intimate association with root.

Answer

Q. No.	Answer	Q. No.	Answer
1	False	6	False
2	True	7	True
3	True	8	True
4	True	9	False
5	True	10	True

B. Descriptive questions

a. Short answer questions

1. Define PGPR.
2. Write down the ten PGPR members.
3. Write down the characteristic of the perfect PGPR.
4. Explain the interactions between PGPR, plants, and pathogens.
5. Write down the benefits of PGPR inoculation to plants
6. What is Soil Solarization?

b. Long answer questions

1. What is Plant growth promoting bacteria? Write down the role of PGPR in plant disease management.
2. How PGPR works ? –Explain in detail
3. Write down the roadmap to the commercialization of PGPR-based biofertilizers and biopesticides.
4. What do you understand by PGPR. List the commercial formulation available in India. Give suitable examples where PGPR has been incorporated in the integrated disease management program.
5. What do you understand by PGPR? Discuss with suitable examples, biological control of soil borne plant pathogens with PGPR.

25

Heritage of Plant Protection in Natural Farming

Historical Background

Since the dawn of human civilization, India has been the source of old farming knowledge. It is stated in our epic history writings that the enlightened Rishis reside in forests in order to comprehend the cycles of the natural world. Parashara Maharishi was one of these great rishis, teaching the science and art of cultivation according to Vedic science. A number of organic inputs are mentioned in passing in some of our ancient literatures, including the Kautilya Arthasashthra, the Ramayana, the Rigveda, and the Mahabharata. The foundation of natural and organic agriculture is found in long-standing agricultural methods that have developed over millennia. Early plant and animal domestication led to the start of systematic Indian agriculture around 9000 BC. Domesticated wheat, barley, and jujube were discovered in Indian subcontinent by 9000 BC. It was said that the farmers of the Indus valley civilization cultivated dates, sesame, and peas. As early as 5440 BC, domesticated wild rice was discovered in Southeast Asia. Thanks to innovation, agricultural prosperity began to increase.

There was no knowledge of plant protection among the Indians when they were learning about farming, harvesting, storing, and managing agriculture in the past. The crop could only be protected by mantras and prayers. It was thought that if the mantra was inscribed in red lac dye and fastened to the crop and, the crop would be safeguarded. However, it cannot be said that the inhabitants of that era were ignorant of the harm that insects and other pests may cause. Among the bugs that harmed crops were gandhi, Shankhi, pandarmundi, dhuli, and shringari (in Sanskrit). The term "gandhi" refers to the foul odor emitted by the gandhi bug, which is currently identified as *Leptocorisa varicornis* F.; "shankhi" indicates that a snail (Pila sp.) is involved; and "pandarmundi" denotes the characteristic white head that appears when the rice stem borer (*Tryporyza incertulus* Walker) attacks. They were undoubtedly aware of the rice stem borer and its symptoms. Since the word "dhuli" means "powder," it seems plausible that it was once used to refer to a powdery mildew of wheat and barley. In Sanskrit, the word "shringari" means "adorned with red color," and it's conceivable that the phrase was once applied to rust-related diseases.

Goats, rats, pigs, deer, wild boars, parrots, and sparrows were also listed as crop-destroying animals in addition to these pests. In fact, they may have turned their attention away from developing protective technology and toward plant protection when the harm caused by various pests to crops reached the point of economic injury. It is noteworthy since at the time, people believed that humans and plants shared a similar physiology. As a result, they separated plant diseases into two groups: internal diseases and exterior diseases.

Insects, birds, and weather were the causes of external diseases, whilst "vata," "pitta," and "kafa" were the causes of internal diseases. Insects, non-insect pests, frost, waterlogging, and drought are examples of external diseases, while nematodes, bacteria, viruses, and fungi are responsible for internal diseases.

Information contained in Surapala's Vrikshayurveda, related to kinds of internal disorders observed in trees, causes and symptoms attributed, and remedies suggested.

Cause given	Symptoms	Cause elaborated	Possible causes
Vata	Slender and crooked trunk; leaves and/or trunk knots; hard fruits that are less juicy and sweet; slow defoliation; flower and fruit drop; yellowing of leaves and fruits in general.	An arid land due to an abundance of dry and smelly substances,	Mechanical barrier beneath the surface, insects that eat leaves, fungus or nematodes that infect roots, viruses, and alkaline or saline soils
Pitta	Yellowing of leaves; early drop; deterioration of fruits and flowers.	Occur near the end of summer if trees receive an excessive amount of water that contains harsh, bitter, sour, and salty substances.	Fruit decay from fungal/bacterial infections; salt in irrigation water; predisposition to blossom blight;Viral diseases
Kapha	Fruit bearing is delayed, and when it does, it ripens too soon and loses its flavor, leaking without injury.	Appear in the winter and spring when trees are overwatered with elements that are sweet, greasy, sour, or chilly.	Fungal gummosis or rot; toxicities or dietary deficits; overwatering

Even though Integrated Pest Management (IPM) is thought of as a relatively new method for protecting plants, it was developed and used for millennia in India. The following lists a few of the customs that were followed back then.

Seed treatment: In the past, seed treatment—which is now seen as a crucial part of integrated pest management (IPM)—was highly valued. Milk, mustard, sesame ash, and cowdung were applied to the seed to improve germination and provide defense against insect pests.

Fumigation: While the idea of fumigation was widely accepted back then, the techniques for it were not as advanced as they are now. For instance, cucurbit infections were managed by smoking cow and dog bones along with cat excrement.

Field application: At that time, hand-dusting different materials and sprinkling aqueous suspension were the methods of plant protection; modern plant protection equipment were not yet invented. There are a number of traditional advice available for the management of insect pest. Here are a few of them:

A mixture of white mustard, black pepper, asafoetida, vidanga (*Embelia ribes*), vaca (*Zingiber zerumbet*), and water infused with beef, buffalo horn, pigeon flesh, and bhillata powder (*Semecarpus anacardium*) can be used to reomove insects that are infesting trees.

- An oilcake-and-water mixture sprayed on creepers will help control the infestation of insects.
- Brick dust and cowdung ash sprinkling will kill insects that feed on leaves.
- To get rid of insects from the branches and roots of trees, cold water is applied for seven days.
- If anointed with a mixture of vidanga, sesame, cow's urine, ghee, and mustard, an insect wound will heal when covered with milk.

Eco-friendly pesticides is a freshly coined word. This word is emphasized more in IPM, and botanical pesticides are employed in place of chemical ones. This is not, in fact, novel. Many years ago, Surapala found and suggested a number of plants and other materials with biocidal qualities to control plant diseases. The well-known "panchamula," or the roots of five plants, had antifungal, antiviral, antibacterial, and antifeeding qualities and was widely utilized at the period. Similarly, mustard has been used to treat a wide range of diseases brought on by "kafa." It is now known that mustard has nematicidal action, antifungal properties, and induces antibiosis in insects.

Some important products used in pest management during Ancient and Medieval periods in India.

Material	Author/Period	Properties
Root of vasika (*Justicia adhatodaa*)	Varahamihira (505–587 AD)	Insecticidal, antifungal, antibacterial, anthelmintic, and soothing action.
Leaves and branches of *Hiptage enghalensis*, or atimuktaka	Varahamihira (505–587 AD)	Insecticidal leaf juice; tannins and hiptagin-containing glycosides found in bark
Mustard (*Brassica alba / Sinapis alba*)	*Brassica alb(*a) Surapala (1000AD)	Insect antixenosis and antibiosis, antifungal, nematicidal and acaricidal,
Bidanga (Embelia Ribes)	Surapala (1000AD) Someshwara Deva (1126 AD)	Antibacterial, Insecticidal, and Anthemintic

Material	Author/Period	Properties
Ash	Someshwara Deva (1126 AD)	Provides micronutrients; desiccates insect eggs on seeds; accelerates germination by softening the seed coat with a small amount of alkalinity
Sesame (*Sesamum indicum)*	Surapala (1000 AD)	Insecticidal; insect repellant; allelopathic to rice
Mahua (*Madhuca* spp.)	Surapala (1000 AD)	oil that kills insects, piscicide, and antibacterial
Kushta (Sussurea lappa)	Surapala (1000 AD)	Insecticidal (repellents, anti-feedant); antiseptic
Bhillata (Semecarpus anarcardium)	Surapala (1000 AD)	Insecticidal; antiseptic' termite repellent; mildew moth-proofing of cloth anthelmintic; antibacterial

Rich Heritage of plant protection in natural farming addresses to plant protection found in Vedas, Buddhist literature, Kautilya's Artha-sastra, Krishi-Parashara, Sangam literature of Tamils, Agni Purana, Brhat Samhita of Varahamihira, Surapala's Vrikshayurveda, Viswavallabha of Chakrapani Misra, and some documents of the mediaeval and pre-modern period. Mainly cover the plant protection and treatment practices recommended in Vrikshayurveda, involving methods of applications that are followed even today, but with more efficacious appliances viz balanced nutrition using spraying of kunapajala, seed dressing (dry/wet), root-dipping, wound dressing, soil drenching, fumigation (smoking), crude spraying (sprinkling), and dusting.

Identification of disorders; Surapala's Vrikshayurveda describes various plant diseases based on Tridosha (Vata, Pitta, Kapha Siddhanta of Ayurveda along with symptoms and remedial measures.

Plant protection practices; Knowledge of plant protection generated by our ancient and mediaeval scholars are applicable to the crop production in modern India, especially at the small farmers level. The use of natural materials available with farmers to control crop disorders has been mentioned since Kautilya's Artha-sastra (c. 300 BC) *viz.* Cut ends of sugarcane setts meant for planting were plastered with a mixture of honey, ghee, fat of hogs, and cow dung. Plant protection practices verified by Surapala (c.1000 AD) viz sprinkling a decoction made out of panchamula [roots of *Clerodendrum phlomidis* L. f., *Aegle marmelos* (L.) Corr., *Stereospermum suaveolens* DC., *Gmelina arborea* L., and *Oroxylum indicum* (L.) Vent.]; (b) drenching tree base by decoction of milk, honey, licorice (*Glycyrrhiza glabra*, and *Madhuca indica;* (c) drenching tree base with decoction of triphala (dried fruits of *Terminalia chebula* Retz., *Terminalia bellirica* (Gaertn.) Roxb., and *Emblica officinalis* Gaertn.), ghee, and honey; (d) dressing tree wounds with a paste made from the barks of banyan and cluster fig trees, cow dung, honey, mustard, and ghee; the same treatment for trees that are oozing, (e) applying paste of vidanga (*Embelia ribes* Burm. f.) and thick mud. Most of these recommendations and use of materials such as botanicals, cow dung, honey, animal fat, cow horn, etc. will be discussed. Varahamihira suggested using milk, ghee, and cow dung for dressing seeds and smoking them by burning animal flesh or turmeric before sowing, sprinkling seeds with a mixture of flour of cereals, legumes, and sesame It has been reported that plant materials contain biocidal qualities and that honey, mustard, and licorice have antibacterial qualities. Antiseptic and biological control properties of milk, and cow dung, which usually gets mixed with urine have also be reported.

Kunapajala

First fermented liquid natural manure of the world. This covers original form of kunap jal, a fermented liquid natural manure, as mentioned in Vrikshayurveda, written by a physician Surapala about one thousand years ago, in verses 101 to 105 by Sarangadhara (1283–1301 AD) a scholar in the court of King Hammira of the present-day Bundelkhand in the verses 171–174 of Sarangadhara's Upvan Vinoda and by Chakrapani Mishra (1577 AD) in "Vishwavallabha" and by Chavundaraya a poet in Karnataka in southern India in 1025 AD in Lokopakara. It also addresses mainly the variants of natural fermented concoctions used for plant nutrition viz herbal kunao jal; Panchagavya, Sasyagavya, Dhanyagavya and Indsafari.

Model Practice Questions

A.Objective Questions

a.Multiple Choice Questions

1.Who was one of the great rishis, taught the science and art of cultivation according to Vedic science.

(a) Parashara Maharishi (b) Tulsidas
(c) Kabirdas (d) Surdas

2. First fermented liquid natural manure of the world is

(a) Kunapajala (b) Jivamrita
(c) Bijamrita (d) Acchadana

3. The characteristic white head that appears when the rice stem borer attacks rice in ancient time

(e) Gandhi (b) Shankhi
(c) Pandarmundi (d) Dhuli

4. What were the causes of internal diseases in plants in ancient time?

(a) Vata (b) Pitta
(c) Kafa (d) All of the above

5. What were the causes of external diseases in plants in ancient time ?
   (a) Insects (b) Birds
   (c) Weather (d) All of the above
6. Who has described a unique fermented liquid fertiliser-cum-plant protection material called Kunapajala?
   (a) Surdas (b) Tulsidas
   (c) Kabirdas (d) Surapala
7. Who found and suggested a number of plants and other materials with biocidal qualities to control plant diseases in ancient time
   (a) Parashara Maharishi (b) Surapala
   (c) Kabirdas (d) Surdas
8. The roots of five plants, had antifungal, antiviral, antibacterial, and antifeeding qualities and was widely utilized at the ancient period
   (a) Panchamula (b) Shankhi,
   (c) Pandarmundi (d) Dhuli,
9.Rich Heritage of plant protection in natural farming addresses to plant protection found in
   (a) Brhat Samhita (b) Vrikshayurveda
   (c)Viswavallabha (d) All of the above
10. Which document of the mediaeval and pre-modern period mainly cover the plant protection and treatment practices recommended
   (a) Brhat Samhita (b) Vrikshayurveda
   (c) Viswavallabha (d) All of the above

Answer

Q. No.	Answer	Q.No.	Answer
1	(a) Parashara Maharishi	6	(d)Surapala
2	(a) Kunapajala	7	(b) Surapala
3	(c) Pandarmundi	8	(a) Panchamula
4	(d) All of the above	9	(d) All of the above
5	(d) All of the above	10	(b) Vrikshayurveda

b.True/False

1. The crop could only be protected by mantras and prayers in ancient agriculture.
2. Cucurbit infections were managed by smoking cow and dog bones along with cat excrement in ancient agriculture.
3. Mustard has been used to treat a wide range of diseases brought on by "kafa in ancient agriculture.
4. Surapala's Vrikshayurveda describes various plant diseases based on Tridosha Siddhanta of Ayurveda along with symptoms and remedial measures.

5. The foul odor emitted by the gandhi bug in ancient time is called Gandhi.
6. Kunapajala, according to Vrikshayurveda, is a fermented liquid natural manure.
7. Vrikshayurveda was written by a physician Surapala .
8. The understanding of plant protection that our ancient and medieval scholars have produced can be applied to the production of crops in contemporary India, particularly at the level of small farmers.
9. The use of readily available natural resources by farmers to control crop diseases has been advocated since Kautilya's Artha-sastra.
10. An insect infestation can be managed by spraying creepers with a mixture of oilcake and water.

Answer

Q. No.	Answer	Q. No.	Answer
1	True	6	True
2	True	7	True
3	True	8	True
4	True	9	True
5	True	10	True

B. Descriptive questions

a. Short answer questions

1. Discuss about the bugs that harmed the crop in ancient time.
2. Write down the causative agent of external and internal diseases of plants in ancient time.
3. Discuss about kunapajala
4. What is the concept of plant protection in natural farming.
5. How does crop rotation contribute to plant protection in natural farming.
6. Why is maintaining soil health important for plant protection.

b. Long answer questions

1. Explain in detail about the information contained in Surapala's Vrikshayurveda, related to kinds of internal disorders observed in trees, causes and symptoms attributed, and remedies suggested.
2. Discuss some important products used in pest management during ancient and medieval periods in India.
3. Discuss about the rich heritage of plant protection in natural farming .
4. Write down the lists a of the customs of plant disease management that were followed at ancient time.
5. How does traditional knowledge contribute to plant protection in natural farming systems?

6. What role does biodiversity play in the natural farming approach to plant protection?
7. Explain the use of biological control agents in natural farming and how they contribute to sustainable plant protection.
8. What are some traditional plant protection methods used in natural farming that have proven effective over time?

26

Plant Protection Through Botanicals

Botanicals

Botanicals are plants and plant products which are safer in protecting plants against diseases and can serve as an alternative to chemical pesticides. The plant product may be leaves, root, stem or even the whole plant parts that contain anti-microbial properties because of the presence of bio-active ingredients in them. Plant products have been found to have fungicidal, bactericidal, and antiviral properties. It is known that 346 plant species have fungicidal, 92 have bactericidal, and 90 have antiviral properties. This proves conclusively that the plant kingdom is home to a sizable stockpile of chemicals that can control a range of plant diseases. Because many of them employ many modes of action and plant compounds are safer for species other than the targeted, there is also less chance of resistance developing Botanical pesticides are cheap, readily biodegradable, effective against a variety of crop pests, and have little toxicity to creatures that are not their intended targets. They also have a variety of modes of action. The phytochemical makeup of various plants is thought to be responsible for their various ways of action. Thus, they can support sustainable agricultural output by being included into integrated pest management systems. One of the direct benefits of this for local farmers is the availability of different plant materials in their different locations, which they may take advantage of to protect their own plants. Additionally, by lowering the amount of money spent on pesticides, this will raise the overall income of the nearby farms.However, botanical pesticides have not been fully adopted due to challenges in formulation and commercialization which are attributed to lack of chemical data and positive controls.

Advantages

- Effective in controlling many agricultural pests.
- Affordable and easily biodegradable.
- Diverse modes of action.
- Accessible with ease.
- Less expensive than chemicals.
- Low toxicity to non target organisms.
- Used as component of integrated disease management.
- Contribute to the development of sustainable agriculture.

- Contribute to a decrease in the cost of pesticides.
- Increasing local farmers' overall income.
- A tool used in natural and organic farming.

Disadvantages

- Extraction techniques are not standardised.
- Rapid degradation.
- Promising results from labs and greenhouses are rarely observed in the field, with very few exceptions.
- Development of the formulation is required.
- Some chemical compounds are dangerous for people.
- Plants are less efficient.
- Formulations with lower availability.
- Bioactive molecule degradation and volatilization

Antimicrobial Secondary Metabolites Produced by Plants

- Several bioactive compounds that defend against pests and diseases and restrict the growth of other plant pathogens are produced by plant secondary metabolism.
- Essential oils and plant extracts contain a variety of bioactive substances such as alkaloids, cyanogenic glycosides, glucosinolates, lipids, phenols, terpenes, polyacetylenes, and polythienyls that protect against insects and pathogens and restrict the growth of other plant pathogens.
- These compounds' diversity and applications in integrated pest and disease control have been studied by scientists.
- To cure plant diseases, products based on plant extracts and essential oils are available in a number of nations.
- Nevertheless, there aren't many plant-based solutions on the market, despite the high potential of plant compounds in the agrochemical industry. This is especially true given the rising need for environmentally friendly alternatives for managing disease and pests in agriculture.
- Six broad chemical groups that plant contains are as follows.
  - ➢ Flavonoides and isoflavonoides
  - ➢ Saponins
  - ➢ Steroides
  - ➢ Tannins
  - ➢ Phenolic and phenolic acids—Chlrogenic acid, Protocatechuic acid, Ferulic acid, Caffeic acid
  - ➢ Coumarins and Pyrones

Commonly used botanicals

1	Plant extracts	• Neem (*Azadirachta indica, A. Juss*), • Garlic (*Allium sativum*, Linn.), • Eucalyptus (*Eucalyptus globulus*, Labil)l., • Turmeric (*Curcuma Longa*, Linn)., • Tobacco (*Nicotiana tabacum*, Linn.), • Ginger (*Zingiber officinale*, Rosc.)
2	Essential oils	• Nettle oil (Urtica spp.), • Thyme oil (*Thymus vulgaris*, Linn.), • Eucalyptus oil (*Eucalyptus globulus*, Labill). • Rue oil (*Ruta graveolens*, Linn.), • Lemon grass oil (*Cymbopogon flexuosus* (Steud.) Wats. • Tea tree oil (*Melaleuca alternifoliia).*
3	Gel and latex	• Aloe vera (Tourn. Ex Linn.).

Volatile oils

- Low molecular weight organic molecules with noticeable vapour pressures at optimal temperatures are called volatiles.
- Contribute to defence systems.
- Prevent the growth of plant pathogenic organisms.
- Generally regarded as safe.
- Isolated using steam or hydro distillation.
- Terpinoides—monoterpenes (C10), ses quiterpenes (C15), diterpines (C20)
- Examples:
    - ➢ Black pepper (*Piper nigrum* Linn).
    - ➢ Clove (*Syzygium aromaticum* Linn.) Merr. & Perry
    - ➢ Nutmeg (*Myristica fragrans* Houtt.)
    - ➢ Oregano (*Origanum vulgare* Linn.)
    - ➢ Thyme (*Thymus vulgaris* Linn.)

Essential oils

- Potential essential oils used against post-harvest pathogens.
- Fungistatic.
- Terpenoids and aromatic compounds.
- Components – Carvacrol, Thymol, Cymene, Terpine, Phenylpropene Derivatives, Eucalyptol, Anisole.
- Examples:
    - Nettle oil (*Urtica* spp.)
    - Thyme oil (*Thymus vulgais* Linn.)
    - Eucalyptus oil (*Eucalyptus globules* La-bill)
    - Rue oil (*Ruta graveolens* Linn.)

- Lemon grass oil (*Cymbopogon flexuosus* (Steud.) Wats.)
- Tea tree oil (*Melaleuca alternifolia* Maiden & Betche).

Neem Products

Among the plant products that have been shown effective in treating a number of diseases are the derivatives of neem. Known variously as the "paradise tree," "china berry," "crackjack," "Nim," "Indian lilac," and "Margosa," the neem tree (*Azadirachta indica*) contains a number of active principles in various regions. There are several important active ingredients having antifungal and insecticidal properties, including Azadirachtin, Nimbin, Nimbidin, Nimbinene, Nimbridic acid, and Azadirone.

(i) Neem Seed Kernel Extract (NSKE)

It is prepared by soaking 5 kg of powdered neem seed kernel in 100 litres of water for eight hours (in a gunny bag). After a thorough shaking, the gunny bag is taken out. Before spraying, 100 cc of teepol is properly mixed. 500 litres of extract are needed to cover one hectare.

(ii) Neem oil solution

First, 100 litres of water are combined with 100 cc of teepol.Then, while continuously shaking, 3 litres of neem oil are gradually added to this mixture. Sprayable, the milky fluid that has created. 500 litres/ha are used for the spray.

(iii) Neem cake extract

100 litres of water are used to soak 10 kg of powdered neem cake for 8 hours. The gunny bag is taken out after a good shake. 100 ml of sticker is then added and well blended. 500 litres of spray fluid per hectare are needed.

(iv) Neem cake

Neem cake powder is applied to the land right away during the final plowing. The amount applied is 150 kg/ha.

Diseases controlled by neem products

(a) Paddy: Tungro virus (Nophotettix virescens is the vector).

Neem cake is applied at a basal dosage of 150 kg/ha. Additionally, 5% NSKE or 3% neem oil can be sprayed at a rate of 500 l/ha. if a single jassid in a plant is identified. At intervals of 15 days, three sprays must be administered.

(b) Rice: Sheath rot (*Acrocyfindrium oryzae*)

At the time of grain emergence, 5% NSKE or 3% neem oil can be sprayed @ 500 lit/ha.

(c) Paddy: Blast (*Pyricularia oryzae*) . Neem oil 5% spray is efficient.

(d) Paddy: Sheath blight (*Rhizoctonia solani*). Neem cake application of 150 kg per hectare.

(e) Groundnut: Rust (*Puccinia arachidis*)

500 lit/ha of 3% neem oil application. The first spray should be administered as soon as the symptom is noticed, and the second 15 days later.

(f) Groundnut: Foot rot (*Sclerotium rolfsii*)

Neem oil application at 1% is successful.

(g) Coconut: Wilt (*Ganoderma lucidum*)

During the wet season, apply 5 kg of neem cake per tree every year.

(h) Black gram: Powdery mildew (*Erysiphe polygoni*)

Effective treatment involves two sprays of 3% neem oil or 5% NSKE, the second of which should be used 15 days after the first.

(i) Black gram: Root rot (*Macrophomina phaseolina*),

a Neem cake application at a rate of 150 kg/ha

(j) Black gram ; Yeliow mosaic (Virus).

Neem oil application at 3% is efficient.

(k) Soybean:Root rot (*Macrophomina phaseolina*)

Application of neem cake @ 150 kg/ha.

Other Plant Products

In addition to the neem components, compounds from several other plant species have also been shown to be helpful in the management of diseases. Paddy brown spot (*Helminthosporium oryzae*), has been effectively managed using an extract from the leaves of tuisi (*Ocimum sanctum*). Early blight of tomato (*Alternaria solani*) and purple blotch of onion (*Alternaria porri*) were greatly reduced by vilvam (*Aegle marmolos*) leaf and pollen extracts. *Alternaria solani* is effectively inhibited by periwinkle flower extract (*Catheranthus roseus*) and garlic bulb extract (*Allium sativum*).

Rice discolouration caused by *Drechslera oryzae* is significantly reduced by extract from mint leaves (*Mentha piperita*). Another advantage of garlic bulb extract is the decrease in paddy blast (*Pyricularia oryzae*) and finger millet leaf blight (*H. nodulosum*). The pathogen that causes Thanjavur wilt of the coconut, *Ganoderma lucidum*, is well combated by both banana rhizome extract and kolinji root exudates. *Puccinia arachidis*, which causes groundnut rust, may be efficiently treated with pinnai seed oil (*Calophyllum inophyllum*). Nochi leaf extract (*Vitex negundo*) efficiently reduced Rice tungro viruses by managing the vector, *Nephotettix virescens*.

Table: Botanicals produced by plants having antimicrobial activity.

Common name	Scientific name	Compound	Class	Effective against
Apple	*Malus pumila* Mill.	Phloretin	Flavonoid derivative	General
Ashwagandha	*Withania somnifera* Dunal.	Withafarin A	Lactone	Bacteria, fungi
Bael tree	*Aegle marmelos* Linn.	Essential oil	Terpenoid	Fungi
Blue gum tree	*Eucalyptus globulus* Labill.	Tannin	Polyphenol	Fungi, Bacteria, Viruses
Onion	*Allium cepa Linn.*	Allicin	Sulfoxide	Fungi, Bacteria
Thyme	*Thymus vulgaris* Linn	Caffeic acid	Terpenoid	Fungi, Bacteria, viruses
Turmeric	*Curcuma longa* Linn.	Curcumin	Terpenoids	Fungi, Bacteria, protozoa
Thorn apple	*Datura stramonium* Linn.	Hyoscymine Scopolamine	Alkaloids	Fungi
Black pepper	*Piper nigrum* Linn.	Piperine	Alkaloid	Fungi
Castorbean	*Ricinus communis* Linn.	Ricinine Ricininoleic	Alkaloids	Fungi
Neem/ Margosa tree	*Azadirachta indica A. Juss.*	Azadirachtin	Terpenoides	Fungi, Bacteria
Garlic	*Allium sativum* Linn.	Allicin	Solfoxide	Fungi, Bacteria

Mode of action of botanical pesticides

Studies on the mechanism of disease control by botanicals suggest that the active ingredients contained there in

1. **Act directly on the pathogen**: The majority of plant extracts that work in *in vitro* testing show mild to significant effects on conidial germination, indicating a fungicidal or fungistatic mechanism of action. Researchers found that *Phytophthora drechsleri f.sp. cajani*, which causes pigeon pea blight, eliminated at a concentration of 24 micro g/ml, and that Achook and Reptlin were fungicidal at a concentration of 48 micron g/ml. Similar to this, *Aspergillus niger* conidiophores are eliminated when *Aspergillus niger* hyphal walls are treated with essential oils obtained from the pericarp of *Citrus sinensis*.
2. **Induce systemic resistance in the host plant:** The active components in botanicals produce systemic resistance in the host plant, according to studies on the mechanism of disease control by them, which suggests a decrease in disease incidence. When applied to apple seedlings, certain plant species' extracts have an altogether different impact, easing symptoms rather than preventing spore germination on slides (*Venturia inaequalis*). We have seen that plant extracts, such, can activate apple seedlings' innate defences, acting as resistance triggers.

Table: Mode of action of phenolic compounds or heterocyclic compound.

S.No	Compound	Remarks
1	Flavonoids	Bind to adhesions
2	Quinones	Bind to adhesions link to cell wall, enzyme inactivation
3	Tannins	Bind to proteins, bind to adhesions, membrane disruption, enzyme inhibition, substrate deprivation and metal ions complexation
4	Phenol	Substrate deprivation
5	Phenolic acids	Membrane disruption

Model Practice Questions

A. Objective Questions

a. Multiple Choice Questions

1. Plant products have been found to have
   (a) Fungicidal properties. (b) Bactericidal properties.
   (c) Antiviral properties. (d) All of the above
2. Botanical pesticides are
   (a) Cheap (b) Readily biodegradable
   (c) Effective against a variety of crop pests (d) All of the above
3. An alkaloid found in turmeric plant is useful against bacterial and fungal plant diseases.
   (a) Allicin (b) Curcumin
   (c) Piperine (d) Tannin
4. Low molecular weight organic molecules produced by plants with noticeable vapour pressures at optimal temperatures are called
   (a) Volatiles (b) Phytoalexins
   (c) PR proteins (d) Phenols
5. Volatile oils
   (a) Contribute to defence systems of plant
   (b) Prevent the growth of plant pathogenic organisms
   (c) Generally regarded as safe to plant
   (d) All of the above
6. Neem plants several important active ingredients having antifungal and insecticidal properties, including
   (a) Azadirachtin (b) Nimbidin
   (c) Nimbridic acid (d) All of the above
7. Garlic contains and alkaloid which is effective against fungal and bacterial plant pathogens
   (a) Allicin (b) Azadirachtin
   (c) Piperine (d) Tannin

8. An alkaloid found in Blue gum tree is useful against bacterial, fungal and viral plant diseases.
   (a) Allicin (b) Azadirachtin
   (c) Piperine (d) Tannin
9. An alkaloid found in Ashwagandha tree is useful against bacterial and fungal plant diseases.
   (a) Withafarin A (b) Azadirachtin
   (c) Piperine (d) Tannin
10. An alkaloid found in Thyme plant is useful against bacterial, fungal and viral plant diseases.
   (a) Allicin (b) Caffeic acid
   (c) Piperine (d) Tannin

Answer

Q. No.	Answer	Q. No.	Answer
1	(d) All of the above	6	(d) All of the above
2	(b) All of the above	7	(a) Allicin
3	(b) Curcumin	8	(d) Tannin
4	(a) Volatiles	9	(a) Withafarin A
5	(d) All of the above	10	(b) Caffeic acid

b. True/False

1. Plant kingdom is home to a sizable stockpile of chemicals that can control a range of plant diseases.
2. Botanical pesticides have not been fully adopted due to challenges in formulation and commercialization.
3. The active components in botanicals produce systemic resistance in the host plant.
4. An alkaloid allicin found in onion plant is useful against bacterial and fungal plant diseases.
5. An alkaloid azadirachtin found in neem plant is useful against bacterial and fungal plant diseases.
6. Rice discolouration caused by *Drechslera oryzae* is significantly reduced by extract from mint leaves (*Mentha piperita)*.
7. Several bioactive compounds that defend against pests and diseases and restrict the growth of other plant pathogens are produced by plant secondary metabolism.
8. Paddy brown spot (*Helminthosporium oryzae*), has been effectively managed using an extract from the leaves of tuisi (*Ocimum sanctum*).
9. An alkaloid piperine found in black pepper plant is useful against bacterial and fungal plant diseases.

10. *Alternaria solani* is effectively inhibited by periwinkle flower extract (*Catheranthus roseus*) and garlic bulb extract (*Allium sativum*).

Answer

Q. No.	Answer	Q. No.	Answer
1	True	6	True
2	True	7	True
3	True	8	True
4	True	9	True
5	True	10	True

B. Descriptive questions

a. Short answer questions

1. What are botanicals? Write down the characteristics of ideal botanical to be used in plant disease management..
2. What are the advantages and disadvantages of using botanicals in plant disease management?
3. Discuss about use of volatile oils in plant disease management
4. What are botanicals in plant protection?
5. How do botanicals help in pest control?
6. What is an example of a botanical pesticide?
7. Are botanicals safe for humans and animals?

b. Long answer questions

1. How do botanicals contribute to integrated pest management (IPM) in sustainable agriculture?
2. Discuss about potential essential oils used against post-harvest pathogens.
3. Discuss the efficacy and safety of common botanical pesticides in controlling major agricultural diseases.
4. Discuss about the mode of action of botanical pesticides in management of plant diseases with suitable examples.
5. What are the advantages and limitations of using botanicals for plant protection in organic farming?
6. Explain the role of essential oils in plant protection and their potential as natural pesticides
7. How do botanicals compare to synthetic pesticides in terms of environmental impact and resistance development?

27

ITKs Related to Management of Plant Diseases

Indigenous Technical Knowledge (ITK)

Indigenous Technical Knowledge (ITK) focuses on the practical application of local people's ways of thinking in various agricultural and related operations. Indigenous knowledge is the distinct, customary, local knowledge that has grown up around the special circumstances of women and men who are indigenous to a given region. Local knowledge, or knowledge specific to a particular culture or community, is known as indigenous knowledge. IK is in opposition to the global knowledge system produced by academic institutions, commercial enterprises, and research centers. It serves as the foundation for local decisions made about a variety of issues pertaining to rural areas, including agriculture, healthcare, food preparation, education, and natural resource management. Indigenous knowledge serves as a society's foundation of information, facilitating communication and decision-making.

Materials and practices that need our early attention

Milk and milk products: For generations, people have used milk and ghee. Buttermilk even proved to be beneficial. Glutamate, leucine, and proline account for about 40% of the total amino acids in milk. It is said that plant growth stimulants are present in milk. According to a recent investigation, milk sprays caused chillies to develop systemic acquired resistance to the viruses that caused leaf curl. A 10% aqueous suspension of milk has also been successfully used to control powdery mildew. Additionally, milk works wonders as a sticker spreader.

Plants have been reported to develop resistance systemically when exposed to the amino acid proline. The synthesis of antibacterial phenolics is stimulated. Cytokinin and auxin contents are elevated in response to high levels of endogenous proline. Proline can be found in animal connective tissues, including fish, in addition to milk.

Application of cowdung: Since the time of Kautilya (c. 300 B(C), cowdung has been used for dressing wounds, sprinkling diluted suspension on plants, plastering cut ends of vegetatively propagating units like sugarcane setts, dressing seeds, and adding to soil. Indian farmers still use cow dung in a variety of ways, although its application for purposes other than manure has been disregarded by agricultural specialists.

In a nutshell, cowdung from the cattleshed is a mixture of urine and dung, usually in a 3:1 ratio. Crude fiber, crude protein, and materials obtainable in either and nitrogen-free extracts are the components of cowdung. The majority of crude fiber is composed of cellulose and lignin, with the presence of hemicellulose and pentosans, which are polysaccharides derived from pentose sugars. There are micronutrients in cowdung too. Only traces of phosphorus are found in the urine section of cowdung, which also contains nitrogen, potash, and sulfur. The nitrogenous substances expelled in feces are made up of two parts: metabolic nitrogen and undigested or unabsorbed dietary nit. The metabolic fraction is made up of materials that are naturally occurring in the body, including bacterial remains alimentary tract epithelial cells, and leftovers of bile and other digestive secretions. Fae-cal residues are essentially undigested fiber, mucus, intestinal bacteria, some expelled compounds generated from bile (such as colors), and debris from sloughed-off intestinal epithelium. The rumen of a cow contains about 100 types of protozoa and over 60 kinds of bacteria. Most of the bacteria are fermenters of cellulose, hemicellulose, and pectin. Bile pigments, bile acids, and bile salts are the components of bile. Bile salts function as emulsifying agents by giving hydrophilic coats to droplets that would otherwise be hydrophobic.

Since bile salts are reabsorbed through the colon and returned to the bile, there shouldn't be any in the dung. Nevertheless, a tiny portion of each cycle (enterohepatic circulation) containing bile salts is lost due to bacterial breakdown in the feces as dyslysin, the slimy substance. Bile salts possess antibacterial properties. The two main bile pigments are biliverdin (green) and bilirubin (reddish/golden yellow). It is the biliverdin (C33H36N4O8) that mostly occurs in herbivorous animals and is responsible for the greenish hue of their dung.

Materials recommended by Surapala to control tree disorders and their currently known properties.

Materials	Properties
Plant species	
Acorus calamus L.	Antibacterial
Black pepper, (*Piper nigrum* L).	Insecticidal alkaloid piperin is antibacterial and antifungal, while oleoresin is both.
Ficus glomerata Roxb.	Latex; 14% tannin in bark; some species of Ficus have antimicrobial properties.
Banyan, (*Ficus benghalensis L).*	Good sealing qualities in latex; tannin
Triphala (*Emblica officinalis*) Gaertn.	Along with the other two triphala species, Anthelmintic.
Burma's (*Embelia ribes* F)	Insecticidal, Anthelmintic, and Antibacterial (Embelinbenzoquinone)
Curcuma longa, Curcuma domestica Val. (Koenig non L.)	Antioxidative Curcuminoids; Antimicrobial

Materials	Properties
Brassica alba L. *Sapiens alba* L., sometimes known as white mustard,	Insecticidal, Antifungal, and Antixenosis; Nematicidal; "Anti-insect" glucosinolate sinalbin and Allyl isothiosinate, a "anti-nematode" Antifungal
Sesame plant, *Sesamum indicum L.*	Insecticidal and repulsive; oil that works in concert with pyrethrums; seed lignins that are antioxidants; 17% protein; 800 mg of calcium, phosphorus, and potassium per 100g; the greatest amount of iron (ash), at 14%.
Animal products and other materials	
Liquid manure (kunap(a)	The benefits would be high yields of high-quality produce; a healthy crop or tree; crop resistance to abiotic stresses like heat, frost, and other environmental conditions; and so on.
Honey	Antimicrobial; prevents plant and animal wounds; Proline is present,honeybee peptide apidecin has antibacterial properties.
Ghee	Same as animal fat
Fish meal	Rich in protein; releases aminoacids including proline.
Cowdung	Urine has antiseptic properties, is full of bacteria that fight diseases, works well as a medium for biocontrol chemicals, and is advantageous to Rhizobium and Azotobacter.
Ash	Ash potassium reacts with surface lipids; hygroscopic particles collect moisture from insect eggs and spores; disrupt insect eating.

ITK and Plant Disease Management

Indigenous Technical Knowledge is an aspect of regional knowledge specific to one particular society or culture. It's the opposite of the worldwide knowledge system that private companies, research centers, and colleges have built. It provides the basis for local decision-making in rural areas on agriculture, healthcare, food preparation, education, the management of natural resources, and many other areas. An organization's information base, or ITK, facilitates communication and decision-making. The idea of sustainable agriculture was first introduced to the Indian agricultural arena in the late 1980s, and it has since acquired importance in indigenous technical knowledge (ITK), which covers the utilization of natural resources to address issues in agriculture and allied fields. Indigenous technological knowledge plays a significant role in the history and culture of a society . It is the wisdom of indigenous peoples from all over the world, each with their own techniques, traditions, abilities, and cultures . ITK stands for specialized, traditional, and regional knowledge that has emerged inside and around certain areas an original to a given place. We need to learn from the local population in order to enhance the development process . ITK refers to the skills and practises that indigenous cultures have developed over time as a result of gaining knowledge and experience, and which are then passed down from generation to generation .

ITK is a wealth of knowledge, expertise, and management techniques for farming systems. They are transmitted through the sharing of cultural and customary knowledge. ITK has a long history because it was developed through years of observation and experience. They are essential tools for development because ITK are farmer-friendly, facilitate the creation of site-specific crop management strategies, save natural resources, increase climate change resilience, and ensure food security. In farming communities, it is important to support and develop traditional agricultural practises because they not only provide access to local knowledge but also enable resource-efficient agricultural management.

Farmers' conventional crop security methods were based on cultural practices that were environmentally friendly and widely accepted, prudent crop rotation, and knowledge of insect life cycles that they acquired from their extensive interactions with the land, its wildlife, and its flora.

Some Famous ITKs developed by progressive organic farmers which have been scientifically validated by universities/ organizations

1. BIJAMRUT

Ingredients

1. Cow dung- 5kg
2. Cow urine- 5L
3. Cow milk- 1L
4. Lime- 250g
5. Water- 100L

Method of Use

- Sprinkled over the seeds before they are sown as seed treatment.

2. JIVAMRUT

Ingredients

1. Cow dung- 10 kg
2. Cow urine- 10L
3. Jaggery- 2kg
4. Flour of gram (Tur, Moong, Cowpea)
5. Urad) – 2kg Live soil (Healthy)
6. Soil- 1 kg
7. Water- 200L

Uses

- Promoting growth and flowering along with acting as a yield enhancer (@5-10% spray with water.
- Soil fertility enhancer (applied along with irrigation water.

3. SANJIVAK

Ingredients

1. Cow urine- 100L
2. Cow dung- 100-200kg
3. Jaggery- 500g
4. Water- 300L
5. Kept for 10 days (Fermentation)

Method of Application (Diluted 20 times before use)

- Along drip irrigation.
- Foliar spray.
- To enrich soil with microorganisms for quick residue decomposition.

4. PANCHAGAVYA

Panchagavya is an organic product having the potential for promoting growth and providing immunity in plant system.

Ingredients

- Cow dung slurry- 4kg
- Fresh cow dung- 1kg
- Cow urine- 3L
- Cow milk- 2L
- Curd- 2L
- Cow butter oil- 1kg

Method of Application 3l Pachagavya diluted in 100L water.

Uses

- Seed and seedling treatment.
- As a soil fertility enhancer by applying through irrigation water.

5. ANDA-ARKH

Ingredients

1. Juice of 25 lemons (Approx.) (can be collected as waste lemon fromØ lemon field.
2. Eggs-5
3. Jaggery- 250g
4. Close for 10 days and use 11th day onwards
5. 10 ml - 50ml of this solution in 10L of water diluted before use.

Uses

- Growth and flowering enhancer (Spray).
- Spray for disease control (10-50 ml/10 litre water).

6. FISH-ARKH

Ingredients

1. Desi Fish- 1kg (depends on availability)
2. Jaggery- 1kg
3. Clean stick (to mix)
4. Clean cloth (to cover)
5. Earthern pot/ Plastic drum

Uses Spray for disease control (3 ltr/100 litre water)

7. Coconut- Buttermilk Ghol

Ingredients

1. Cow buttermilk (chaanch) – 5
2. Coconut water- 1L
3. Fruit juice- 1L
4. Turmeric- 100g
5. Hing- 20g

It has pesticidal actions.

Method of application

- 1L of this solution diluted with 10L of water before spray.
- Used as a tool for plant protection against fungal disease and insects.

8. Khatti Lassi (buttermilk) spray

The Khatti lassi would fight the fungal rot, So let's see what is Khatti Lassi / Buttermilk. Well this is how its made:

1. Take one litre of milk and make curd from it.
2. Dilute the curd to make 10L Lassi. Put the Lassi in a plastic drum. Keep the content for one week.
3. After one week, add about one foot long copper strip or about one meter long copper wire (after making a spiral). Keep the content for at least 5-days (can be kept even for 7 days).
4. Lassi will become greenish-blue, and is ready for use.
5. Dilute the 10L Lassi to about 100L by adding water. Spray the diluted Lassi on crop in one acre area.

This is how our Khatti Lassi looked and we happily sprayed it all over our potatoes, with beneficial results.

Some plant disease control methods are

1. A small amount of ash dusting and spraying with liquid tanned leather waste in tribal areas to treat chilies with bunchy top disease.
2. Farmers used to spray fresh cow dung in the collar area of the chilli plant to treat fungal diseases like damping off and dieback.

3. Farmers treated ginger and turmeric seeds for disease management and improved germination by using fresh cow dung slurry (1 kilogramme of cow dung in 5 liters of water).
4. Farmers in betelvine-growing areas employed neem cake, sesame, and mustard to control soil-borne diseases.
5. In order to manage the chickpea wilt disease, farmers in the Sagar district combined 30 kg of chickpea seed with 200 gm of salt and 0.5 mg of heeng in one liter of buttermilk. To enhance development and guard against soil-borne fungi, cow urine is sprinkled on pulse seeds.
6. Castor cake, Karanja cake, and neem cake are used to control root rot and collar rot when the disease is soil-borne.
7. Water has been boiled with 20 kg of *Casuarina equisetifolia* leaves for 20 minutes. After cooling, the solution needs to be filtered. After that, the extract is diluted with water and applied topically to treat fungal and bacterial infections.
8. 2 kilogram of fresh papaya leaves should be combined with 3–4 litres of water to make a solution, which should be left overnight. Following filtration, the mixture is diluted with 50–60 liters of water and 250 milliliters of soap solution is added in order to manage rice brown spot disease.
9. Cultivating marigolds and then solanaceous vegetable crops together effectively reduces bacterial infections.
10. Rice with brown spot disease can be controlled by placing khair (*Acacia catechu*) leaves in a water cannel.

Over the millennia, Indian farmers have developed the skills necessary to grow food and endure difficult conditions. Their agricultural practices have been influenced by the rich legacy of ITK. It would be difficult to improve the quality of life for Indians, who mostly reside in and depend on agricultural production systems, if we were to ignore the rich heritage of ITK. Indigenous technological knowledge is distinct in a number of ways. It is a compilation of experiences made by members of those communities, making it "local" in the sense that it is rooted in a particular community and situated within larger cultural traditions.

Changes in farming operations brought about by a move away from old methods and toward the newest ones, as well as changes in the socioeconomic circumstances of rural residents. The main obstacles to broader ITK implementation include a lack of documentation for ITK, a lack of comprehension of its scientific justifications, and a sluggish pace of research. Acknowledging the inventiveness of customary societies is essential for safeguarding biodiversity and preserving the diverse knowledge of the community's farmers for a future generation .

Model Practice Questions

A. Objective Questions

a. Multiple Choice Questions

1. The acronym ITK denotes

(a) Indigenous Technical Knowledge (b) Indigenous Technology Knowledge

(c) Indigenous Technological Knowledge.(d) Indian Technical Knowledge

2. An aspect of regional knowledge specific to one particular society or culture is called

(a) Indigenous Technical Knowledge (b) Indigenous Technology Knowledge

(c) Indigenous Technological Knowledge.(d) Indian Technical Knowledge

3. The idea of sustainable agriculture was first introduced to the Indian agricultural arena in the late

(a) 1960s (b) 1970s

(c) 1980s (d) 1990s

4. ITK stands for which knowledge that has emerged inside and around certain areas an original to a given place.

(a) Specialized (b) Traditional

(c) Regional (d) All of the above

5. ITK are farmer-friendly, facilitate the

(a) Creation of site-specific crop management strategies,

(b) Save natural resources,

(c) Prevent the growth of plant pathogenic organisms

(d) All of the above

Answer

Q. No.	Answer	Q. No.	Answer
1	(a) Indigenous Technical Knowledge	4	(d) All of the above
2	(a) Indigenous Technical Knowledge	5	(a) All of the above
3	(c) 1980s		

b. True/False

1. Indigenous Technical Knowledge is the opposite of the worldwide knowledge system that private companies, research centers, and colleges have built.
2. ITK provides the basis for local decision-making in rural areas on agriculture.

3. ITK, is an organization's information base, facilitates communication and decision-making.
4. ITK stands for specialized, traditional, and regional knowledge that has emerged inside and around certain areas an original to a given place.
5. Farmers in betelvine-growing areas employed neem cake, sesame, and mustard to control soil-borne diseases.
6. ITK refers to the skills and practises that indigenous cultures have developed over time as a result of gaining knowledge and experience, and which are then passed down from generation to generation .
7. Cultivating marigolds and then solanaceous vegetable crops together effectively reduces bacterial infections.
8. ITK is a wealth of knowledge, expertise, and management techniques for farming systems.
9. ITK are farmer-friendly, facilitate the creation of site-specific crop management strategies, save natural resources, increase climate change resilience, and ensure food security .
10. Farmers used to spray fresh cow dung in the collar area of the chilli plant to treat fungal diseases like damping off and dieback.

Answer

Q. No.	Answer	Q. No.	Answer
1	True	6	True
2	True	7	True
3	True	8	True
4	True	9	True
5	True	10	True

B. Descriptive questions

1. Define ITK. How ITK play important role in management of plant disease under natural farming condition.
2. Dicuss about materials recommended by Surapala to control tree disorders and their currently known properties.
3. Write the name, constituent and use of some famous ITKs developed by progressive organic farmers for plant disease management which have been scientifically validated by universities/ organizations.

28

Major Diseases of Rice Crop and Management

S. No	Major Diseases	Causal Organism
1	Blast	*Pyricularia oryzae*
2	Brown spot	*Helminthosporium oryzae*
3	Sheath blight	*Rhizoctonia solani*
4	False smut	*Ustilaginoidea virens*
5	Bacterial leaf blight	*Xanthomonas oryzae pv. oryzae*
6	Bacterial leaf treak	*Xanthomonas oryzae pv. oryzicola*
7	Tungro	*Rice tungro bacilliform virus* and *Rice tungro spherical virus*
8	Khaira	Zn Deficiency

1. Blast

Diagnostic Symptoms

- Pathogen attacks the crop at all stages of crop growth.
- Symptoms appear on leaves, nodes, rachis, and glumes. On the leaves, the lesions appear as small bluish green flecks, which enlarge under moist weather to form the characteristic spindle shaped spots with grey centre and dark brown margin **(Leaf blast).**
- Spots coalesce as the disease progresses and large areas of the leaves dry up and wither. Severely infected nursery and field appear as burnt.
- Black lesions appear on nodes girdling them. The affected nodes may break up and all the plant parts above the infected nodes may die **(nodal blast).**
- During flower emergence, the fungus attacks the peduncle and the lesion turns to brownish-black which is referred to as rotten neck / neck rot / panicle blast (**neck blast**).
- In early neck infection, grain filling does not occur while in late infection, partial grainfilling occurs.

Etiology

Scientific Classification

Kingdom: Fungi

Division: Eumycota

Sub-**division:** Deuteromycotina

Class: Hyphomycetes

Order: Hyphomycetales (Moniliales)

Family: Dematiaceae

Genus: *Pyricularia*

Species: *oryzae*

- Anamorphic stage:*Pyricularia oryzae*
- Telemorphic stage: *Magnaporthe grise*
- Mycelium is hyaline to olivaceous and septate.
- Conidia are produced in clusters on long septate, olivaceous conidiophores.
- Conidia are pyriform to ellipsoid, attached at the broader base by a hilum.
- Conidia are hyaline to pale olive green, usually 3 celled.
- Perfect state of the pathogen is *M. grisea* producing perithecia.
- Ascospores are hyaline, fusiform, 4 celled and slightly curved.

Disease Cycle

- Primary infections generally result in a few widely scattered spots on leaves. Mycelium and conidia in the infected straw and seeds are major sources of inoculums.
- Pathogen also survives on collateral hosts *viz., Brachiaria mutica, Leersia hexandra, Panicum repens, Digitaria marginata,* and *Echinochloa crusgalli.*
- Spores arising from the primary infections are capable of causing many more infections. This cycling is called secondary spread.
- Secondary spread is responsible for the severe epidemics of blast in fields and localized areas.
- Disease spreads primarily through airborne conidia since spores of the fungus present throughout the year.

Management

- Grow resistant to moderately resistant varieties ADT36, ADT39, CO47, IR 20, ASD 18 and IR64.
- Remove and destroy the weed hosts in the field bunds and channels.
- Treat the seeds with *Pseudomonas fluorescens* @ 10g/kg of seed.
- Early sowing of seeds after the onset of the rainy season is more advisable..
- Excessive use of fertilizer should be avoided as it increases the incidence of blast. Nitrogen should be applied in small increments.

- Water management practices in rain fed areas lessen the likelihood of stress, which also aid in blast control.

2.Brown Spot

Diagnostic Symptoms

- Initial symptoms appear as minute circular or oval, brown lesions or dots on the coleoptiles, leaf blade, leaf sheath and glumes, being most prominent on the leaf blade and glumes.
- These lesions may girdle the coleoptiles and cause distortion of the primary and secondary leaves leading to seedling blight.
- Spots have a light brown to gray centre, surrounded by a reddish brown margin.
- Several spots coalesce and the leaf dries up.
- Infected seedlings become stunted or die.
- Affected nurseries can be often recognized from a distance by their brownish scorched appearance.
- Velvety appearance of lesions is visible on infected glumes under severe conditions.
- Infected grains show black discoloration or brown lesions.
- Disease-causing fungi can also penetrate grains, causing *'pecky rice'*, a term used to describe spotting and discoloration of grains.
- Disease appears from seedling to milking stage.

Etiology

Scientific Classification

Kingdom: Fungi

Division: Eumycota

Sub-**division:** Deuteromycotina

Class: Hyphomycetes

Order: Hyphomycetales (Moniliales)

Family: Dematiaceae

Genus: *Drechslera*

Species: *oryzae*

- Anamorphic stage: *Drechslera oryzae*
- Telemorphic stage: *Cochliobolus miyabeanus*
- Pathogen grows both intra and intercellularly inside the host.
- Mycelium brown in color and produces conidiophores that emerge in group by rupturing epidermis or through stomatal openings.
- Conidiophores arise singly or in small groups, geniculate and brown in colour.

- Conidia are usually curved with a bulged center and tapered ends.They are pale to golden brown in colour and are 6-14 septate.
- Conidia germinate readily producing germ tube mostly from cells of both ends.
- Perithecia with asci containing 6-15 septate, filamentous or long cylindrical, hyaline to pale olive green ascospores.

Disease Cycle

- Pathogen perennates by infected seeds and collateral hosts growing to the adjacent to paddy fields.
- Fungus also survives on collateral hosts like *Leersia hexandra* and *Echinochloa colonum, Cynodon dactylon etc.*
- Infected seeds and stubbles are the most common source of primary infection.
- When such seeds are sown in nurseries, the pathogen results in seedling blight.
- Similarly,condia produced on infected collateral hosts get disseminated by air and causes primary infection on seedlings.
- Secondary infection takes place with air and water borne conidia, infect the plants both in nursery and in main field.

Management

- Removal of collateral hosts and infected debris from the field.
- Use of slow release nitrogenous fertilizers.
- Use disease free seeds.
- Seed treatment with hot water (53−54°(C) for 10−12 minutes before planting, to control primary infection at the seedling stage. To increase effectivity of treatment, pre-soak seeds in cold water for 8 hours.

3. Sheath Blight

Diagnostic Symptoms

- Initial symptoms are noticed on leaf sheaths near water level.
- On the leaf sheath oval or elliptical or irregular greenish grey spots are formed.
- As the spots enlarge, the centre becomes greyish white with an irregular blackish brown or purple brown border.
- Lesions on the upper parts of plants extend rapidly coalesing with each other to cover entire tillers from the water line to the flag leaf.
- Presence of several large lesions on a leaf sheath usually causes death of the whole leaf, and in severe cases all the leaves of a plant may be blighted.
- Infection extends to the inner sheaths resulting in death of the entire plant.

- Plants heavily infected in the early heading and grain filling growth stages produce poorly filled grain, especially in the lower part of the panicle.
- Pathogen affects the crop from tillering to heading stage.

Etiology

Scientific Classification

Kingdom: Fungi

Division: Eumycota

Sub-**division:** Deuteromycotina

Class: Hyphomycetes

Order: Agonomycetales

Family: Agonomycetaceae

Genus: *Rhizoctoniza*

Species: *solani*

- Anamorphic stage: *Rhizoctonia solani* Kuhn
- Teleomorphic: *Thanetophorus cucumeris*
- In India, sheath blight of rice is caused by anastomosis group-1 (AG-I-1) of the fungus having 3-16 nuclei per cell.
- Mycelial stage producing sclerotia is the commonly encountered stage in India.
- Pathogen produces a fast growing stout mycelium with septate hyphae.
- Young colonies may look white but older mycelium is invariably some shade of brown.
- Sclerotia may or may not formed. In some isolates the sclerotia are light brown while in some they are so dark as to look black.
- In some isolates the mycelium produces barrel shaped cells which have been termed as chlaymydospores.
- Cells are as a rule multinucleate (in other species of Rhizoctonia they are binucleate).
- Sclerotia of the fungus are distinct from many other fungi in that there is no differentiation of the sclerotial tissue into a rind and internal medulla although outer cells may be darker and thicker walled.
- Basidial stage is mostly saprophytic stage of the pathogen. It appears as a flaky pellicle on the surface of the leaves near the ground level or dead parts of the stem.
- Basidia are barrel shaped, obpyriform or obclavate or clavate and measure 9-25x5-12 µm, bear 4 sterigmata.
- Basidiospores are ellipsoid or oblong-ellipsoid, flattened on one side, hyaline, truncate, and measure 5-15x 4-8 µm.

Disease Cycle

- Pathogen is seed borne. In addition to the weed hosts and infected seed, the pathogen mainly survives as sclerotia and /or mycelia in diseased plant debris left in the field.
- Sclerotia and mycelia which survive in plant debris are brought up on the soil surface during puddling, leveling, and other operations.
- They come in contact with new planted seedlings, germinate and cause infection.

Management

- Deep ploughing in summer and burning of stubbles.
- Grow resistant varieties.
- High seeding rate or dense plant spacing should be discouraged .
- Need-based or balanced application of nitrogen fertilizers in the fields.
- Sanitation, specifically removing of weeds, infected stubbles or crop residues from the field .
- Irrigation should be restricted in the affected crop and avoid flow of irrigation water from infected fields to healthy fields.

False Smut

Diagnostic Symptoms

- False smut is one of the emerging fungal grain diseases of rice.
- Symptoms are visible after flowering only, where the fungus infects the young ovary of individual kernel and transform them into large, velvety green balls (smut balls).
- Initially the smut balls are small and remain confined between glumes.They gradually enlarge and enclose the floral parts.
- Smut balls are initially yellow in colour and are covered by membrane which later bursts and the colour changes to orange, yellowish green, green, olive green and finally greenish black(Fig.-4).
- Only few grains in a panicle are usually infected and the rest are normal.

Etiology

Scientific Classification

Kingdom: Fungi

Division: Eumycota

Sub-**division:** Ascomycotina

Class: Plectomycetes

Order: Clavicipitales

Family: Clavicipitaceae

Genus: *Claviceps*

Species: *oryzae – sativae*

- Anamorphic stage*:* *Ustilaginoidea virens*
- Teleomorphic stage: *Claviceps oryzae - sativa*
- Mycella filling in the ovary are united, fine and colorless and they develop conidial. Pseudosclerotial and sclerotial stages.
- Conidia are spherical, echinulate, olivaceous in color, and measure 4-6x25μm.
- Conidia germinate in water giving rise to short germ tube, 1-3 secondary conidia are produced on each germ tube usually after 12-24 hours.
- Secondary conidia are subglobose or globose to oblong, hyaline, granulate, and measure 4-8x2-5μm.
- Late in the season, the conidial masses in smut balls harden and form pseudosclerotia.
- Sclerotia are formed in each green ball.The sclerotia are hard, variously shaped, and 5-13x2-5 mm in size.It germinate to produce stipes bearing perithecial stroma.
- Perithecia are ovate to pyriform measuring 150.5 to 430x86-193.5μm.
- They bear hyaline, elongated,cylindrical and each containing eight ascospores.
- Ascospores are hyaline, unicellular; germinate in water giving rise to 1-2 germ tubes, which bear secondary conidia.

Disease Cycle

- Sclerotia are reported to be the perennating structures.
- Sclerotia are considered to be the major source of primary inoculum
- Ascospores produced inside the perithecia embedded in sclerotial stroma are disseminated and brought on to the host surface whereupon they germinate and infect the host just before the appearance of ears.
- Primary inoculums also come from the collateral hosts.
- It is considered however, that conidia play an important role in causing secondary infection during the growing season.

Management

- Use healthy disease free seeds for sowing.
- Seeds should not be taken from false smut affected fields.
- Early planting can help to escape heavy disease incidence.
- Infected plants should be removed and destroyed at the time of harvesting,
- Field bunds and irrigation channels should be kept clean.

Bacterial Leaf Blight

Diagnostic Symptoms

The symptoms of this disease occur in two phases viz., leaf blight phase and seedling blight or 'kresek' phase.

Leaf Blight Phase

- Initially water-soaked lesions formed on the tip of the leaves which increase in length and width downwards and turn yellowish to straw colour stripes with a wavy margin.
- These lesions may develop at one or both edges of the leaves or along the mid rib.
- There is an appearance of bacterial ooze that looks like milky or opaque dewdrop on young lesions early in the morning.
- Lesions turn yellow to white as the disease progresses.
- Severely infected leaves tend to dry quickly.

Kresek Phase or Seedling Wilt

- Green water-soaked layer along the cut portion or leaf tip of leaves are early symptoms.
- Leaves wilt and roll up and become greyish green to yellow.
- Entire plant wilts completely.
- Infected leaves of mature plant turn yellow to pale yellow.
- Youngest leaf is uniform pale yellow or has broad yellow stripe.
- Under severe conditions., panicles remain sterile and unfilled but not stunted.
- Kresek phase is observed 1-3 weeks after transplanting.
- The kresek infected tillers may be confused with the stem borer injury but the latter can be easily pulled out while it is not so with the kresekinfected tillers.

Quick diagnosis

- Cut a young lesion across and place in a transparent glass container with clear water.
- After a few minutes, hold the container against light and observe for thick or turbid liquid coming from the cut end of the leaf.

Etiology

Scientific Classification

Kingdom: Prokaryotae

Division: Gracilicutes

Class: Proteobacteria

Family: Psudomonadaceae

Genus: *Xanthomonas*

Species: *campestris pv. oryzae*

- Causal organism*: Xanthomonas oryzae pv. oryzae*
- Rod shaped cells of the bacterium measure 0.5-0.8 x 1.0-2.0μ.
- Single and not formed capsule.
- Gram negative.
- Motile by a single polar flagellum.
- Colonies are slow growing, mucoid and straw colored to yellow.

Disease Cycle

- Pathogen survives in soil, seed, infected stubbles and collateral hosts.
- Seed from infected crop, diseased wild rice growing in ponds and Irrigation water contaminated with the bacterium flowing through field to field a provides the primary inoculum.
- Secondary spread is through wounds and stomata by bacterial cells disseminated by wind borne raindrop splashes, by irrigation water or rainwater coming from infested fields and by contact between diseased and healthy leaves.

Management

- Use of healthy disease free certified seed.
- Grow resistant cultivars like IR 20 and TKM 6.
- Hot water seed treatment (53–54°(C) for 30 minute.
- Application of stable bleaching powder @12.5 kg/ha .
- Moderate level of 60 – 80 kg N /ha with required potassium may be recommended in endemic area during wet season. The nitrogen should be applied in 3–4 splits.
- Practicing field sanitation such as removing weed hosts, infected plants and crop debris, rice straws, and volunteer seedlings.
- Maintaining shallow water in nursery beds, providing good drainage during severe flooding.
- Movement of irrigation water from affected field to the adjacent rice field should be restricted.
- Rouge out the infected plants from the field very carefully.

Tungro

Diagnostic Symptoms

- Plants are markedly stunted.
- Leaves show yellow to orange discoloration and interveinal chlorosis.
- Young leaves are sometimes mottled while rusty spots appear on older leaves.

- Tillering is reduced with poor root system.
- Panicles not formed in very early infection, if formed, remain small with few, deformed and chaffy grains.

Etiology

- Causal organism: *Rice tungro bacilliform virus* (RTBV) and *Rice tungro spherical virus* (RTSV)
- Two morphologically unrelated viruses present in phloem cells.
- Rice tungro bacilliform virus (RTBV), bacilliform capsid, circular ds DNA genome.
- Rice tungro spherical virus (RTSV) isometric capsid, ss RNA genome.

Disease Cycle

- Tungro disease viruses are transmitted from one plant to another by leafhoppers that feed on tungro-infected plants.
- Most efficient vector is the green leafhopper.
- Leafhoppers can acquire the viruses from any part of the infected plant by feeding on it, even for a short time.
- It can, then, immediately transmit the viruses to other plants within 5–7 days.
- Viruses do not remain in the leafhopper's body unless it feeds again on an infected plant and re-acquires the viruses.
- Tungro incidence depends on the availability of the virus sources and vector population.
- Other than infected rice plants in the farmer's field, other primary sources for tungro, include: stubble of previous crops, new growth from infected stubbles that had not been properly plowed under and harrowed effectively, volunteer rice, infected plants in nearby rice fields
- Seedlings raised in nurseries or seedbeds can also be infected with Tungro prior to transplanting and can be a primary source of virus.
- Transplanting seedlings from nurseries in tungro-infected areas has also shown to increase infection rates in the field, particularly, in cases where seedbed is in a tungro-endemic area or when the nursery duration is 5–6 weeks.

Management

- Field sanitation, removal of weed hosts of the virus and vectors.
- Grow disease tolerant cultivars like Pankhari 203, BM66, BM68, Latisail, Ambemohar 102, Kamod 253, IR50 and Co45.
- Set up light traps to monitor the vector population.
- Adjust planting times to when green leafhopper are not in season or abundant, if known plow infected stubbles immediately after harvest to

reduce inoculum sources and destroy the eggs and breeding sites of green leaf hopper.

Khaira

Zinc deficiency

Diagnostic Symptoms

- The leaves of affected plants show chlorosis at the base.
- Large number of small, brown or bronze spots appears on the lamina surface, which coalesce to form bigger spots and ultimately the entire leaf becomes bronze coloured and dries up.
- The growth of diseased plant remains stunted.
- Root growth is also restricted and usually the main roots turn brown.
- The finer roots are destroyed.
- The plants fails to grow further and produce no panicles in severe conditions but sometimes there is natural recovery after 45 days of transplanting.
- The disease appears in nursery but may also appear in patches after 10–15 days of transplanting.

Management

- Spray mixture of zinc sulphate (5 kg) and lime (2.5 kg) in 500 litres of waters after 10 days of sowing in the nursery followed by a second spray as above after 25 days of sowing in the nursery.
- Spray can be repeated after 10–15 days of transplanting if symptoms appear in the main field.
- Zinc sulphate (5 kg) and urea (10 kg) in 500 litres of water/ ha can also be sprayed for control of the disease.

Model Question Paper

A. Objective Type Questions

a. Choose the correct answer from the following:

1. Which disease is related to the Bengal famine?

(a) Blast of rice (b) Brown spot of rice

(c)False smut of rice (d) None of the above

2. What leads to false smut of rice:

(a) *Ustilaginoidea virens* (b) *Helminthosporium oryzae*

(c) *Tilletia indica* (d) *Clavibacter tritici*

3. The first place in India where blast disease was initially recorded to have occurred in 1918.

(a) Tamil Nadu (b) Uttar Pradesh

(c) Bihar (d) None of the above

4. The perfect stage of rice blast is ?
   (a) *Pyricularia oryzae* (b) *Magnaporthe grisea*
   (c) *Ustilaginoidea virens* (d) *Clavibacter tritici*
5. *Magneporthe grisea*produces:
   (a) Perithecium (b) Apothecium
   (c) *Cleistothecium* (d) Pycnidium
6. Mucoid and straw-colored, colonies of bacteria *Xanthomonas oryzae pv. oryzae* grow slowly and eventually turn
   (a) Yellow. (b) Brown
   (c) White (d) Blue
7. Bcterial blight of rice is managed by seeds treatment with hot water for 30 minutes at temperature
   (a) 50^{0}C (b) 54^{0}C
   (c) 58^{0}C (d) 64^{0}C
8. Stable bleaching powder application in the rice control which disease.
   (a) Blast of rice (b) Brown spot of rice
   (c) False smut of rice (d) Bacterial blight of rice
9. Some isolates of *Rhizocotnia solani* cusing sheath blight of rice the mycelium produces barrel shaped cells which have been termed as
   (a) Chlaymydospore (b) Basidiospores
   (c) Pycnidiospore (d) Ascospore
10. Uyeda and Ishiyama provided a detailed description of which disease and its causative organism.
   (a) Blast of rice (b) Brown spot of rice
   (c) False smut of rice (d) Bacterial blight of rice

Answer

S. No	Answer	S. No	Answer
1	b. Brown spot of rice	6	a.Yellow
2	a. *Ustilaginoidea virens*	7	b. 54^{0}C
3	a.Tamil Nadu	8	d. Bacterial blight of rice
4	b. *Magnaporthe grisea*	9	a. Chlaymydospore
5	a. Perithecium	10	d. Bacterial blight of rice

(b) State whether the following statements are *True* or *False:*

1. In India, the stage that produces sclerotia most frequently occurs during the mycelial stage of Rhizoctonia solani..
2. Rice bacterial blight can be effectively controlled by treating seeds with streptocycline.
3. Xanthomonas oryzae pv oryzae is a gram-negative bacterium.

4. The *Rhizoctonia solani* causing sheath blight of rice primarily survives as sclerotia and mycelia in diseased plant debris left in the field.
5. To control rice bacterial blight, it is preferable to restrict the irrigation water flow from the impacted field to the adjacent rice crop.
6. Airborne conidia are primarily responsible for the blast disease's spread.
7. Khaira disease first manifests in nurseries but can also show up in patches ten to fifteen days after rice is transplanted into the main field.
8. Drechslera oryzae, another name for *Helminthosporium oryzae*.
9. Foliar spray the crop rice with Edifenphos (0.1%) manage brown spot of rice.
10. A typical term for brown spot on rice is fungal sesame leaf spot.

Answer

S. No	Answer	S. No	Answer
1	True	6	True
2	True	7	True
3	True	8	True
4	True	9	True
5	True	10	True

B. Descriptive Questions

a. Long answer question

1. Briefly describe the diagnostic symptoms, causal organism, disease cycle and management of the following plant disease
   (a) Blast of rice
   (b) Brown spot of rice
   (c) Sheath blight of rice.
2. Describe in detail the symptoms, causal organism, disease-cycle and management of bacterial blight of rice.
3. Describe in detail the symptoms, causal organism, disease-cycle and management of bacterial streak of rice.
4. Describe in detail the symptoms, causal agent, and management of khaira disease of rice.
5. Write the causal organism and major symptoms of any three of the following plant disease:
   (i) Brown spot of rice
   (ii) Sheth rot of rice
   (iii) False smut of rice.
6. Describe the symptoms and disease management of any two plant diseases caused by the following pathogens:
   (i) *Pyricularia oryzae*
   (ii) *Rhizoctonia solani*
   (iii) *Helminthosporium oryzae*

7. Distinguish between
   a. Leaf blast and Nodal blast
   b. Bengal famine and Irish famine
   c. Epidemiology of Blast and Brown leaf spot.

b. Short answer Questions

1. Differentiate between sheath blight and sheath rot of rice.
2. How would you identify the disease in the field: (a) False smut of rice (b) Bacterial blight rice
3. Write the epidemiological condition of brown spot of rice.
4. Write short notes on integrated management of bacterial blight of rice and Etiology of false smut.
5. Describe epidemiological conditions for development of shealth blight of rice.
6. How would you identify the khaira disease of rice in field.

29

Major Diseases of Wheat Crop & Management

S. No	Major Diseases	Causal Organism
1	Black or stem rust	*Puccinia graminis tritici* Pers
2	Brown or leaf rust	*Puccinia recondita* f. sp. *tritici* Eriks. & Henn
3	Yellow or stripe rust	*Puccinia striiformis var. striiformis* Westend
4	Loose smut	*Ustilago tritici* (Pers.) E. Rostr.
7	Karnal bunt	*Tilletia indica* Mitra

Rust

The rust of wheat has been studied more than any other disease in the world. K.C. Mehta and colleagues conducted exceptional research on the disease in India. There are three recognized wheat rusts. These include *Puccinia graminis tritici*, which causes black stem rust; *Puccinia recondita*, which causes brown rust (leaf rust); and *P. striiformis*, which causes yellow rust. Considering their later seasonal arrival, brown and yellow rusts inflict more damages than black rust.

1. Black or Stem Rust

Diagnostic Symptoms

- Symptoms are produced on almost all aerial parts of the wheat plant but are most common on stem, leaf sheaths.
- Brown pustules of uredina appear on lower surface of the leaves, leaf sheaths and the stems, and on the spikes. which give them a rusty appearance.
- Uredina form urediniospores, which are the repeating spores.
- Uredina gradually start producing teliospores and finally transform into telia.
- Telia are black and crust like, formed abundantly on the stem.
- For this reason, the rust is also called black stem rust.
- Othe stymptoms are stunted growth of the plants, and poor tilleirng, which in severe infections, are not forme(Fig. 2).

Etiology

Systemic position

Phylum: Basidiomycotina

Class: Teliomyetes
Order: Uridinales
Family: Pucciniaceae
Genus: *Puccinia*
Species: *graminis tritici*

- Pathogen ;*Puccinia graminis tritici.*
- Pathogen is an obligate biotroph and has a complex life cycle featuring alternation of generations.
- Pathogen is macrocyclic, heteroecious, requiring two hosts to complete its life cycle - the wheat and barberry.
- Wheat is the primary host and barberry is the alternate host.
- There are many species in *Berberis* and *Mahonia* that are susceptible to stem rust, but the common barberry is considered to be the most important alternate host.
- *Puccinia graminis* can complete its life cycle either with or without barberry (the alternate host).
- *Puccinia graminis* is macrocyclic (exhibits all five of the spore types that are known for rust fungi).
- Characteristic rust color on stems and leaves is typical of a general stem rust as well as any variation of this type of fungus.

Disease Cycle

On Wheat

- Due to its cyclical nature, there is no true 'start point' for this process. Here, the production of urediniospores is arbitrarily chosen as a start point.
- Urediniospores are formed in structures called uredinia, which are produced by fungal mycelia on the cereal host 1–2 weeks after infection.
- Urediniospores are dikaryotic (contain two un-fused, haploid nuclei in one cell) and are formed on individual stalks within the uredinium. They are spiny and brick-red.
- Urediniospores are the only type of spores in the rust fungus life cycle which are capable of infecting the host on which they are produced and this is therefore referred to as the 'repeating stage' of the life cycle.
- It is the spread of urediniospores which allows infection to spread from one cereal plant to another. This phase can rapidly spread the infection over a wide area.
- Towards the end of the cereal host's growing season, the mycelia produce structures called telia.
- Telia produce a type of spore called teliospores.These black, thick-walled spores are dikaryotic.
- Teliospores are the only form in which *Puccinia graminis* is able to overwinter independently of a host.

- Each teliospore undergoes karyogamy (fusion of nuclei) and meiosis to form four haploid spores called basidiospores. This is an important source of genetic recombination in the life cycle.
- Basidiospores are thin-walled and colourless.

Barberry

- Basidiospores cannot infect the cereal host, but can infect the alternative host (Usually barberry).They are usually carried to the alternative host by wind.
- Once basidiospores arrive on a leaf of the alternative host, they germinate to produce a haploid mycelium which directly penetrates the epidermis and colonises the leaf.
- Once inside the leaf the mycelium produces specialised infection structures called pycnia.
- Pycnia produce two types of haploid gametes, the pycniospores and the receptive hyphae.
- Pycniospores are produced in a sticky honeydew which attracts insects.The insects carry pycniospores from one leaf to another.
- Splashing raindrops can also spread pycniospores.
- A pycniospore can fertilise a receptive hypha of the opposite mating type, leading to the production of a dikaryotic mycelium.
- This is the sexual stage of the life cycle and cross-fertilisation provides an important source of genetic recombination.
- This dikaryotic mycelium then forms structures called aecia, which produced a type of dikaryotic spores called aeciospores.
- These have a worty appearance and are formed in chains - unlike the urediniospores which are spiny and are produced on individual stalks.
- Chains of aeciospores are surrounded by a bell-like enclosure of fungal cells.
- Aeciospores are able to germinate on the cereal host but not on the alternative host (they are produced on the alternative host, which is usually barberry).
- They are carried by wind to the cereal host where they germinate and the germ tubes penetrate into the plant.
- Fungus grows inside the plant as a dikaryotic_mycelium. Within 1–2 weeks the mycelium produces uredinia and the cycle is complete.

Disease cycle without barberry

- Since the urediniospores are produced on the cereal host and can infect the cereal host.
- It is possible for the infection to pass from one year's crop to the next without infecting the alternative host (barberry).For instance, infected volunteer wheat plants can serve as a bridge from one growing season to another.

- In other cases the fungus passes between winter wheat and spring wheat, meaning that it has a cereal host all year round.
- Since the urediniospores are wind dispersed, this can occur over large distances.
- This cycle consists simply of vegetative propagation - urediniospores infect one wheat plant, leading to the production of more urediniospores which then infect other wheat plants.

Management

- Use of resistant varieties like HP2278, HW741, WL614, Sonara 63,64 etc. Mostly,the resistant cultivar used in India contain the Sr31 gene, which, unfortunately is susceptible to the fast spreading UG-99 race of the pathogen. The race originally came up in Uganda in 1999, and is fast spreading.It may soon enter South Asia, and knock at our door.The wheat breeders have been altered by GRI (Global Rust Initiative) regarding the need for developing cultivars resistant to the UG-99.
- Mixed cropping of wheat and barley with suitable crop.
- Reduction in proportion of nitrogen in the N.P.K in a fertilizer.

2. Brown or Leaf Rust

- This is the most cmmon rust in North and South India.
- It appears earlier than the black rust.

Diagnostic Symptoms

- Most common site for symptoms is on upper leaf blades, however, sheaths, glumes and awns may occasionally become infected and exhibit symptoms.
- Symptoms begin as small, circular to oval yellow spots on infected tissue of the upper leaf surface.
- As the disease progresses, the spots develop into orange colored pustules which may be surrounded by a yellow halo .
- The pustules produce a large number of spores that are easily dislodged from the pustule resulting in an "orange dust" on the leaf surface or on clothes, hands and equipment.
- As the disease progresses, black spores may be produced resulting in a mixture of orange and black lesions on the same leaf.
- Tiny orange lesions may be present on seed heads, but these lesions do not develop into erumpent pustules.
- Yield loss often occurs

Etiology

Systemic position

Kingdom: Fungi

Phylum: Basidiomyotina

Class: Teliomycetes

Order: Uredianles

Family: Pucciniaceae

Genus: *Puccinia*

Species: *recondita*

- Pathogen *Puccinia triticina (P. recondita).*
- Pathogen is a biotroph and heteroecious.
- Primary host is wheat and the secondary host is *Thalictrum.*Leaf rust is spreads via airborne spores.
- Five types of spores are formed in the life cycle.
- Uredospores, teleutospores, and basidiospores develop on wheat plants and pycnidiospores and aeciospores develop on the alternate hosts.
- Uredospores are brown, round or oblong, echinulate, stalked and have 7-10 germ pores.These spores are dispersed in the air and cause infection to other plants.
- Teliospores are bicelled with flattened top of the upper cell, brown and smooth.They are produced by teliosori, which are small, oval to linear, black in color, and covered by epidermis.
- Teliospores germinate to produce basidia and basidiospores, which normally cause infection in alternate host.
- Functional alternate host, which has not been reported from India, is *Isopyrumfumarioides.*
- In India, the non functional host is *Thalictrum,* which commonly occurs in hills.
- Spermogonial and aecial stages of the pathogen have not been reported.

Disease Cycle

- Pathogen over-summers in low and mid altitudes of Himalayas and Nilgiris.
- Wind blown urediniospores from hills, cause th primary infection.
- They germinate and penetrate through closed stomata on the lower surface, and form dikaryotic mycelium.
- Uredinia are formed and urediniospores cause secondary infections, till end of the season when teliospores are formed.
- The pycnial and aecial stages are not reported in India.

Management

- Collect and destroy crop debris.
- Provide irrigation at critical stages of the crop.
- Avoid water logging.
- Avoid water stress during flowering stage.
- Balance application of NPK.
- Follow mixed cropping and crop rotation.
- Avoid excess application of "N".

3. Yellow or Stripe Rust

- Yellow rust does considerable damage in Punjab, Haryana and Uttar Pradesh, which are the main wheat growing areas.
- It is rare in South India because of the high heat during the crop season.

Diagnostic Symptoms

- Mainly occur on leaves than the leaf sheaths and stem.
- Bright yellow pustules (Uredia) appear on leaves at early stage of crop and pustules are arranged in linear rows as stripes.
- Stripes are yellow to orange yellow and this gives the name 'yellow' rust to the disease
- In severe infections, all parts of the plant are infected, and the stripes disappear due to the crowding of uredinia.
- Teliospores are also arranged in long stripes, dull black in colour.
- Telia appear at the end of the season, as black streaks, on the lower surface.

Etiology

Systemic position

Phylum: Basidiomycotina
Class: Teliomyetes
Order: Uredinales
Family: Pucciniaceae
Genus: *Puccinia*
Species: *striiformis*

- Pathogen;*Puccinia striiformis*
- Pathogen is biotroph and its primary host is wheat.
- Uredospores of rust pathogen are almost round or oval in shape and bright orange in colour.
- Each uredospores possesses 6-10 scattered germ pores.
- Teliospores are bright organge to dark brown, two celled and flattened at the top.
- Sterile paraphyses are also present at the end of sorus.

- No intervening hosts for pycnial and aecial stages of the pathogen have been discovered.

Disease Cycle

- Inoculum survives in the form of uredospores /teliospores in the hills during off season on self sown crop or volunteer hosts, which provide anexcellent source of inoculum.
- About ninety weeds serve as collateral hosts, which can provide the urediniospores for primary infection.
- Wind borne uredospores from hills are lifted due to cyclonic winds cause primary infection in the plains during crop season.
- In India, role of alternate host (Barberis) is not there in completing the life cycle.
- Urediniospores spread the disease until the end of the season, when telia are formed.
- Teliospores have no role in disease causation.

Management

- Grow resistant varieties like PBW 343, PBW 550, PBW 17.
- Collect and destroy crop debris.
- Provide irrigation at critical stages of the crop.
- Avoid water logging.
- Avoid water stress during flowering stage.
- Balance application of NPK.
- Follow mixed cropping and crop rotation.
- Avoid excess application of "N".

Loose Smut

Diagnostic Symptoms

- Infected plants is very difficult to detect in the field until heading.
- Infected heads emerge earlier than normal heads.
- Entire inflorescence is commonly affected and appears as a mass of olive-black spores, initially covered by a thin gray membrane.
- Once themembrane ruptures, the head appears powdery.
- Spores are dislodged, leaving only the rachis intact.
- In some cases remnants of glumes and awns may be present on the exposed rachis.
- Smutted heads are shorter than healthy heads due to a reduction in the length of the rachis and peduncle.
- While infected heads are shorter, the rest of the plantis slightly taller than healthy plants.

Etiology

Systemic position

Kingdom: Fungi

Division: Eumycota

Phylum: Basidiomycotina

Class: Teliomyetes

Order: Ustilaginales

Family: Tilletiaceae

Genus: *Ustilago*

Species: *tritici*

- Pathogen: *Ustilago nuda tritici.*
- Mycelium is hyaline primarily but turns to brown near maturity.It is septate, dikaryotic, and grows systemically inside the plant.
- Mycelia cells get transformed into brown, spherical, echinulate teleutospores.
- Teleutospores measure 5-9 μm in diameter, germinate readily producing promycelium or basidium consisting of four uninucleate cells.
- Basidium, unlike other smut causing fungi, do not produces basidiospores.
- Its cells germinates and give rise to uninucleate hyphae, the primary hyphae; a pair of sexually compatible uninucleate primary hyphae fuse and results in dikaryotic hypha often called infection hypha.

Disease Cycle

- The pathogen perennates as dormant mycelium lying inside the kernel (seed), which has been infected by the infection hypha.
- Mycelium lies dormant within the seed until next season when they are sown.
- On seed germination and growth of the seedling, the mycelium becomes active, starts developing, and moves systemically along the growing seedling.
- Mycelium is hyaline during its growth through the plant, but it turns to brown at maturity.
- When the head emerges, the hyphae accumulate in the floral parts.Their cells are transformed into teleutospores, which completely fill in the spikelets.
- They first remain covered by adelicate silvery membrane, which soon bursts spreading almost all teleutospores to atmosphere.
- They are now wind blown leaving a bare rachis behind. This is usually the time when the healthy heads of adjacent plants are in flowering stage.
- Disseminated teleutospores lodge between the glumes and reach the feathery stigma germinate thereupon in moist stigmatic fluid giving out promycelium or basidium consisting of four cells.

- The basidium produces no basidiospores, but its cells germinate and produce short uninucleate, sexually compatible primary hyphae that fuse in pairs and give rise to dikaryotic mycelium the infection thread.
- Infection hypha penetrates the flower through the stigma or the young ovary wall and become established in the pericarp, integuments, and in the tissues of the embryo before the kernels become mature.
- Mycelium then becomes inactive and remains dormant, primarily in scutellum of the kernel.
- Fungal presence in no way happers the grain formation .
- When such infected kernels are sown in the next season the hyphae become active and show further course of their action.

Management

- Use of resistant varieties like NP710, 718,761,770, Bansipali 808, Bansi 224 etc.
- **Hot water treatment**: The seeds are soaked in water for 4-6 hours at 68-86^{0}F and then placed in water for 2 minutes at 120^{0}F, and finally taken out, dried and sown.
- **Solar energy treatment**: The seeds are soaked into water for four hours (8 am-12 noon), then taken out and dried in sun for four hours (12 noon-4 pm), and then sown.
- In the standing crop, the plants showing yellowing of the boot leaf tip normally are the ones which will give smutted ear heads on emergence. Uproot such plants before ear emergence to reduce the infestation of healthy seeds at later stage.
- Use disease free seeds in the healthy field. For seed production, disease free field/areas to be identified for having crop without considerable inoculum load.
- Burry the infected ear heads in the soil, so that secondary spread is avoided.

Karnal Bunt

Diagnostic Symptoms

- Symptoms of karnal bunt are often difficult to distinguish in the field owing to the fact that incidence of infected kernels on a given head is low.
- Symptoms are most readily detected on seed after harvest.
- Black sorus, containing dusty spores is evident on part of the seed, commonly occurring along the groove.
- Heavily infected seed is fragile and the pericarp ruptures easily.
- Foul, fishy odor associated with karnal bunt.
- Odor is causedby the production of trimethylamine by the pathogen.
- Seed that is not extensively infected maygerminate and produce healthy plants

Etiology

Scientific Classification

Kingdom: Fungi

Division: Eumoycota

Sub-Division: Basidiomyotina

Class: Teliomycetes

Order: Ustilaginales

Family: Tilletiaceae

Genus: *Neovosia*

Species: *indica*

- Pathogen: *Neovosia indica*
- Telitospores: spherical to oval with reticulation looking as curved spines on the episore, dark brown and measure 20-49 μ in diameter.
- Telitospore wall consists of three layers; perisporium, episporium and endosporium.
- On germination, a teleutospore gives rise to a promycelium that bears long, sickle shaped sporidia in a cluster at its tip; each cluster or whorl generally consists of 60-185 sporidia called primary sporidia and are unicellular.
- Primary sporidia are incompatible with each other, so produce another crop of sporidia called secondary sporidia.
- The compatible primary sporidia and secondary sporidia fuse and develop into the dikaryotic hyphae or sporidia.
- The infection is caused by these dikaryotic hyphae or sporidia.

Disease Cycle

- Disease is soil borne and pathogen perennates through teleutospores, which reach the soil either directly falling on the ground or through sticking on the surface of infected seeds and survive until flowering time of the next wheat crop.
- When favourable conditions become available at the time of flowering of the wheat crop, the perennating teleutospores germinate producing short, stout promycelia which give rise to primary sporidia in whorl.
- Primary sporidia produce secondary sporidia.
- Sporidia are wind or water splash disseminated to the flowers whereupon the compatible primary and secondary sporidia fuse producing dikaryotic hyphae and or sporidia.
- These are the dikaryotic hyphae and or sporidia which cause primary infection of individual florets.

- Secondary spread of the pathogen within and between spikelets takes place through secondary sporidia and dikaryotic mycelia produced on the primary infected spikelets.

Management

- Grow resistant / tolerant varieties.
- Low-lying areas of the field accumulate water and are more prone to kernal bunt. Effective land leveling and drainage can reduce disease incidence.
- Decrease seed rate during sowing.
- Increase row spacing during sowing.
- Delayed sowing.
- Avoiding irrigation during the period of awn emergence and end of flowering may hinder disease development.
- Avoid lodging by using balance dose of Nitrogen and Potash.
- Plastic mulching or solarization reduces the chance of teliospore germination.
- Burn the stubble after harvesting.
- Crop rotation with non-host crop.

Model Question Paper

A. Objective Type Questions

a. Choose the correct answer from the following

1. Which of the following diseases is conveyed internally seed borne?

   (a) Powdery mildew of wheat (b) Brown leaf rust of wheat

   (c) Loose smut of wheat (d) None of the above

2. The cause of karnal bunt of wheat is:

   (a) *Puccinia graminis tritici* (b) *Erysiphe graminis tritici*

   (c) *Tilletia indica* (d) *Clavibacter tritici*

3. Which of the following rust appears earliest on wheat ?

   (a) Brown or Leaf rust of wheat(b)Black or Stem rust of wheat

   (c) Yellow or Stripe rust of Wheat (d) None of the Above

4. Who studies the detailed disease cycle of cereal rusts in India?

   (a) R.S. Singh (b) K.C. Mehta

   (c) B.B. Mundkur (d) E.J. Butler

5. *Clavibacter tritici* is the causal organism of:

   (a) Flag smut of wheat (b) Black or Stem rust of wheat

   (c) *Tundu* or Yellow ear rot of wheat (d) Leaf blight of wheat

6. *Neovosia indica* is a soil borne pathogen perennates through

   (a) Teleutospores (b) Uredeospores

   (c) Basidiospore (d) Aeciospore

7. The *Neovosia indica* belongs to which class.
   (a) Teleiomycetes (b) Oomycetes
   (c) Plectomycetes (d) Bsidiomycetes
8. Foul, fishy odor associated with which disease of wheat.
   (a) Powdery mildew (b) Brown leaf rust
   (c) Loose smut (d) Karnal bunt
9. Foul, fishy odor associated with karnal bunt disease of wheat due to the production of chemical.
   (a) Trimethylamine (b) Uric acid
   (c) Hydrocynic acid (d) Succinic acid
10. The mycelia cells of *Ustilago tritici* get transformed into brown, spherical, echinulate
   (a) Teleutospores (b) Uredeospores
   (c) Basidiospore (d) Aeciospore

Answer

S. No	Answer	S. No	Answer
1	c. Loose smut of wheat	6	a.Teleutospores
2	c. *Tilletia indica*	7	a.Teleiomycetes
3	a. Brown or Leaf rust of wheat	8	d. Karnal bunt
4	b. K.C. Mehta	9	a. Trimethylamine
5	c. Tundu or Yellow ear rot of wheat	10	a.Teleutospores

b. State whether the following statements are *True* or *False:*

1. K.C. Mehta described the wheat rust disease cycle in India.
2. Loose smut infected plants is very difficult to detect in the wheat field until heading.
3. The fungus responsible for the karnal bunt infection produces trimethylamine, which is the source of the rotting fish odor in the wheat fields.
4. The rust fungus *Paccinia graminis tritici* is autoecious.
5. Powdery mildew of wheatis caused by *Erysiphe graminis tritici.*
6. Sporids or dikaryotic hyphae are the source of the *Neovosia indica* infection in wheat.
7. In wheat crops, the teleutospore of *Neovosia indica* germinates to form a promycelium that contains long, sickle-shaped sporidia.
8. After harvest, wheat seed is the best place to find the signs of karnal bunt.
9. Because the incidence of infected kernels on a given head is low, it might be challenging to detect the symptoms of karnal bunt in a wheat field.
10. It can be somewhat challenging to identify loose smut-infected plants in a wheat field prior to heading.

Answer

S. No	Answer	S. No	Answer
1	True	6	True
2	True	7	True
3	True	8	True
4	True	9	True
5	True	10	True

B. Descriptive Questions

a. Long Answer Questions

1. How do the three rusts of wheat differ? Describe in brief the different stages of any heteroecious rust occurring on wheat.
2. Describe the present views or recent work on annual recurrenc of wheat rusts in plants of India.
3. Describe the loose smut of wheat under heads, diagnostic symptoms, causal organism, disease cycle and disease management.
4. Describe in detail the symptoms, causal organism, disease-cycle and management of Karnal bunt of wheat.
5. Briefly describe the diagnostic symptoms, causal organism, disease cycle and disease management of the following plant disease

   (a) Brown or Leaf rust of wheat (b) Loose smut of wheat

   (c) Powdery mildew of wheat.
6. Write the causal organism and major symptoms of any three of the following plant disease:

   (i) Yellow or Stripe rust of wheat (ii) Seed gall of wheat

   (iii) Altermeria leaf blight of wheat.

b. Short answer Questions

1. Explain solar energy treatment of wheat seeds.
2. Give the different spore stages of any heteroecious rust.
3. Differentiate between smut and rust.
4. Name the three rust diseases of wheat with their causal organism.
5. Differentiate between yellow rust and brown rust of wheat.
6. How would you identify the flowing disease in the field: (a)Loose smut of wheat (b) Black or Stem rust of wheat
7. Give the disease cycle of powdery mildew of wheat.

30

Major Diseases of Maize Crop & Management

S. No	Major Diseases	Causal Organism
1	Banded leaf and sheth blight	*Rhizoctonia solani*
2	Southern leaf blight	*Helminthosporium maydis*
3	Northern leaf blight	*Exserohilum turcicum*
4	Downy mildew	*Peronosclerospora sorghi*
5	Bacterial stalk rot	*Erwinia dissolvens*

Banded Leaf and Sheath Blight

Diagnostic Symptoms

- From the crop's seedling stage until crop maturity, this pathogen can infect the plant at any point in its growth.
- The disease initially appears on the first and second leaf sheaths above the ground, and it eventually spreads to the ears, where it results in ear rot.
- The hallmark symptoms of ear rot include the presence of light brown, cottony mycelium on the plant's ear, small, rounded black sclerotia, early ear drying, and caking of the ear sheath.
- When there is humidity, the infection that first appears on the lower sheath extends to the upper sheaths, causing leaf sheath rot and complete leaf drying.
- When the condition worsens, the majority of the plant's leaves become blighted, and easily detachable sclerotia form on the lesions.
- The disease manifests as a straw-colored, wet, irregular to roundish patch on both leaf surfaces on the basal sheath.
- Mycelium is growing profusely in the space between the stem and the leaf sheath as well as on the leaf sheath affected area.
- Eventually, as the disease worsens, many sclerotial bodies are discovered growing on the plant's afflicted areas.
- All of the maize plant's aerial sections, with the exception of the tassel, exhibit disease symptoms.
- Leaf and sheath blight is one way that the disease shows up on leaves, stalks, ears, and leaf sheaths

- A disease that affects plants 30 to 40 days old during the pre-flowering period of the plant causes severe blights and the apical part of growing maize plants to die.

Etiology

Classification

Sub division: Deuteromycotina

Class: Deuteromycetes

Sub class: Hyphomycetidae

Order: Agonomycetales

Genus: *Rhizoctonia*

Species: *solani* Kuhn

- Anamorph: *Rhizoctonia solani*
- Teleomorph: *Thanatephorus cucumeris*
- *Rhizoctonia solani* immature colonies appear almost white in the media, but as the colonies age, they take on a hint of brown.
- The consistent features that set this fungus apart are its many colors of brown.
- The majority of hyphae have a diameter that ranges from 5 to 14 µm.
- The most characteristic feature of *R. solani* is its dolipore septa.
- Although it's not always accurate, right-angled mycelium branching is sometimes thought to be a unique characteristic of this fungus.
- Two consistent and dependable traits of *R. solani* seem to be the constriction of branch hyphae at the place of origin and the creation of septum at a right angle in the point of origin.
- *R. solani's* sclerotia are initially grey-white in color, but later on they turn to brown to black in colour.
- They are subglobose and in shape they are slightly flattened varying from 0.5-5.0 mm in size

Disease cycle

- The primary sources of inocolum are the sclerotia found in the soil or in the infected host debris, as well as the active mycelium that is developing close to the maize plant in the field.
- Seeds are not thought of as the source of inoculum and may not have a significant function in an outbreak of serious disease.
- Sclerotia that live on plant debris emerge in the soil during land preparation and other operations, come into contact with freshly planted, healthy seedlings, and infect them.
- After 48 to 72 hours, the infection spreads upward in bands, destroying the tissue in its path and reaching the distal half of the leaf lamina, where

patches of buff white mycellial growth appear. In the 30 to 40 day older plants that are in the pre-flowering stage, sclerotia become active.

- When diseased leaves or sheaths come into contact with healthy plants, the disease spreads secondary. The entire plant is affected by the disease up to the mid-dough stage, including the ear that leaves the tassel.
- Affer ear infections happen in the later stages, affecting seed germination and causing seed rot and seedling blight.
- In maize crops, there has never been evidence of a secondary disease spread by basidiospores.

Management

- Select field with good drainage .
- Sanitation and removal of previous crop debris/wheat straw
- Deep ploughing 2 to 3 times at 10 to 15 days interval to destroy crop debris & weeds.
- Proper seed bed preparation, planting seed in warm, fairly moist soil (above 12.8 °C)
- Use of resistant/tolerant hybrids; Pratap Kanchan-2, Pratap Makka-3, Shaktiman-1, Shaktiman-3.
- Adoption of crop rotation .
- Use balanced soil fertility, avoid high level of N and low level of K .
- Stripping of lower leaves along with their sheath.
- Seed treatment with *Trichoderma harzianum* 2.0% WP @ 20 g/kg of seeds.

Southern Leaf Blight

Diagnostic Symptoms

- Disease affects leaves, leaf sheaths, ear, and maize grains.
- **On leaves**, the grayish-red or stramineous long lesions with dark brown center appear along leaf veins, being spindle-shaped or elliptical.
- Length of the lesions is 40 mm, and their width is about 6 mm.
- Lesions can coalesce, causing death of leaves.
- **On sheaths,** the lesions are brown with purple border.
- Length of the lesions is 50 mm.
- **On ears,** the lesions are spindle-shaped, brown with dark border.
- Germ part of seeds becomes dark, and the seeds lose germinating ability.

Etiology

Scientific Classification

Kingdom: Fungi

Division: Eumycota

Sub-division: Deuteromycotina

Class: Hyphomycetes

Order: Hyphomycetales (Moniliales)

Family: Dematiaceae

Genus: *Helminthosporium*

Species: *maydis*

- Pathogen: *Helminthosporium maydis* (Syn: *H. turcicum*)
- Facultative parasite.
- Teleomorph stage is *Cochliobolus heterostrophus* (Drechsler) Drechsler.
- Produces a specific T-toxin causing the same symptoms of disease as those caused by conidia.
- Conidiophores are in group, geniculate, mid dark brown, pale near the apex and smooth.
- Conidia are fusiform, markedly curved, pale to mid dark golden brown with 5-11 septa.
- Size of conidia is 25-115 x 8.5-20.6 microns.
- Race T appeared on maize hybrids with Texas male sterile cytoplasm.
- Disease development was related to the growing of maize with T-cytoplasm on large areas.
- Since 1990, the use of cultivars with T male sterile cytoplasm has been forbidden in selection.
- Presently, the cultivars with M and C types of sterility are used.

Disease Cycle

- Seed-borne.
- Infected seeds and corn residues are sources of the primary infection.
- Secondary infection through wind borne conidia.
- Wind spreads the spores long distance.
- Also infects sorghum, wheat, barley, oats, sugarcane etc.

Management

- Use of disease-resistant hybrids of maize.
- Removal of infected maize residues.
- Crop rotation.
- Removal of crop residues that contain the infection.

Northern Leaf Blight

Diagnostic Symptoms

- Characteristic symptum is the one-to-six inch long cigar-shaped gray-green to tan-colored lesions on the lower leaves.
- As the disease develops, the lesions spread to all leafy structures, including the husks.

- Lesions may become so numerous that the leaves are eventually destroyed causing major reductions in yield due to lack of carbohydrates available to fill the grain.
- Leaves then become grayish-green and brittle, resembling leaves killed by frost.
- Yield losses can reach as high as 30-50% if the disease establishes itself before tasseling.

Etiology

Scientific Classification

Kingdom: Fungi

Division: Eumycota

Sub-division: Deuteromycotina

Class: Hyphomycetes

Order: Hyphomycetales (Moniliales)

Family: Dematiaceae

Genus: *Helminthosporium*

Species: *turcicum*

- Pathogen: *Exserohilum turcicum.*
- Mycelium is localized in the infected lesion.
- Conidiophores emerge through stomata and are simple, olivaceous, septate and geniculate.
- Conidia are olivaceous brown, 3-8 septate and thick walled.

Disease Cycle

- Pathogen overwinters as mycelia and conidia on corn residues left on the soil surface.
- Conidia are transformed into thick-walled resting spores called chlamydospores.
- New conidia are produced on the old corn residue, and the conidia are carried by the wind or rain to lower leaves of young corn plants.
- Secondary spread within fields occurs by conidia produced on the leaf tissues.

Management

- Planting resistant hybrids.
- A one- to two-year rotation away from corn.
- Destruction of old corn residues by tillage.

Downy mildew/Crazy top

- Sorghum downy mildew - *Peronosclerospora sorghi.*

- Phlippine downy mildew - *Peronosclerospora philippinensis.*
- Crazy top - *Sclerophthora macrospora.*

Diagnostic Symptoms

- Most characteristic symptom is the development of chlorotic stripes along almost the whole length of the leaves.
- Stripes are yellowing white in the start, continue as such, and turn darker quite late.
- Plants exhibit a stunted and bushy appearance owing to shortening of the internodes.
- White downy growth is seen on the lower surface of leaf.
- Downy growth furthermore occurs on bracts of green unopened male flowers in the tassel.
- Small to large leaves are noticed in the tassel.
- Proliferation of auxillary buds on the stalk of tassel and the cobs is common (Crazy top).

Etiology

Scientific Classification

Kingdom: Fungi

Division: Eumycota

Class: Oomycetes

Order: Peronosporales

Family: Peronosporaceae

Genus: *Peronosclerospora*

Species: *sorghi, philippinensis* and *sacchari*

- Hemibiotroph
- Pathogen grows as white downy growth on both surface of the leaves, consisting of sporangiophores and sporangia.
- Sporangiophores are quite short and stout, branch profusely into series of pointed sterigmata which bear hyaline, oblong or ovoid sporangia (conidi(a).
- Sporangia germinate directly and infect the plants.
- In advanced stages, oospores are formed which are spherical, thick walled and deep brown.

Disease Cycle

- Primary source of infection is through oospores in soil or dormant mycelium present in the infected maize seeds or conidia produced in over wintering crop debris and infected neighboring plants.
- Oospores in the soil germinate in response to root exudates from susceptible maize seedlings.

- Germ tube infects the underground sections of maize plants leading to characteristic symptoms of systemic infection including extensive chlorosis and stunted growth.
- Whole plants show symptoms, if the pathogen is seed borne.
- Once the pathogen has colonized host tissue, sporangiophores emerge from stomata and produce sporangia which are wind and rain splash disseminated and initiate secondary infections.
- Sporangia are always produced in the night, fragile and cannot be disseminated more than a few hundred meters and do not remain viable for more than a few hours.
- Germination of sporangia is dependent on the availability of free water on the leaf surface.
- Oospores are produced in large numbers,as the crop approaches senescence.

Management

- Deep ploughing.
- Grow resistant varieties and hybrids *viz.* CO1, COH1and COH2.
- Crop rotation with pulses.
- Rogue out infected plants.

Model Question Paper

A.Objective Type Questions

a.Choose the correct answer from the following

1. A 40 mm diameter and 6 mm long spindle-shaped or elliptical lesion on the leaf is a diagnostic indicator of a maize disease.

(a)	Powdery mildew	(b)	Crazy top
(c)	Southern leaf blight	(d)	Northern leaf blight

2. Teleomorph stage of pathogen *Helminthosporium maydis* is

(a)	*Helminthosporium turcicum*	(b)	*Erysiphe graminis*
(c)	*Tilletia indica*	(d)	*Cochliobolus heterostrophus*

3. Helminthosporium maydis is a member of the order

(a)	Hyphomycetales	(b)	Oomycetales
(c)	Albuginales	(d)	None of the Above

4. Downy mildew of maize is caused by

(a)	*Helminthosporium turcicum*	(b)	*Erysiphe graminis*
(c)	*Peronosclerospora sorghi*	(d)	*Cochliobolus heterostrophus*

5. Southern leaf blight disease development was related to the growing of maize with
   (a) T-cytoplasm (b) C-cytoplasm
   (c) M-cytoplasm (d) K-cytoplasm
6. The one-to-six inch long cigar-shaped gray-green to tan-colored lesions on the lower leaves of maize is characteristic symptom of
   (a) Powdery mildew of maize (b) Crazy top
   (c) Northern leaf blight. (d) Southern leaf blight
7. The conidia of *Helminthosporium turcicum* are transformed into thick-walled resting spores under adverse condition called
   (a) Chlamydospore (b) Uredospore
   (c) Macroconidia (d) Microconidia
8. Proliferation of auxillary buds on the stalk of tassel and the cobs of maize infected with downy mildew is called
   (a) Low top (b) Crazy top
   (c) High top (d) Wavy top
9. Teleomorph stage of pathogen *Rhizoctonia solani* is
   (a) *Helminthosporium turcicum* (b) *Thanatephorus cucumeris*
   (c) *Tilletia indica* (d) *Cochliobolus heterostrophus*
10. The most characteristic feature of *R. solani* is its
   (a) Dolipore septa (b) Clamp connection
   (c) Ascospore formation (d) Basidiospore formation

S. No	Answer	S. No	Answer
1	Southern leaf blight	6	Northern leaf blight
2	*Cochliobolus heterostrophus*	7	Chlamydospores
3	Hyphomycetales	8	Crazy top
4	*Peronosclerospora sorghi*	9	b. *Thanatephorus cucumeris*
5	T-cytoplasm	10	a. Dolipore septa

(b) State whether the following statements are True or False

1. Race T of *Helminthosporium maydis* appeared on maize hybrids with Texas male sterile cytoplasm..
2. Foliar spray with Metalaxyl + Mancozeb @ 0.2% manage downy mildew of maize.
3. Infected seeds and corn residues are sources of the primary infection of southern leaf blight disease.
4. The pathogen *Peronosclerospora sorghi* is hemibiotroph.
5. Primary source of downy mildew infection in maize is oospores in soil or dormant mycelium present in the infected seeds .

6. Oospores are produced in large numbers in the mize infected with downy mildew,as the crop approaches senescence.
7. Young plants are highly susceptible to downy mildew than older plants.
8. Most characteristic symptom downy mildew of mize is the development of chlorotic stripes along almost the whole length of the leaves.
9. *Helminthosporium maydis is* facultative parasite in nature .
10. *Helminthosporium maydis* belong to family dematiaceae.

Answer

S. No	Answer	S. No	Answer
1	True	7	True
2	True	8	True
3	True	9	True
4	True	10	True
5	True		
6	True		

B. Descriptive Questions

a. Long Answer

1. Dfferentiate between Southern and Northern blight disese of maize on the basis of the causal organisms, symptomatology, etiology, disease cycle and management.
2. Describe in detail the occurrence, importance, symptoms, etiology, disease cycle and management of banded leaf and sheath blight of maize.
3. Give the diagrammatic representation of disease cycle of the following diseases of maize.

   i. Downy mildew, ii Banded Leaf and Sheath blight, iii. Powdery mildew
4. Briefly describe the diagnostic symptoms and integrated management of any two of the following maize diseases:

   (i) Southern blight (ii) Banded leaf and Sheath blight

   (iii) Northern blight
5. Illustrate the downy mildew of maize in the following headings:

   (i) Pathogen (ii) Symptoms

   (iii) Disease cycle (iv) Disease management

b. Short Answer Questions

1. How will you differentiate Southern and Northern blight of maize?
2. Why the diseases is called Southern and Northern blight of maize?
3. Which fungal pathogen causes downy mildew of maize? Give its systematic position.
4. Write a short note on Race-T of *Helminthosporium maydis.*
5. Describe the mode of parasitism of *Peronosclerosporasorghi.*

31

Major Diseases of Ground Nut Crop & Management

S. No	Major Diseases	Causal Organism
1	Early leaf spot	*Cercopora arachidicola*
2	Late leaf spot	*Phaeoisariopsis personata*
3	Rust	*Puccinia arachidis*
4	Wilt	*Fusarium oxysporum* and *F. solani*
5	Collar rot or Seedling blight or Crown rot	*Aspergillus niger and A. pulverulentum*

Tikka or Leaf Spot

Cercospora arachidicola/ Phaeoisariopsis personatum

Diagnostic Symptoms

- Disease is caused by two different pathogens, namely, *Cercospora arachidicola and Phaeoisariopsis personatum.*
- The two pathogens mostly occur on the same leaves.
- The *Cercospora arachidicola* appears about a month earlier than the *Phaeoisariopsis personatum.*
- The two pathogens cause symptoms differing each other.

Early leaf spot

- Sub-circular dark brown spots having 1 to over 10 mm diameter are produced on the upper leaflet surface.
- Spots are of lighter shade of brown on the lower side of the leaflets.
- Yellow halo is seen around the brown spots.
- Dark brown in colour on the upper leaflet surface where most sporulation occurs and a lighter shade of brown on the lower leaflet surface.
- Distribution of fruiting structures is random on upper leaflet surface.
- Oval to elongate spots are also seen on stems, petioles and pegs.
- These are oval to elongate in shape and have more distinct margins than the leaflet lesions.
- On severity of the diseases, leaflets become chlorotic, then necrotic, lesions coalesce and leaflets are shed.

Late leaf spot

- Late leaf spots are nearly circular and darker than early leaf spots.
- Lesions circular to sub-circular in shape with 1 to 6 mm diameter.
- All lesions are dark brown to black in colour.
- On the lower leaflet surface where most sporulation occurs the lesions are black in colour and slightly rough in appearance.
- Fruiting structures are in concentric rings on the lower leaflet surface.
- Lesions on other parts and effects on disease severity are similar to that of early leaf spot.

Etiology

Scientific Classification

Kingdom: Fungi

Division: Eumycota

Sub-division: Deuteromycotina

Class: Hyphomycetes

Order: Hyphomycetales (Moniliales)

Family: Dematiaceae

Genus: *Cercospora/Phaeoisariopsis*

Species: *arachidicola/personatum*

Character	*Cercospora arachidicola*	*Phaeoisariopsis personatum*
Mycelium	Primarily intercellular but it becomes intracellular in later stages when the host cells die	Septate, entirely internal, intercellular
Haustoria	No haustoria are produced	Haustoria produced
Conidiophores	Light olivaceous brown, unbranched, geniculate, septate, and are produced on small stromata.	Uniformally olivaceous brown, small unbranched, septate and geniculate (knee shaped),develop on a stroma emerge in groups by rupturing the epidermis
Conidia	Sub hyaline to yellow brown in color, obclavato-cylindric in shape, 35 X 110-2.5 X5.4μm in size and septate (4-12 septa)	Obclavate to obclavato cylindric, colored, 18 X 60-6 X 10μm in size spetate (1-7 septa)
Perfect stage produces	Perithecia and Ascostromata	Perithecia and Ascostromata
No of ascospore produced	08	08
Ascospore	Hyaline, slightly curved, 2 celled, apical cell larger than lower cell	2 celled and constricted at septum and hyaline

Disease Cycle

- Pathogen survives for long period of time in the infected plant debris through conidia, dormant mycelium and perithecia in soil.
- Volunteer groundnut plants also harbour the pathogen.
- Primary infection is by ascospores or conidia from infected plant debris or infected seeds.
- Secondary spread is by wind borne conidia .
- Rain splash also helps in the spread of conidia.
- Conidia germinate by germ tubes and cause infection directly or through stomata.
- Incubation period is 8-15 days.

Management

- Grow tolerant varieties like ICGV 89104, ICGV 91114 (EM), ICGV 920920, ICGV 92093 (MM).
- Intercropping pearl millet or sorghum with groundnut (1: 3) .
- Follow cereal-cereal- groundnut crop rotation.
- Deep burying of crop residues in the soil, and removal of volunteer groundnut plants .
- Foliar application of aqueous Neem leaf extract (2-5%) or 5% Neem seed kernel extract at 2 week's interval 3 times starting from 4 weeks after planting.

Rust

Diagnostic Symptoms

- Every aerial portion of the plant is affected by the disease. Usually, the disease is found when the plants are roughly six weeks old.
- On the underside of leaves, tiny, dusty, brown to chestnut-colored pustules called uredosori develop.
- A powdery mass of uredospores is visible when the epidermis ruptures.
- The upper surface of leaves develops tiny, necrotic dark patches that correspond to the sori.
- On the petioles and stem, rust pustules are visible.
- Brown teliosori, which resemble dark pustules, emerge among the necrotic areas towards the end of the season.
- Lower leaves that are severely infected dry out and drop early. Small, shriveled seeds are produced as a result of the serious disease(Fig-3).

Etiology

- Pathogen: *Puccinia arachidis*
- There are uredial and telial phases produced by the pathogen.

- In groundnuts, uredial stages are produced in large quantities, but telia production is restricted.
- Uredospores are yellow, circular, oval, or pedicellate, unicellular, and echinulated with two or three germpores.
- Teliospores have two cells and a dark brown color.
- No records exist for pycnial or aecial stages, and knowledge regarding the function of an alternate host is nonexistent.

Disease Cycle

- On volunteer groundnut plants, the pathogen survives as uredospores.
- The fungus can also thrive in soil that has been infected by plant debris.
- Uredospores are mostly dispersed via wind-borne inoculum.
- Additionally, the uredospores propagate through contaminating seeds and pods. Dissemination aids include rainsplash and tools.
- Additionally, the fungus thrives on secondary hosts such as *Arachis marginata*, *A. nambyquarae*, and *A. prostrate.*

Management

- Avoid monoculturing of groundnut.
- Remove volunteer groundnut plants and reservoir hosts.
- Grow moderatelyresistant varieties like ALR 1.

Wilt

Fusarium oxysporum

Diagnostic Symptoms

- Germinating seeds are attacked by the pathogens shortly before emergence.
- General tissue disintegration and the surface of the seedling is covered with sporulating mycelium.
- Damping off symptoms characterized by brown to dark brown water soaked sunken lesions on the hypocotyl which later encircle the stem and extend above the soil level.
- Roots are also attacked, especially the apical portions.
- Affected seedlings become yellow and wilted.
- Leaves turn greyish green and the plants dry up and die.
- Roots and stems show internal vascular browning and discoloration.
- Pathogens are also commonly associated with pod rot.

Etiology

Scientific Classification

Kingdom: Fungi

Division: Eumycota

Sub-division: Deuteromycotina

Class: Hyphomycetes

Order: Hyphomycetales (Moniliales)

Family: Dematiaceae

Genus: *Fusarium*

Species: *oxysporum*

- Aerial mycelium first appears white, and then may change to a variety of colors.
- Pathogen produces three types of asexual spores; macroconidia,microconidia and chlamydospores.
- Macroconidia are nearly straight, slender and thin-walled. They usually have three or four speta, a foot-shaped basal cell and a curved and tapered apical cell. They are generally produced from phialides in false heads by basipetal division.
- Microconidia are ellipsoidal and have either a single septum or none at all.
- They are formed from phialides in false heads by basipetal division.
- Chlamydospores are globose with thick walls. They are either formed from hyphae or by the modification of hyphal cells.

Disease Cycle

- Pathogen survives in the soil between crop cycles in infected plant debris.
- The pathogen can survive either as mycelium, or as any of its three different spore types viz., macroconidia, microconidia and chlamydospores.
- Pathogen can invade a plant either with its sporangial germ tube or mycelium by invading the plant's roots.
- Roots can be infected directly through the root tips, through wounds in the roots, or at the formation point of lateral roots, Once inside the plant, the mycelium grows through the root cortex intercellulary.
- When the mycelium reaches the xylem, it invades the vessels through the xylem's pits. At this point, the mycelium remains in the vessels, where it usually advances upwards toward the stem and crown of the plant.
- As it grows, the mycelium branches and produces microconidia, which are carried upward within the vessel by way of the plant's sap stream. When the microconidia germinate, the mycelium can penetrate the upper wall of the xylem vessel, enabling more microconidia to be produced in the next vessel.
- Due to the growth of the pathogen within the plant's vascular tissue, the plant's water supply is greatly affected.
- This lack of water induces the leaves' stomata to close, the leaves wilt, and the plant eventually dies.

Management

- Avoid contamination of clean fields with soil or crop debris from infested fields.
- Adequate irrigation should be applied.
- Harvest early where a large number of plants are infected early in the season.
- Avoid crop rotations with susceptible crops such as cotton and potato.
- Long rotations with non-hosts such as corn, grain sorghum, and sudan grass.
- Seed treatment with spores and mycellial fragments of *Trichoderma viride* .

Model Question Paper

A.Objective Type Questions

a. Choose the correct answer from the following:

1. *Phaeoisariopsi personatum* belongs to order.
   - (a) Agonomycetales (b) Peronosporales
   - (c) Hyphomycetales (d) Moniliales
2. Early leaf spot of groundnut is caused by:
   - (a) *Cercospora arachidicola* (b) *Rhizoctonia bataticola*
   - (c) *Sphacelotheca sorghi* (d) *Sphacelotheca cruenta*
3. Late leaf spot of groundnut is caused by:
   - (a) *Cercospora arachidicola* (b) *Phaeoisariopsis personatum*
   - (c) *Sphacelotheca sorghi* (d) *Sphacelotheca cruenta*
4. The symptom of which disease remain restricted to the lower leaf surface and cushion of conidiophores can be seen in concentric rings
   - (a) Early leaf spot (b) Late leaf spot
   - (c) Wilt (d) None of the above
5. Secondary spread of leaf spot disease of groundnut is by.
   - (a) Soil borne conidia (b) Air borne conidia
   - (c) Externally seed borne conidia (d) Enternally seed borne conidia
6. Spots are surrounded by yellow halo is characterisitic symptom of which disease of groundnut
   - (a) Early leaf spot (b) Late leaf spot
   - (c) Wilt (d) None of the Above
7. The teleomorphic stage of *Cercospora arachidicola*is
   - (a) *Mycosphaerella arachidis* (b) *Sphacelotheca cruenta*
   - (c) *Rhizoctonia bataticola* (d) None of the Above

8. The teleomorphic stage of *Phaeoisariopsis personatum*
   (a) *Mycosphaerella arachidis* (b) *Mycosphaerella berkeleii*
   (c) *Rhizoctonia bataticola* (d) None of the above

Answer

S.No	Answer	S. No	Answer
1	c. Hyphomycetales	5	b. Air borne conidia
2	a.*Cercospora arachidicola*	6	a. Early leaf spot
3	b. *Phaeoisariopsis personatum*	7	a. *Mycosphaerella arachidis*
4	b. Late leaf spot	8	b. *Mycosphaerella berkeleii*

b. State whether the following statements are *True* or *False:*

1. The groundnut leaf spot pathogen lives for a considerable amount of time in the diseased plant waste through conidia, latent mycelium, and perithecia in the soil.
2. Ascospores or conidia from contaminated plant debris or seeds cause the primary infection of groundnut leaf spot disease.
3. Wind-borne conidia are responsible for the secondary transmission of groundnut leaf spot disease.
4. Three different asexual spore forms are produced by *Fusarium oxysporum*: macroconidia, microconidia, and chlamydospores.
5. The establishment of the disease *Fusrium oxysporum* within the groundnut vascular tissue has a significant impact on the plant's water supply. As a result, the leaves wilt and the plant eventually dies.
6. In groundnuts, the *Phaeoisariopsis personatum* appears approximately one month later than the *Cercospora arachidicola.*
7. The early phases of groundnut late leaf spot development do not exhibit a yellow halo surrounding the spots.
8. Early leaf spot on groundnuts has spots with a yellow halo surrounding them.
9. The lower leaf surface of groundnuts is still home to *Phaeoisariopsis personatum* conidia, which are visible as a cushion of conidiophores arranged in concentric circles.
10. *Phaeoisariopsis personatum* uses its haustoria to absorb nutrients from its host.

Answer

S. No	Answer	S. No	Answer
1	True	6	True
2	True	7	True
3	True	8	True
4	True	9	True
5	True	10	True

B. Descriptive Questions

a. Long Answer Questions

1. Describe the late leaf spot of ground nut under heads, diagnostic symptoms, causal organism, disease cycle and disease management.
2. Which are the important fungal diseases of groundnut ? Describe any one under heads, diagnostic symptoms, causal organism, disease cycle and disease management.
3. Describe in detail the symptoms, causal organism, etiology, disease-cycle, epidemiology and management of wilt of groundnut.
4. Describe in detail the symptoms, causal organism, etiology, disease-cycle, epidemiology and management of rust of groundnut.
5. Write short notes on
   a. Disese cycle of wilt of groundnut
   b. Management of rust of groundnut
   c. Etiology of *Cercopora arachidicola* cusing early leaf spot of groundnut.

b. Short Answer Questions

1. Write down the etiology of *Phaeoisariopsis personata*.
2. Which fungal pathogen cause early leaf spot of groundnut? Give its systematic position.
3. Elaborate disease cycle of late leaf spot of groundnut in pictorial form .
4. Describe in brief about mangment of leaf spot of groundnut.
5. Differentiate between early and late leaf spot of groundnut.
6. How would you identify the flowing disease in the field: (a) wilt of groundnut(b) rust of groundnut.
7. Write the epidemiological condition for wilt of groundnut

32

Major Diseases of Gram Crop & Management

S. No	Name of Diseases	Causal Organism
1	Ascochyta blight	*Ascochyta rabiei* Pass Labr
2	Wilt	*Fusarium oxysporum*f.sp. *ciceris*(Padwick)Matuoet. K Sato
3	Gray mold	*Botrytis cinerea* Pers.Ex.Fr
4	Rust	*Uromycesciceris-arietini* Grogn.Jacz.&Beyer
5	Collar rot	*Sclerotium rolfsii* Sacc
6	Dry root rot	*Rhizoctonia bataticola*Taub Butler
7	Wet root rot	*Rhizoctonia solani*

Ascochyta Blight

Diagnostic Symptoms

- All above ground parts of the plant are infected.
- Round or elongated lesions, bearing irregularly depressed brown spot and surrounded by a brownish red margin on leaf.
- Similar spots may appear on the stem and pods.
- Spots on the stem and pods have pycnidia arranged in concentric circles as minute block dots.
- When the lesions girdle the stem, the portion above the point of attack rapidly dies.
- If the main stem is girdles at the collar region, the whole plant dies (Fig.1).

Etiology

Scientific Classification

Kingdom: Fungi

Division: Eumycota

Sub-division: Deuteromycotina

Class: Coelomycetes

Order: Sphaeropsidales

Family: Sphaeropsidaceae

Genus: *Aschochya*

Species: *rabiei*

- Pathogen: *Ascochyta rabiei*
- Pathogen occurs as both an anamorph (nonsexual state) and teleomorph (sexual state).
- Anamorph: *Ascochyta rabiei*
- Mycelium: hyaline to brown, septate.
- Pycnidia: fruiting body formed on stems, leaves and pods, round or flattened, globose, dark brown, immersed in the tissues, measuring 100 to 200 µm in diameter, with a prominent ostiole are spherical to sub-globose.
- Conidia: formed inside pycnididum,hyaline, oval to oblong, straight or slightly curved, 1-septate, some 0-septate, slightly or not constricted at the septum, rounded at each end and measure 10-16x3.5 µm in size.
- Teleomorph:*Didymella rabiei(Mycosphaerella rabiei)*
- In perfect stage, the dark colored, globose pseudothecia are formed, each measuring 100 to 140 µm.
- Pseudothecia are characterized by eight ascospore asci.The ascospores are bicelled with one cell larger than the other,measure 12.5-19.0x6.7 -7.6 µm.

Disease Cycle

- Pathogen survives in the infected plant debris as pycnidia or in seed.
- Primary infection is from inoculums produced by pathogen from seed-borne pycnidia and plant debris in the soil.
- Secondary infection is mainly through air-borne conidia.
- Rain splash also helps in the spread of the disease.

Management

- Use only certified seed.
- Use resistant varieties *viz.* Him chana -1, Gaurav, Vardan, Samrat, PBG 1 and BG 261
- Exposure of seed at 40-50°C reduced the survival of *A. rabiei*by about 40-70 per cent.
- Destroy blight-infested crop residues and volunteer chickpea plants.
- Choose crop rotations that permit 3 to 4 years between successive chickpea crops.

Botrytis Gray Mold

Diagnostic Symptoms

- Disease is usually seen at flowering time when the crop canopy is fully developed.
- Lack of pod setting is the first indication of the disease.
- Under favorable conditions foliage shows clear symptoms and plants often die in patches.

- More severe on portions of the plant hidden under the canopy.
- Shed flowers and leaves, covered with the spore mass, can be seen on the ground under the plants.
- When humidity is very high, the symptoms appear on stems, leaves, flowers and pods as gray or dark brown lesions covered with moldy sporophores.
- Lesions on stem girdle the stem completely .
- Tender branches break off at the point where grey mold has caused rotting.
- Affected leaves and flowers turn into a rotting mass.
- Lesions on the pod are water soaked and irregular.
- Grayish white mycelium may be seen on the infected seeds.

Etiology

- Pathogen: *Botrytis cinerea* pers. Ex. Fr.
- Mycelium: septate, brown, hyphae being 8-16μ m wide.
- Conidiophores: light brown, septate, erect and their tips are slightly enlarged bearing small and pointed sterigmata.
- Conidia: hyaline, 1-celled, oval and are borne in clusters on short sterigmata. Conidia in mass are ash-grey in colour, and measure 4-16 x 4-20μ m.

Disease Cycle

- Pathogen survives in the infected plant debris. However, infected seed is thought to be the primary source of infection.
- Pathogen is a facultative parasite with a very wide host range in the temperate and sub-tropical region.
- It causes gray mould disease in a number of crop plants, such as strawberry, grapevine, apple, cabbage, carrot, cucumber, eggplant, lettuce, pepper, squash, tomato and several ornamentals.
- Pathogen also produces highly resistant sclerotia as survival structures in older cultures. It overwinters as sclerotia or intact mycelia, both of which germinate in spring to produce conidiophores.
- Spores of *B. cinerea* are released from conidiophores and then dispersed by air currents or rain splash and cause new infections.

Management

- Grow resistant varieties: BG 276, GL 90159, GL 91040, GL 91071, and GL 92162.
- Adopting wider spacing.
- Intercropping with linseed.
- Avoiding excessive vegetative growth.
- Avoiding excessive irrigation.
- Growing resistant varieties.

Wilt

Diagnostic Symptoms

- At first green chlorosis, typically appear on lower leaves and extending up the plant. Leaves eventually take on a dull-yellow colour, wilt and the plant collapses and dies.
- Adult plants wilt, with their petioles and rachises drooping.In some cases there may be leaf vein clearing before wilt begins.
- Internally, the xylem tissues stain dark brown to almost black.
- Wilting may initially affect only one side of the plant.

Etiology

Scientific Classification

Kingdom: Fungi

Division: Eumycota

Sub-division: Deuteromycotina

Class: Hyphomycetes

Order: Hyphomycetales (Moniliales)

Family: Dematiaceae

Genus:*Fusarium*

Species:*oxysporum*f sp *ciceri*

- Pathogen: *Fusarium oxysporum*f.sp. *ciceris.*
- Mycelium: Produces profusely branched septate hyphae, hyaline to light brown in color.
- Macroconidia:1-7 (mostly 3-5) septate, 25-55 to 2.5-6 μm, fusiform, curved, with a tapering, pointed, sometimes hooked, apical cell and distinctly pedicellate basal cell. In some strains macroconidia are sparse, in others, abundantly produced in pinkish sporodochial conidiomata.
- Microconidia: Borne on simple short conidiophores arising laterally on the hyphae. These are oval to cylindrical, straight to curved, and 5-15 to 2-5 μm, collecting in small, slimy droplets.
- Chlamydospores: Globose, hyaline 7-15 μm diameter, usually forming abundantly in mature colonies, terminal or intercalary, single or in small groups or chains.

Disease Cycle

- Pathogen is mainly soil borne as well as seed borne and also it is a facultative saprophyte.It can survive in the soil upto six years in the absence of susceptible host.
- Primary infection is through chlamydospores in soil, which remain viable upto next crop season.

- Secondary spread is through irrigation water, cultural operations and implements.
- Following infection of host roots, the fungus crosses the cortex and enters the xylem tissues.
- It then spreads rapidly up through the vascular system, becoming systemic in the host tissues, and may directly infect the seed.

Management

- Grow resistant cultures like Avrodhi, Alok Samrat, Pusa-212, JG- 322 GPF-2, Haryanachana-1 and Kabuli chickpea like Pusa-1073 and Pusa-2024, ICCC 42, H82-2
- Seeds treatment with *Trichoderma viride* at 4 g/kg or *Pseudonomas fluorescens* @ 10g/kg of seed.
- Apply heavy doses of organic manure or green manure.

Model Question Paper

A. Objective Type Questions

a. Choose the correct answer from the following

1.*Aschochyta rabiei* belongs to order.

(a) Sphaeropsidales (b) Peronosporales
(c) Hyphomycetales (d) Moniliales

2. Ascochyta blight of gram is caused by:

(a) *Ascochyta rabiei* (b) *Rhizoctonia bataticola*
(c) *Sphacelotheca sorghi* (d) *Sphacelotheca cruenta*

3. The teleomorphic stage of *Aschochyta rabiei* is

(a) *Didymella rabiei* (b) *Sphacelotheca cruenta*
(c) *Rhizoctonia bataticola* (d) None of the above

4. The teleomorphic stage of *Aschochyta rabiei* form dark colored, golobose

(a) Pseudothecia (b) Perithecia
(c) Apothecia (d) Cleistothecia

5. The fruiting body formed by *Aschochyta rabiei* in chickpea is

(a) Pseudothecia (b) Pycnidia
(c) Apothecia (d) Cleistothecia

6. The secondary spreads of *Ascochyta blight* in gram is through

(a) Soil borne conidia (b) Air borne conidia
(c) Seed borne conidia (d) Vector

7. The pathogen *Botrytis cinerea* produces highly resistant surviving structure in in old culture called

(a) Pseudothecia (b) Sclerotia
(c) Apothecia (d) Cleistothecia

8. Wilt of chickpea is caused by:
   (a) *Ascochyta rabiei* (b) *Rhizoctonia bataticola*
   (c) *Fusarium oxysporum f.sp. ciceri* (d) *Sphacelotheca cruenta*
9. *Fusarium oxysporum f.sp. ciceri* produces:
   (a) Mcroconidia (b) Microconidia
   (c) Chlaymydospores (d) All
10. Dry root rot of gram is caused by:
   (a) *Rhizoctonia bataticola* (b) *Rhizoctonia solani*
   (c) *Sphacelotheca sorghi* (d) *Sphacelotheca cruenta*

Answer

S. No	Answer	S.No	Answer
1	a. Sphaeropsidales	6	b. Air borne conidia
2	a. *Ascochyta rabiei*	7	b. Sclerotia
3	a. *Didymella rabiei*	8	c. *Fusarium oxysporum f.sp. ciceri*
4	a. Pseudothecia	9	d. All
5	b. Pycnidia	10	a. *Rhizoctonia bataticola*

b. State whether the following statements are *True* or *False*.

1. In nature, *Fusarium oxysporum* f.sp. *ciceri*, the pathogen that causes chickpea wilt, is primarily carried by soil.
2. *Ascochyta rabiei's* ideal condition, or teleomorph, is *Didymella rabiei.*
3. The earliest sign of the botrytis gray mold disease in grams is the absence of pod setting.
4. Chlamydospores, microconidia, and macroconidia are the three forms of conidia produced by the fungus *Fusarium oxysporum* f.sp.*ciceri*.
5. The application of Trichoderma as a protective agent can help manage chickpea wilt disease.
6. Eight ascospore asci distinguish *Ascochyta rabiei* pseudothecia.
7. Grayish white mycelium found on the infected seeds of chickpea is a diagnostic symptom of botrytis grey mold disease .
8. Chickpea wilt disease-infected xylem tissues discolor from dark brown to nearly black.

Answer

S. No	Answer	S. No	Answer
1	True	5	True
2	True	6	True
3	True	7	True
4	True	8	True

Descriptive Questions

a. Long Answer Questions

1. Describe in detail the most distinguishing symptoms, causal organism, disease cycle and management of botrytis grey mould of chickpea.
2. Describe in detail the symptoms, etiology, disease cycle and management of Ascochyta blight of chickpea.
3. Briefly describe the most distinguishing symptoms and disease management of any two of the following plant disease:
   (i) Collar rot of chickpea (ii) Dry root rot of chickpea
   (ii) Wiltof chickpea
4. Describe the wilt of chickpea in the following headings
   (i) Pathogen (ii) Symptoms
   (ii) Disease cycle (iv) Disease management
5. How would you identify the following diseases in field:
   (i) Collar rot of chickpea (ii) Dry root rot of chickpea
   (ii) Wilt of chickpea (iv) Wet root rot of chickpea
6. Describe the disease cycle and management of the following diseases:
   (i) Wilt of chickpea (ii) Botrytis grey mould of chickpea

b. Short Answer Questions

1. Compare and contrast wet root rot and dry root rot of chickpea giving suitable measures for their management.
2. Mention on diagnostic symptoms and management of wilt of chickpea.
3. Discuss about the disease cycle and epidemiology of botrytis grey mould of chickpea.
4. Give the disease cycle and predisposition of Ascochyta blight of chickpea.
5. Write down an integrated management practices of Botrytis grey mould of chickpea.

33

Major Diseases of Pea Crop & Management

S. No	Name of Disease	Causal Organism
1	Downy mildew	*Peronospora pisi* Sydow
2	Powdery mildew	*Erysiphe pisi* DC. Syn. *E. polygoni* DC.
3	Rust	*Uromyces fabae (Pers.)* Schrot. &Uromyces *fabae DB. f. sp. pisisativae* Hiratsuka
4	Anthracnose	*Colletotrichum pisi*
5	Fusarium wilt	*Fusarium oxysporum f. sp. pisi*

Downy Mildew

Diagnostic Symptoms

- At first grayish white, moldy growth appears on the lower leaf surface, and a yellowish area appears on the opposite side of the leaf.
- Infected leaves can turn yellow and die if weather is cool and damp.
- Stems may be distorted and stunted.
- Brown blotches appear on pods, and mold may grow inside pods.

Etiology

Scientific Classification

Phylum: Heterokontophyta

Class: Oomycota

Order: Peronosporales

Family: Peronosporaceae

Genus: *Peronospora*

Species: *pisi*

- Pathogen: *Peronospora pisi.*
- Pathogen is an obligate parasite.
- Mycelium: coenocytic, hyaline, profusely branched, intercellular sending branched, finger shaped haustoria inside the host cells.
- Conidiophores: unbranched for about two third or more of their length and dichotomously branch at the tip.Branches are long, slender and pointed, and are called sterigmata.A single conidium is produced at the tips of each sterigmata.

- Conidia: oval to elliptical and light yellow in color.Conidia are short lived and germinate readily by germ tube.
- Antheridia &Oogonnia: When the pathogen moves towards the end of the growing season, its hyphae move deep into the host tissue interspaces wherein they produce sex organs (antheridia and oogonia) in close proximity.
- Oospore: After fertilization via fertilization tube, thick walled, green yellow, warty oospores are produced at the rate of one in each oogonium.

Disease Cycle

- Pathogen survives in the soil and on old pea trash through oospores.It can also be seed-borne.
- Oospores perennating in soil germinate by germ tube at the return of favorable conditions.
- The germ tube infects underground parts of young seedlings and the pathogen progresses upwardly systemically causing primary infection..
- These diseased plants provide the wind blown spores that spread the disease in the field and serve as source of secondary infection.
- The oogonial and oospore stages occur later; before harvest, the oospores finally reaches the soil, where they survive one to two years.

Management

- Field sanitation.
- Growing of resistant varieties.
- Deep ploughing of fields during summer.
- Three summer ploughings at 10 days interval reduces diseases level.
- Timely sowing should be done.
- Destroy the alternate host plants .
- Apply manures and fertilizers as per soil test.
- Destruction of diseased plant debris.

Powdery Mildew

Diagnostic Symptoms

- At first symptom appears on leaves.
- White floury patches appear on both surface of leaves as well as tendrils, stem and pod.
- As plant become older, the symptoms alsomt cover the entire plant, become more or less grayish brown and the infected parts impart dirty appearance.
- Fruits do not either set or remain very small.
- Later stages, powdery growth also covers the pods.

Etiology

Scientific Classification

Kingdom: Fungi

Division: Eucomycota

Division: Ascomycotina

Class: Pyrenomycetes

Order: Erysiphales

Family: Erysiphaceae

Genus:*Erysiphe*

Species:*polygoni*

- Pathogen: *Erysiphe polygoni*
- Pathogen: obligate parasite and ectophite, only haustoria enter into the cell.
- Mycelium: septate, hyaline, profusely branched, and superficial forming a white web like coating over the surface of the infected part.It sends finger like haustoria into the epidermal cells of the host to obtain nutrients.After maturity, many hyphal branches give rise to erect, stout, conidiophores, which produce conidia in chain.
- Conidiophores:bears spores in chain, septate and their cells appear like conidia in shape.
- Conidia: is elliptical or barrel shaped, hyaline, single celled, and measures 25-35x13-16 µm.
- Cleistothecia: black, minute bodies provided with a number of appendages, and remain scattered in the mycelia web, contains usually 2-8 asci.
- Ascospores: 3-8, hyaline, one celled, elliptical and unicellular, measuring 19-25x9-14µm.

Disease Cycle

- Pathogen survives in soil and plant debris in the form of cleistotheica and may survive in seed in the form of dormant mycelium.
- Primary infection occurs by asci and ascospores releases after breakdown of cleistotheica or conidia from collateral hosts.
- Secondary spread takes plae by wind borne conidia produced as a result of primary infection.

Management

- Field sanitation.
- Growing of early maturing and resistant varieties like Rachna, Pant P5, DMR 11, HUP 2, JP 885, KFP 103, Ambika, Shubhra, Aparna, Azad P4, Pusa Panna
- Control volunteer field peas.
- Avoid sowing field pea crops adjacent to last season's stubble.

- Burn infected pea stubble soon after harvest.
- Control volunteer field peas which can support disease.

Rust

Diagnostic Symptoms

- Disease appears usually at the beginning of flowering.
- Leaves of infected plants exhibit many small, orange-brown pustules usually at the lower surface.
- Severely infected leaves wither and may drop from the plant.
- Larger pustules occur on the stems and isolated pustules may be found on the pods.

Etiology

Phylum: Basidiomycotina
Class: Teliomyetes
Order: Uridinales
Family: Pucciniaceae
Genus: *Uromyces*
Species: *pisi*

- Pathogen: *Uromyces pisi* (Pers.) Schrot.
- Autoecious rust:Pathogen passes all stages on peas completing its life cycle on pea.
- Aecial stage: appear first on the lower surface of leaves and on the stems and petioles, where the spermogonia also occur mixed with aecia.
- Aeciospores:are elliptical, yellowish brown, measure 14-22μ in diameter and have finely warty cell walls.
- Uredial stage: which is repeated several times during the season, produces spiny light brown uredospores.
- Urediniospores: are rounded, single, spinose, light brown, 21-30 x 18-26 microns.Pathogens produce several generations of urediniospores on peas.
- Telial stage: are not powdering, dark-brown, almost black.
- Teliospores: are unicellular, rounded, glabrous, positioned on colorless leg, 25-40 x 18-28 microns in size.It germinate to produce four celled basidia.
- Basidiospores: on which four sporidia are formed.
- Spores are spread by means of wind.

Disease Cycle

- Pathogen survives in soil and plant debris.
- Primary infection on the newly growing crop is considered to be caused by the germination of teleutospores present in soil, seed and collateral hosts in the month of January.
- Secondary infection by aeciospores or secondary aecia.

Management

- Field sanitation.
- Growing of resistant varieties.
- Early sowing.
- Destruction of diseased plant debris.
- Crop rotation.

Model Question Paper

A. Objective Type Questions

a. Choose the correct answer from the following

1. *Peronospora pisi* is an order member.
   (a) Sphaeropsidales (b) Peronosporales
   (c) Hyphomycetales (d) Moniliales
2. Downy mildew of pea is caused by:
   (a) *Peronospora pisi* (b) *Rhizoctonia bataticola*
   (c) *Sphacelotheca sorghi* (d) *Sphacelotheca cruenta*
3. The *Peronospora pisi* responsible for pea downy mildew is
   (a) Facultative parasite (b) Obligate parasite
   (c) Obligate saprophyte (d) Fcultative saprophyte
4. The pathogen *Peronospora pisi* survives in the soil and on old pea trash through
   (a) Oospore (b) Perithecia
   (c) Apothecia (d) Cleistothecia
5. The fruiting body formed by *Erysiphe polygoni* in pea is
   (a) Pseudothecia (b) Pycnidia
   (c) Apothecia (d) Cleistothecia
6. The secondary spreads of powdery mildew in pea is through
   (a) Soil borne conidia (b) Air borne conidia
   (c) Seed borne conidia (d) Vector
7. Wetable sulfur is a chemical that has been discovered to be very effective for controlling
   (a) Downy mildew (b) Powdery mildew
   (c) Rust (d) Late blight
8. Which pathogen is an obligate parasite and ectophite, only haustoria enter into the cell
   (a) *Ascochyta rabiei* (b) *Rhizoctonia bataticola*
   (c) *Erysiphe polygoni* (d) *Sphacelotheca cruenta*
9. *Uromyces pisi* is a:
   (a) Autocious rust (b) Heterocious rust
   (c) Dioecious rust (d) All

10. Rust of pea is caused by:
   (a) *Uromyces pisi* (b) *Rhizoctonia bataticola*
   (c) *Sphacelotheca sorghi* (d) *Sphacelotheca cruenta*

Answer

S. No	Answer	S. No	Answer
1	a. Sphaeropsidales	7	b. Sclerotia
2	a. *Ascochyta rabiei*	8	c. *Fusarium oxysporum f.sp. ciceri*
3	a. *Didymella rabiei*	9	d. All
4	a. Pseudothecia	10	a. *Rhizoctonia bataticola*
5	b. Pycnidia	11	a. *Rhizoctonia solani*
6	b. Air borne conidia	12	b. *Sclerotium rolfsii*

b. State whether the following statements are True or False

1. The *parasite* Erysiphe *polygoni* is an obligatory one.
2. To control pea downy mildew, use a folar spray containing 0.2% of Metalaxyl and Mancozeb.
3. The white floury patches indication of powdery mildew disease appears on both sides of the pea leaves.
4. Three different kinds of conidia—macroconidia, microconidia, and chlamydospores—are produced by the fungus *Fusarium oxysporum* f.sp. *pisi*.
5. Applying Trichoderma to the soil can help control pea wilt disease.
6. Coenocytic mycelium is present in the fungus *Peronospora pisi.*
7. An obligatory parasite is the fungus *Peronospora pisi.*
8. Near the conclusion of the pea season, the disease *Peronospora pisi* produces oogonia and antheridia, which are sex organs.
9. Asci and ascospores released following the breakdown of cleistothecia or conidia from collateral hosts cause the primary infection of powdery mildew in pea plants.
10. Conidia released by the main infection and carried by the wind cause the secondary spread of powdery mildew.

Answer

S. No	Answer	S. No	Answer
1	True	6	True
2	True	7	True
3	True	8	True
4	True	9	True
5	True	10	True

B. Descriptive Questions

a. Long Answer Questions

1. Describe in detail the most distinguishing symptoms, causal organism, disease cycle and management of anthracnose of pea.
2. Describe the causal organisms and mode of penetration of powdery mildew and downy mildew diseases of pea.
3. Describe in detail the occurrence importance, symptoms, etiology, disease cycle and management of powdery mildew of pea.
4. Give the diagrammatic representation of disease cycle of the following diseases of pea

   i. Rust   ii. Downy mildew

   iii. Anthracnose   iv. Powdery mildew
5. Briefly describe the diagnostic symptoms and integrated management of any two of the following pea diseases:

   (i) Powdery mildew (ii)Anthraconse (iii) Wilt
6. Illustrate the powdery mildew of pea in the following headings:

   (i) Pathogen   (ii) Symptoms

   (iii) Disease cycle   (iv) Disease management

b. Short Answer Questions

1. How will you differentiate Erysiphe from Sphaerotheca ?
2. Why the diseases caused by Erysiphe are called powdery mildew?
3. Which fungal pathogen causes powdery mildew of pea? Give its systematic position.
4. Give the characteristic of the oospore of *Peronospora pisi.*
5. What is cleistothecia.

34

Major Diseases of Sugarcane Crop & Management

S.No	Major Diseases	Causal Organism
1	Red rot	*Colletotrichum falcatum*
2	Smut	*Ustilago scitaminea*
3	Grassy shoot	*Phytoplasma*
4	Ratoon stunting	*Clavibacterium xyli* sub sp. *xyli*
5	Pokkahboeng	*Fusarium moniliforme* and *Fusarium subglufinans*

Red Rot

Diagnostic Symptoms

- Practically all above ground parts are affected, but stalks are the main targets of stack.
- Spindle leaves (third or fourth)starts drying.
- Further stalks becomes discolored and hollow.
- Diagnostic symptom is the formation of red streaks, interrupted by white patches, inside the stalk.This can be seen by splitting the stalks longitudinally.
- White patches are characteristics of disease, not found in other stalk rots.
- After splitting opened the diseased stalk, a sour smell due to conversion of glucose to alcohol.
- Red longitudinal lesions are also seen on the mid ribs of the leaves on which black color fruiting body (Acervuli) are formed.

Etiology

Phylum: Deuteromycotina
Class: Hyphomycetes
Order: Melanconiales
Family: Melanconiaceae
Genus: *Colletotrichum*
Species: *falcaturm*

- Anamorphic stage:*Colletotrichum falcatum.*
- Perfect stage: *Physalospora tucumanensis.*
- Hyphae: profusely branched septate, hyphae containing oil droplets.
- Acervuli:black, minute velvettyacervuli with long, rigid bristle-like, septate setae.

- Conidiophores: a closely packed inside the acervulus, which are short, hyaline and single celled.
- Conidia: single celled, hyaline, falcate, granular and guttulate.
- Perithecia: dark brown to black with a papillate ostiole.
- Asci: clavate, unitunicate and eight-spored.
- Ascospore: hyaline, straight or slightly curved, ellipsoid or fusoid and unicellular which measure 18-22 µm x 7-8µm.

Disease Cycle

- Pathogen can survive as chlaymydospores in crop debris or in soil.
- Pathogen produces acervuli and conidia in the soil that serve as primary inoculums and infect the setts. Diseased sets also give rise to infected plants.
- Secondary infection that spreads the disease is brought about by conidia formed on the mid rib lesions of the leaves.
- Dissemination of conidia is through wind, rin, heavy dew and irrigation waters.

Management

- Grow resistant and moderately resistant varieties CO 62198, CO 7704, CO 8001, CO8201.
- Select the setts from the disease free fields or disease free areas.
- Setts can be treated with by moist hot air at 54°C for 2 hours and by aerated steam at 52 °C for 4 to 5 hours.
- Adopt crop rotation by including rice and green manure crops.
- Aviod ratooning of the diseased crop.

Smut

Diagnostic Symptoms

- Production of long whip-like structure from the gwoing point of of affected cane.
- Whip may be few millimeters to 20 mm in diameter.
- It is unbranched and made up of a fairly hard core of parenchyma and fibrovascular element in which the fungal hyphae ramify with the telioospore.
- Whip enclosed at first in a thin silvery sheath enclosing mass of black powdery spores.
- Initally thin canes with elongated internodes later become reduced in length.
- Profuse sporuting of lateral buds with narrow erect leaves especially in ratton crop.

Etiology

Phylum: Basidiomycotina
Class: Teliomyetes
Order: Ustilaginales
Family: Tilletiaceae
Genus: *Ustilago*
Species: *scitaminea*

- Pathogen: *Ustilago scitaminea*
- Pathogen is true culmicolous (stem infectiong) smut.
- Fungal hyphae are primarily intercellular and collect as a dense mass between the vascular bundles of host cell and produce tiny black spores.
- Teliospores are spherical light brown, and 10.5 to 17.04 micro meter in diameter.
- Teliospores germinate to produce 3-4 celled, hyaline promycelium and produce 4 sporidia (basidiospore) which are hyaline and oval shaped with pointed ends.
- Dikaryotic phase is established through fusion between sporidia, between cells of promycelium or between cells of hyphae developing from the germinating teliospores.
- Dikaryotic infection hypha invades the host tissue.
- When the whip is fully emerged the membrane ruptures releasing teliospores in the air.

Disease Cycle

- Pathogen survives as smut spores and dormant mycelium present in or on the infected setts and in the infected plant materials in the soil.
- Primary spread of the disease is through diseased seed-pieces (setts).
- Secondary spread in the field is mainly through the wind borne smut spores.
- Disease spread centrifugally in the field.

Management

- Plant healthy setts taken from disease free area.
- Grow resistant varieties like Co 7704, COC 85061 and COC 8201.
- Physical or chemical treatment of sets before planting.
- Physical method include hot water treatment, the sets are dipped in water at 55^{0}C to 60 ^{0}C for 10 minute, or exposure of sets to moist hot air at 54^{0}C for eight hours .
- Discourage ratooning of the diseased crops having more than 10 per cent infection.

- Remove and destory the smutted clump (collect the whips in a thick cloth bag/polythene bag and immerse in boiling water for 1 hr to kill the spores).
- Follow crop rotation with green manure crops or dry fallowing.

Grassy Shoot

Diagnostic Symptoms

- Characterized by proliferation of vegetative buds from the base of the cane giving rise to crowded bunch of tillers bearing narrow leaves.
- Leaves become pale yellow to completely chlorotic, thin and narrow.
- Plants appear bushy and 'grass-like' due to reduction in the length of internodes premature and continuous tillering.
- Cane formation rarely occurs in the affected clumps, if formed, thin with shorter internodes having aerial roots at the lower nodes.

Etiology

- Pathogen: *Phytoplasma*
- Two types of bodies are seen in ultrathin sections of phloem cells of infected plants.
- Spherical bodies of 300-400 nm diameter and filamentous bodies of 30-53 mm diameter in size.

Disease Cycle

- Pathogen is perpetuated through crop rattoning.
- Primary transmission isthrough infected seed material.
- Secondary spread is by aphid,grass hopper and dodder.

Management

- Select setts from diseased free area.
- Grow resistant varieties viz., Co 86249, CoG 93076 and Coc 22
- Pre-treating the healthy setts with hot water at 52°C for 1 hour before planting.
- Avoid ratooning if GSD incidence is more than 15 % in the plant crop.

Model Question Paper

A. Objective Type Questions

a. Choose the correct answer from the following:

1. The cause of grassy shoot disease is:

   (a) Phytoplasma (b) Viroid

   (c) Prions (d) Virus

2. The perfect stage of *Thanatephorus cucumaris* is:

   (a) *Colletotrichum falcatum* (b) *Rhizoctonia solani*

   (c) *Rhizoctonia bataticola* (d) *Fusarium oxysporum* f. sp. *vasinfectum*

3. *Ustilago scitaminea* causes:
   (a) Smut of Sugarcane (b) Wilt of Cotton
   (c) Damping off of Tobacco (d) Wilt of Sugarcane
4. Red rot is a destructive disease of:
   (a) Cotton (b) Tobacco
   (c) Sugarcane (d) Potato
5. The perfect state or teleomorph of *Colletotrichum falcatum* is:
   (a) *Physalsospora tucumanensis* (b) *Glomerella tucumanensis*
   (c) *Thanatephorus cucumeris* (d) *Fusarium oxysporum*
6. Ratoon stunting disease is caused by:
   (a) Phytoplasma (b) Viroid
   (c) Bacteria (d) Virus
7. Red longitudinal lesions are also seen on the mid ribs of the leaves of sugarcane on which black color fruiting bod are formed
   (a) Acervuli (b) Pycnidia
   (c) Sporodochia (d) Cleistothecia
8. Diagnostic symptom of which disease is formation of red streaks, interrupted by white patches, inside the stalk that can be seen by splitting the stalks longitudinally.
   (a) Red rot of Sugarcane (b) Smut of Sugarcane
   (c) Grassy shoot of Sugarcane (d) Wilt of Sugarcane
9. Production of long whip-like structure from the gwoing tip of of damaged sugarcane is a diagnostic sign of
   (a) Red rot (b) Smut
   (c) Grassy shoot (d) Wilt
10. Which disese of sugarcane is characteriszed by proliferation of vegetative buds from the base of the cane giving rise to crowded bunch of tillers bearing narrow leaves
   (a) Red rot (b) Smut
   (c) Grassy shoot (d) Wilt

S. No	Answer	S. No	Answer
1	a. Phytoplasma	6	Bacteria
2	b. *Rhizoctonia solani*	7	a. Acervuli
3	a. Smut of Sugarcane	8	a. Red rot of sugarcane
4	c. Sugarcane	9	a. Red rot
5	a. *Physalospora tucumanensis*	10	c. Grassy shoot

b. True /Flse

1. Sugarcane infected with red rot disease produces small, velvety, black acervuli with long, stiff bristle-like setae that are septate in nature.
2. After sugarcane was infected with the pathogen *Colletotrichum falcatum*, dark brown to black perithecia with papillate ostioles appeared.
3. Conidia that develop on the midrib lesions of the leaves cause a secondary infection that spreads the red rot disease in sugarcane.
4. Sugarcane monoculturing encourages the disease red rot.
5. To prevent red rot disease in sugarcane, soak the setts in a Triademefon (0.05%) solution for 15 minutes before to planting.
6. Discourage ratooning of the smut diseased sugarcane crops having more than 10 per cent infection.
7. *Clavibactor xyli* is a slow growing, gram positive, rod shaped bacteria.
8. Top-rot phase is the most serious stage of Pokkahboeng disease.
9. Pokkahboeng of sugarcane is a air-borne disease.
10. Knife-cut Phase is an exaggerated stage of a typical ladder lesion of a Pokkahboeng disease.

Answer

S. No	Answer	S. No	Answer
1	True	6	True
2	True	7	True
3	True	8	True
4	True	9	True
5	True	10	True

B. Decriptive Questions

a. Long Answer Questions

1. Describe in detail the most distinguishing symptoms, causal organism, etiology, disease cycle and management of grassy shoot disease of sugarcane.
2. Describe in brief the most distinguishing symptoms, transmission and disease management of ratoon stunting.
3. Describe the disease Pokkahboeng giving symptoms, pathogen, disease cycle and management.
4. Describe in detail the diagnostic symptoms, etiology, disease cycle and management of red rot of sugarcane.
5.

(a) Briefly describe the most distinguishing symptoms, causal organism and management of smut of sugarcane.

(b) Briefly describe the most distinguishing symptoms, causal organism and management Pokahboeng of sugarcane.

b. Short Answer Questions

1. Describe in brief about disease management of red rot of sugarcane.
2. Explain the symptom and management of smut of sugarcane.
3. Explain disease cycle, epidemiological condition and management of pokkahboeng disese of sugarcane
4. Give digramatic representation of disease cycle of smut of sugarcane and their management.
5. Explain the symptom and management of grassy shoot of sugarcane.

35

Major Diseases of Mustard Crop & Management

S. No.	Major Diseases	Causal Organism
1	Alternaria blight	*Alternaria brassicae*, *A. brassicicola*, and *A.raphani*
2	White rust	*Albugo candida*
3	Downy mildew	*Peronospora parasitica*
4	Sclerotinia stem rot	*Sclerotinia sclerotiorum*
5	Powdery mildew	*Erysiphe cruciferarum*

1. Alternaria blight

Diagnostic Symptoms

- Symptom first appear on the lower leaves as small circular brown necrotic spots which slowly increase in size.
- Many concentric spots coalesce to cover large patches showing blightening and defoliation in severe cases.
- Circular to linear, dark brown lesions also develop on stems and pods, which are elongated at later stage.
- Infected pods produce small, discolored and shriveled seeds.

Etiology

Scientific Classification

Kingdom: Fungi

Phylum: Ascomycota

Class: Dothideomycetes

Subclass: Pleosporomycetidae

Order: Pleosporales

Family: Pleosporaceae

Genus: *Alternaria*

Species: *brassicae/brassicicola*

Alternaria brassicae

- Mycelium: Immersed, hyphae branched, septate, hyaline, smooth,4-8μm thick.

- Conidiophores: arising in groups of 2-10 or more from the hyphae which are straight or curved.
- Conidia: light olive coloured with transverse and longitudinal septa. These are around 3-5 septate and conidia are borne in chain.

Alternaira brassicicola

- Mycelium: immersed, hyphae branched, septate, hyaline, olivaceous brown, smooth, upto 70 µm long and 4-8µm thick.
- Conidiophores: arising in groups of 2-12, emerging through stomata, usually simple, erect, straight or curved, smooth upto 70µm long, 5-8 µm thick.
- Conidia: mostly in chains of upto 20 or more, sometimes branched, straight, nearly cylindrical, usually slightly tapering towards the apex, or obclavate, the basal cell rounded, the beak usually almost nonexistent, are light olive coloured with1-11 mostly less than 6 transverse septa.

Disease Cycle

- Pathogens survive between mustard and canola crops in infested crop debris and on seed.
- These pathogens are disseminated as conidia within and among fields by splashing water and wind.
- Pathogens infect their hosts by direct penetration or through wounds andnatural openings.

Management

- Plant high quality seed free from the Alternaria black spot pathogens.
- Practice a three year or longer rotation to non-hosts such as small grains.
- Promptly incorporate crop residues after harvest to hasten their breakdown.
- Eliminate volunteer canola, mustard and cruciferous weeds.
- Early sowing in first fortnight of October .

White Rust

Diagnostic Symptoms

- Intially chlorotic (yellowed) lesions and sometimes galls appeared on the upper leaf surface.
- There are corresponding white blister-like dispersal pustules of sporangia on the underside of the leaf.
- Branches and flower parts deformed and pathogen stimulates hypertrophy and hyperplasia.
- Flower and young fruits may also malformed and malformed pod lack seed development.
- Disease is sometimes called the cancer of rapeseed as one phase of this disease shows grotesque malformation of the young shoots and inflorescence.

Etiology

Scientific Classification

Phylum: Heterokontophyta

Class: Oomycota

Order: Peronosporales

Family: Albuginaceae

Genus: *Albugo*

Species: *candida*

- Pathogen: *Albugo candida.*
- Pathogen is obligate parasite and it needs a living host to grow and reproduce.
- Mycelium: Intercellular producing small globose, knob shaped haustoria in the host cells.
- Sporangiophores: hyaline, clavate, free from each other laterally and are very thick walled especially towards base.
- Sporangia: hyaline, nearly oval to spherical with uniform thin walled and 14-16 x16-20μm in size.It may germinate by means of germ tube or through zoospores.
- Zoospores: Biflagellate, lose their flagella, come to rest, and then germinate by germtube which penetrates the epidermis of the host directly or through the stomata.
- Oogonium: globose, terminal or intercalary and multinucleate.
- Antheridium:clavate, paragynous and contains 6-12 nuclei.
- Oospores: globose, chocolate brown,30-35μm in diameter. It germinate by formation of vesicles in which zoospores are formed or the oospores may germinate by production of one or two simple or branched germ tube by release of zoospores from sessile vesicles.
- Germination of sessile vesicles is most common.

Disease Cycle

- Pathogen reproduce by producing both sexual spores (called oospores) and asexual spores (called sporangia) in a many-stage (polycyclic) disease cycle.
- Thick-walled oospores are the main overwintering structures, but the mycelium can also survive in conditions where all the plant material is not destroyed during the winter.
- In the spring, the oospores germinate and produce sporangia (s. sporangium) on short stalks called sporangiophores that become so tightly packed within the leaf that they rupture the epidermis and are consequently spread by the wind.

- Liberated sporangia can either germinate directly with a germ tube or begin to produce biflagellate motile zoospores.
- Zoospores then swim in a film of water to a suitable site and each one produces a germ tube – like that of the sporangium – that penetrates the stoma.
- When the oomycete has successfully invaded the host plant, it grows and continues to reproduce.

Management

- Sanitation.
- Use of resistant cultivars.
- Early sowing preferably in the first fortnight of October.
- Use the seed from stag-head free plants to avoid carry over of oospore through seeds.
- Destroy the diseased plant debris.
- Crop rotation.

Downy Mildew

Diagnostic Symptoms

- Intially grayish white irregular necrotic patches develop on the lower surface of leaves.
- Later under favorable conditions, brownish white fungal growth may also be seen on the spots.
- Most conspicuous and pronounced symptom is the infection of inflorescence causing hypertrophy of the peduncle of inflorescence and develop stag head structure.

Etiology

Scientific Classification

Kingdom: Fungi

Division: Eumycota

Class: Oomycetes

Order: Peronosporales

Family: Peronosporaceae

Genus: *Peronospora*

Species: *parasitica*

- Pathogen: *Peronospora parasitica*
- Obligate parasite
- Mycelium: strictly intercellular, hyaline with large, finger shaped haustoria.
- Sporangiophores: 100-300μm long and unbranched for a major portion of

their length.Dichotomous branching occurs, 6-8 times, at the tip. They are at acute angle with each other .A single sporangium is borne at the tip of each ultimate branch.

- Sporangia: dispersed by rain splash or wind and may germinate by means of germ tube.
- Oogonia: pale yellow, irregularly round, and swollen into crest like folds.
- Antheridia: tendril like and are produced on separate hyphae.
- A fertilization tube grows from the antheridium through the receptive papilla towards a central body in the ooplasm to discharge a single male nucleus.The two nuclei fuse and initiate the uninucleate oospore.
- Oospores: globose, measuring 26-43μm in diameter and germinate by germ tube.

Disease Cycle

- Pathogen survives between crops in the soil as oospores.
- Infection occurs when soilborne resting structures called oospores germinate and produce sporangia under moist, cool conditions.
- Sporangia are produced on the underside of leaves in the evening and are released during the day as leaves dry.
- Windblown sporangia land on leaves and directly penetrate leaves and flowers.

Management

- Plant resistant or tolerant varieties.
- Seed treatment with biofungicides.
- Sow the crop on normal planting date.
- Destroy diseased plant parts.

Sclerotinia Stem Rot

Diagnostic Symptoms

- Initial symptoms appear as the plant canopy closes over rows during flowering and pod development.
- Lodged stems in contact with the soil develop watery lesions, with snowy white mycelium and black, irregularly shaped sclerotia become apparent as disease progresses.
- When the stem is completely girdled by such lesions, the plants wilt and die.
- When the crop is at seed stage, the plants tend to lodge, touching the siliquae to the soil level.
- Such plant show rotting of the siliquae with fungal growth along with the sclerotial, bodies just above the soil level.

Etiology

Scientific Classification

Kingdom: Fungi

Division: Ascomycota

Class: Leotiomycetes

Order: Helotiales

Family: Sclerotiniaceae

Genus:*Sclerotinia*

Species: *sclerotiorum*

- Pathogen:*Sclerotinia sclerotiorum.*
- Pathogen is characterized by the formation of hard blackish sclerotinia which germinate and produce cup shaped brown color apothecia.
- Sclerotia borne superficially usually on dense white mycelium, globose to cylindrical, but quite variable in shape, 2-15x2-15μm with black outer rind and white inner contex.
- Apothecia arising one to several form a sclerotium ochraceous often darker at the base of the stipe, receptacle 2-8 mm broad applanate to slightly concave, often with control depression frequently with an undulate margin, tapering to form a stipe 3-10 mm long, 1-2 mm wide.

Disease Cycle

- Pathogen survives between crops in the soil as sclerotia.
- Infection occurs when soilborne sclerotia (dormant resting structures) germinate when the soil surface is continuously wet for at least two weeks.
- Germinating sclerotia form small, tan, cup-shaped structures called apothecia, which release millions of airborne ascospores that colonize dead plant parts (such as senescent flowers) and infect host tissues.
- Sclerotia can survive in soil for upto 8 to 10 years.

Management

- Plant high quality seed free from sclerotia.
- Practice a four-year or longer rotation to non-hosts such as corn or small grains.
- Avoid excessive irrigation and fertilization.

Model Question Paper

A.Objective Type Questions

a. Choose the correct answer from the following

1. Sclerotinia stem rot disease is caused by:
   (a) Phytoplasma (b) Vioroid
   (c) Fungi (d) Virus
2. Light olive coloured conidia with transverse and longitudinal septa
   (a) *Colletotrichum falcatum* (b) *Alternaria brassicae*
   (c) *Rhizoctonia bataticola* (d) *Fusarium oxysporum*
3. White blister-like dispersal pustules of sporangia on the underside of the leaf of mustard is characteristic symptom of disesase
   (a) White rust (b) Alternaria blight
   (c) Damping off (d) Downy mildew
4. Which disease is sometimes called the cancer of rapeseed
   (a) White rust (b) Alternaria blight
   (c) Damping off (d) Downy mildew
5. Which pathogen is an obligate parasite and it needs a living host to grow and reproduce.
   (a) *Albugo candida* (b) *Glomerella tucumanensis*
   (c) *Thanatephorus cucumeris* (d) *Fusarium oxysporum*
6. Thick-walled oospores are the main overwintering structures found in
   (a) *Albugo candida* (b) *Glomerella tucumanensis*
   (c) *Thanatephorus cucumeris* (d) *Fusarium oxysporum*
7. *Peronospora parasitica* survives between crops in the soil a
   (a) Acervuli (b) Pycnidia
   (c) *Oospores* (d) Cleistothecia
8. Most conspicuous and pronounced symptom of which disease of mustard is the infection of inflorescence causing hypertrophy of the peduncle of inflorescence and develop stag head structure
   (a) White rust (b) Alternaria blight
   (c) Damping off (d) Downy mildew
9. Foliar spray with Ridomil MZ @ 0.2% manage
   (a) Rust (b) Alternaria blight
   (c) Powdery mildew (d) Downy mildew
10. Mustard lodged stems in contact with the soil develop watery lesions, with snowy white mycelium and black, irregularly shaped sclerotia become apparent as which disease progresses.
   (a) White rust (b) Alternaria blight
   (c) Sclerotinia stem rot (d) Downy mildew

Answer

S. No	Answer	S. No	Answer
1	c Fungi	6	a. *Albugo candida*
2	b. *Alternaria brassicae*	7	c. *Oospores*
3	a. White rust	8	d. Downy mildew
4	a. White rust	9	d. Downy mildew
5	a. *Albugo candida*	10	c. Sclerotinia stem rot

b. State whether the following statements are True or False.

1. The pathogen *Sclerotinia sclerotiorum* persists in the soil as sclerotia in between crops.
2. When the soil surface is consistently moist for at least two weeks, soilborne sclerotia germinate and cause *Sclerotinia sclerotiorum* infection in mustard.
3. The distinctive indication of Alternaria blight on mustard is the existence of a yellow halo and a concentric ring.
4. Hypertrophy and hyperplasia in mustard branches are induced by *Albugo candida.*
5. The germination of sessile vesicles occurs most frequent in *Albugo candida.*
6. The pathogen *Albugo candida* reproduces by producing both sexual and asexual spores at various stages of its disease cycle.
7. *Peronospora parasitica* infection occurs in mustard when oospores germinate and form sporangia under moist and cool circumstances.
8. The hard blackish sclerotinia that germinate and create brown, cup-shaped apothecia are the defining feature of *Sclerotinia sclerotiorum.*
9. Splashing water and wind spread the Alternaria blight disease within and among mustrd fields.
10. Sclerotia that are germinating Apothecia are tiny, tan, cup-shaped structures formed by *Sclerotinia sclerotiorum.*

Answer

S. No	Answer	S. No	Answer
1	True	6	True
2	True	7	True
3	True	8	True
4	True	9	True
5	True	10	True

B.Descriptive Questions

a. Long Answer Questions

1. Describe in detail the symptoms, etiology, disease cycle and management of white rust of mustard.
2. Describe the most distinguishing symptoms, causal organism, disease cycle and management of downy mildew disease of mustard.

3 Briefly describe the most distinguishing symptoms and management of any of the following plant disease:
   (i) Alternaria blight of mustard
   (ii) Sclerotinia stem rot of mustard
4. Describe in detail the symptoms, causal organism, disease cycle and management of powdery mildew of mustard.
5. Describe the symptoms, causal organism, disease cycle and the management of any disease of mustard studied by you.

b. Short Answer Questions

1. Explain the chemical management of white rust of mustard.
2. Distinguish between powdery and downy mildew of mustard.
3. Describe the hypertrophied symptoms of white blisters of mustard.
4. What is etiology? Write down the etiology of *Albugo candida* causing white blister in mustard.
5. Discuss about disease cycle and epidemiology of Sclerotinia stem rot diseases of mustard.

36

Major Diseases of Cotton Crop & Management

S. No	Name of Diseases	Causal Organism
1	Anthracnose	*Colletotrichum capsici*
2	Vascular wilt	*Fusarium oxysporum* f.sp. *vasinfectum* (Akt) Synedar & Hansen
3	Black arm	*Xanthomonas axonopodis* pv. *malvacearum*(Smith) Dye
4	Leaf blight	*Alternaria macrospora* Zimm
5	Leaf Curl	*Cotton leaf curl virus*

Anthracnose

Diagnostic Symptoms

- Pathogen infects the seedlings and produces small reddish circular spots on the cotyledons and primary leaves.
- Lesions develop on the collar region, stem may be girdled, causing seedling to wilt and die.
- In mature plants, the fungus attacks the stem, leading to stem splitting and shredding of bark.
- Most common symptom is boll spotting. Small water soaked, circular, reddish brown depressed spots appear on the bolls.
- The lint is stained to yellow or brown, becomes a solid brittle mass of fiber. The infected bolls cease to grow and burst and dry up prematurely.

Etiology

Scientific Classification

Kingdom: Fungi

Division: Eumycota

Sub-division: Deuteromycotina

Class: Coelomycetes

Order: Melanconiales

Family: Melanconiaceae

Genus:*Colletotrichum*

Species:*capsici*

- Pathogen: *Colletotrichum capsici*
- Pathogen forms large number of acervuli on the infected parts.
- Conidiophores: are slightly curved, short and club shaped.
- Conidia are hyaline and falcate, borne single on the conidiophores.
- Setae: Numerous black colored and thick walled setae are also produced inacervulus.

Disease Cycle

- Pathogen survives as dormant mycelium in the seed or as conidia on the surface of seeds for about a year.
- Pathogen also perpetuates on the rotten bolls and other plant debris in the soil.
- Secondary spread is by air-borne conidia.

Management

- Remove and burn the infected plant debris and bolls in the soil.
- Rogue out the weed hosts.
- Plant high quality seed free from infections.
- Practice a four-year or longer rotation to non-hosts such as corn or small grains.
- Avoid excessive irrigation and fertilization.

Vascular Wilt

Diagnostic Symptoms

- Disease affects the crop at all stages.

Seedling stage

- Cotyledons of the seedling turn yellow and then brown.
- Base of petiole shows brown ring, followed by wilting and drying of the seedlings.

Young and grown up plants

- Symtoms start from the older leaves at the base, followed by younger ones towards the top, finally involving the branches and the whole plant.
- Initial symtoms is yellowing of edges of leaves and area around the veins i.e. discoloration starts from the margin and spreads towards the midrib.
- The leaves loose their turgidity, gradually turn brown, droop and finally drop off.
- Defoliation or wilting may be complete leaving the stem alone standing in the field.
- Sometimes partial wilting occurs; where in only one portion of the plant is affected, the other remaining free.
- Browning or blackening of vascular tissues is the other important symptom.

- Black streaks or stripes may be seen extending upwards to the branches and downwards to lateral roots.
- In severe cases, discolouration may extend throughout the plant starting from roots extending to stem, leaves and even bolls.
- In transverse section, discoloured ring is seen in the woody tissues of stem.
- Plants affected later in the season are stunted with fewer bolls which are very small and open before they mature.

Etiology

Scientific Classification

Kingdom: Fungi

Division: Eumycota

Sub-**division:**Deuteromycotina

Class: Hyphomycetes

Order: Tuberculariales

Family: Tuberculariaceae

Genus:*Fusarium*

Species:*oxysporum* f. sp.*vasinfectum*

- Pathogen:*Fusarium oxysporum* f.sp. *vasinfectum*
- Mycelium: white to grayish white or bluish purple, often form a mat on the collar region of the stem near the ground level.
- Hyphae: both inter as well as intracellular .
- Conidiophores: produced in sporodochia,sometimes on the mycelium, vertically branched.
- Conidia: macrocondia, microconidia and chlamydospores.
- Macroconidia: 1 to 5 septate, hyaline, thin walled, falcate with noculum ends, 40-50x3-4.5µm.
- Microconidia: hyaline, thin walled, spherical or elliptical, unicellular, measuring 5-12x 2-3.5 µm.
- Chlamydospores: dark coloured and thick walled, terminal or intercalary.

Disease Cycle

- It is a soil inhabitant and has good saprophytic ability.
- Pathogen can survive in soil as saprophyte and as chlamydospores for many years.
- Pathogen is both externally and internally seed-borne.
- Primary infection is mainly through conidia produced from dormant hyphae and chlamydospores in the soil.
- Secondary infection in case of this disease is rare as the conidia produced seldom succeed in establishing aerial infection on the plant.

Management

- Deep summer ploughing during June-July.
- Grow disease resistant varieties like Varalakshmi, Vijay Pratap, Jayadhar and Verum.
- Remove and burn the infected plant debris in the soil.
- Apply heavy doses of farm yard manure or other organic manures.
- Follow mixed cropping with non-host plants.
- Delay planting to the end of October.
- Avoid dense planting.
- Retain cotton residues on the surface for 60 days.
- Bare fallow rotation is best.
- Always practice good farm hygiene.

Black Arm/Angular leaf Spot

Diagnostic Symptoms

- Pathogen attacks all stages from seed to harvest.
- Small water soaked spots, angular lesions of 1 to 5 mm across the leaves and bracts, especially on the undersurface of leaves. Hence called angular leaf spot.
- Sometimes extensive dark green, water soaked lesions along the veins known as vein blight.
- Lesions dry and darken with age and leaves may be shed prematurely resulting in extensive defoliation.
- Black lesions on the stem which girdle and spread along the stem or branch known as black arm.
- Dark green, water soaked, greasy, circular lesions of 2 to 10mm across the bolls, especially at the base of the boll under the calyx crown. As the boll matures the lesions dry out and prevent normal boll opening. This phase of symptom is called as "Boll rot"
- The bacterium spreads inside the boll and lint gets stained yellow because of bacterial ooze and loses its appearance and market value. The pathogen also infects the seed and causes reduction in size and viability of the seeds

Etiology

Scientific Classification

Kingdom: Prokaryotae

Division: Gracilicutes

Class: Proteobacteria

Family: Psudomonadaceae

Genus: *Xanthomonas*

Species: *axonopodis* pv. *malvacearum*

- Pathogen: *Xanthomonas axonopodis* pv. *malvacearum.*
- Gram negative.
- Short rod with a single polar flagellum.
- Non-spore forming measures 1.0-1.2 X 0.7-0.9 μm.
- In culture, it forms yellow colony.

Disease Cycle

- Pathogen inoculums may either be present in the field on infected crop residues from a previous season or it may be introduced at planting within infected seed.
- Infected seeds are the source of primary inoculums.They carry the pathogen externally as well as internally.
- Plant debris is another source of inoculums.
- Lesions on cotyledons may be initiated by inoculums within the seed during germination.
- Inoculum from infected crop residues may be splashed onto the foliage and into the growing point of young seedlings where it can survive saprophytic ally on leaf surfaces.
- When environmental conditions are favorable the bacteria enter the plant via the stomata or wounds.
- As lesions develop, bacteria exude out onto the leaf surface for further dispersal through wind driven rain.
- Pathogen is able to enter the seed when mature, open, blight-infected bolls are exposed to wet weather prior to harvest.

Management

- Grow resistant varieties like CRH 71 and Sujatha, 1412
- Delint the cotton seeds with concentrated sulphuric acid at 100ml/kg of seed.
- Remove and destroy the infected plant debris.
- Rogue out the volunteer cotton plants and weed hosts.
- Follow crop rotation with non-host crops.
- Early earthing up with potash.

Model Question Paper

A. Objective Type Questions

a. Choose the correct answer from the following

1. *Colletotrichum capsici* belongs to order.

(a)	Melanconiales	(b)	Peronosporales
(c)	Hyphomycetales	(d)	Moniliales

2. Anthracnose of cotton is caused by:
   (a) *Colletotrichum capsici* (b) *Rhizoctonia bataticola*
   (c) *Sphacelotheca sorghi* (d) *Sphacelotheca cruenta*
3. Angular leaf spot disease of cotton is caused by
   (a) *Xanthomonas axonopodis* pv. *malvacearum* (b) *Sphacelotheca cruenta*
   (c) *Rhizoctonia bataticola* (d) None of the above
4. Black lesions on the stem which girdle and spread along the stem or branch of cotton known as.
   (a) Black arm (b) Angular leaf spot
   (c) Anthracnose (d) None

(5) Infected seeds are the source of primary inoculums and they carry the pathogen.
   (a) Externally (b) Internally
   (c) Externally as well as Internally (d) None

6. *Xanthomonas axonopodis* pv. *malvacearum* is a type of bacteria
   (a) Gram negative (b) Gram positive
   (c) Gram neutral (d) None
7. The pathogen *Fusarium oxysporum* f.sp. *vasinfectum*produces highly resistant surviving structure in soilcalled
   (a) Pseudothecia (b) Sclerotia
   (c) Chlaymydospore (d) Cleistothecia
8. Fusarial wilt of cotton is caused by:
   (a) *Ascochyta rabiei* (b) *Rhizoctonia bataticola*
   (c) *Fusarium oxysporum* f.sp. *vasinfectum* (d) *Sphacelotheca cruenta*
9. *Fusarium oxysporum f.sp. vasinfectum produces:*
   (a) Mcroconidia (b) Microconidia
   (c) Chlaymydospores (d) All
10. *Fusarium oxysporum* f.sp. *vasinfectumbelongs to order.*
   (a) Tuberculariales (b) Peronosporales
   (c) Hyphomycetales (d) Moniliales

(a) Answer

S. No	Answer	S. No	Answer
1	a. Melanconiales	6	a. Gram negative
2	a. *Colletotrichum capsici*	7	c.Chlaymydospore
3	a. *Xanthomonas axonopodis* pv. *malvacearum*	8	c. *F. oxysporum* f.sp. *vasinfectum*
4	a. Black arm	9	d. All
5	c. Externally as well as Internally	10	a.Tuberculariales

b State whether the following statements are True or False.

1. A yellow colony of *Xanthomonas axonopodis* pv. *malvacearum* forms in culture.
2. Cotton angular leaf spot can be controlled by foliar spraying a 100g mixture of streptomycin sulfate and tetracycline with 1.25 kg/ha of copper oxychloride.
3. Black lesions known as "black arm" that girdle and spread down the stem or branch of cotton infected with *Xanthomonas axonopodis* pv. *malvacearum.*
4. Chlamydospores, macroconidia, and microconidia are the three forms of conidia that are produced by the fungus *Fusarium oxysporum* f.sp. *vasinfectum.*
5. Trichoderma can be used to the soil to combat cotton wilt disease.
6. An extensive dark green, water soaked lesions seen along the veins of cotton dueto infection of*Xanthomonas axonopodis* pv. *malvacearum*known as vein blight.
7. A small water soaked spots, angular lesions of 1 to 5 mm across the leaves and bracts, especially on the undersurface of leaves due to infection of *F oxysporum* f.sp. *vasinfectum* called angular leaf spot.
8. *Xanthomonas axonopodis* pv. *malvacearum*is a gram negative, short rod shaped bacteria with a single polar flagellum.
9. *Fusarium oxysporum* f.sp. *vasinfectum*is a soil inhabitant fungi having good saprophytic ability.
10. Yellowing of edges of leaves and area around the veins i.e. discoloration starts from the margin and spreads towards the midrib is a initial symptom of Fusarium wilt disease of cotton.

Answer

S. No	Answer	S. No	Answer
1	True	6	True
2	True	7	True
3	True	8	True
4	True	9	True
5	True	10	True

B. Descriptive Questions

a. Long answer questions

1. Describe in detail the most distinguishing symptoms, causal organism, disease cycle and management of para wilt of cotton.
2. Describe in detail the occurrence, importance, symptoms, etiology, disease cycle and management of wilt of cotton.

3. Give the diagrammatic representation of disease cycle of the following diseases of cotton

i.	Wilt	ii.	Anthracnose
iii.	Bacterial blight	iv.	Leaf curl

4. Briefly describe the diagnostic symptoms and integrated management of any two of the following cotton diseases:

(i)	Collar rot	(ii)	Bacterial blight
(iii)	Wilt		

5.. Illustrate the Angular leaf spot of cotton in the following headings:

(i)	Pathogen	(ii)	Symptoms
(ii)	Disease cycle	(iv)	Disease management

b. Short answer questions

1. Write an integrated management scheduele of angular leaf spot of cotton.
2. Explain the disease cycle and epidemiological factor of vascular wilt of cotton.
3. Write systemic position and etiology of pathogen causing anthracnose of cotton.
4. Compare between symptom of angular leaf spot and black arm of cotton giving suitable measures for their management.
5. Explain the disease cycle and predisposition of bacterial blight of cotton.

37

Major Diseases of Banana Crop & Management

S. No	Name of Disease	Causal Oranism
1.	Sigatoka	*Mycosphaerella gloeosporioides*
2.	Panama wilt	*Fusarium oxysporum* f.sp *cubense*
3.	Bacterial wilt	*Pseudomonas solanacearum / Burkholderia solanacearum*
4.	Bunchy top	*Banana bunchy top virus*
5.	Anthracnose	*Gloeosporium gloeosporioides*

1. Sigatoka

Diagnostic Symptoms

- Early symptoms appear on the young leaves.
- Small spindle shaped spots on foliage with greyish centreand yellowish halo running parallel to veins.
- Spots are mostly seen along the edge of the leaf with defined margin and possess dark brown to black margin. Spots coalesce and whole leaf blade dries up.
- On the upper surface of the spots, fructificationsof the fungus appear as black specks unevenly and individual bananas appear undersized and their flesh develops a buff pinkish colour, and store poorly.

Etiology

Scientific Classification

Kingdom: Fungi

Division: Eumycota

Sub-division: Deuteromycotina

Class: Hyphomycetes

Order: Hyphomycetales (Moniliales)

Family: Dematiaceae

Genus: *Cercospora*

Species: *musae*

- Pathogen: *Mycosphaerella musicola (Cercospora musae*)
- Mycelium is hyaline, septate and branched.

- Conidia are elongated, narrow and multiseptate.
- Perithecia are dark brown to black and ostiolate.
- Asci are oblong and clavate.
- Ascospores are hyaline, two celled, obtuse to ellipsoid.

Disease Cycle

- Survives on dry infected leaves on the field soil and primary infection takes place through ascospores in the infected plant debris.
- Secondary spread through wind borne conidia and ascospores.
- Infection takes place through stomata on the lower surface of young leaves.
- Surface moisture is necessary for release of both conidia and ascospores. Hence the disease is severe in moist weather.

Management

- Use resistant varieties
- Plant tissue culture saplings
- Maintain proper drainage
- Avoid water logging
- Control weeds
- Remove diseased suckers
- Adopt correct spacing
- Maintain field sanitation

2. Panama wilt

Diagnostic Symptoms

- Symptoms are most clear on the plants that are at least five month old.
- Sudden wilting of the plant or individual leaves is a characteristic symptom of the disease.
- Yellowing of the lower most leaves starting from margin to midrib of the leaves.
- Yellowing extends upwards and finally heart leaf alone remains green for some time and it is also affected.
- The leaves break near the base and hang down around pseudostem.
- Discolored vascular strands varying from light yellow to dark brown are the distinguishing internal symptoms.
- Usually the discoloration manifests first in the outer or oldest leaf sheath and extends upto the pseudostem.

Etiology

Scientific Classification

Kingdom: Fungi

Division: Eumycota

Sub-division: Deuteromycotina

Class: Hyphomycetes

Order: Hyphomycetales (Moniliales)

Family: Dematiaceae

Genus: *Fusarium*

Species: *oxysporum* f. sp *cubense*

- Pathogen:*Fusarium oxysporum* f. sp *cubense.*
- Mycelium is septate, hyaline and branched.
- Fungus produces micro, macro conidia and also chlamydospores.
- Micro conidia - Single celled or rarely one septate hyaline elliptical or oval.
- Macro conidia - Sickle shaped hyaline, 3-5 septate and tapering at both ends.
- Chalamydospores - Thick walled, spherical to oval, hyaline to slightly yellowish in color.

Disease Cycle

- The disease is soil borne.
- The pathogens survives in the soil mainly as chlamydospores and invades the soft roots or rhizomes, very often through the injuries.
- The pathogen can also survive in the diseased rhizomes and other plant parts saprophytically for a long period.
- Primary infection always takes place through injured roots.
- Deep wounding exposing the xylem help in easy infection.
- When the germ tube of chlaymydospores enters into the root, hyphae proceed internally along the root to the rhizome where they develop extensively in vascular tissues before passing up the vascular system into the pseudostem and the older leaf petioles.
- The secondary infection takes place by means of conidia.
- The disease can easily spread by contact of root system of adjacent healthy plants with conidia released by the diseased plants.
- Flood also help in the local dispersal of the secondary inoculums.

Management

- Use of disease free suckers for planting.
- The planting material should be dipped in allyl alcohol.
- Infected plant should be eradicated by weed killer.
- Resistant cultivar should be cultivated. Cavendesh, Poovan, Moongil, Peyladen, Rajabale etc.
- Avoid ill drained soils, and prefer slightly alkaline soils (7-7.5 pH) for cultivation.

- Flood fallowing for 6 to 24 months or crop rotation with rice.
- Application of lime (1-2 kg/pit) to the infected pits after chopping of the plants parts.

2. Bacteiral Wilt

Diagnostic Symptoms

- Symptoms start on rapidly growing young plants.
- The youngest three to four leaves turn pale green or yellow and collapse near the junction of lamina and petiole.
- Characteristic discoloration of vascular strands, wilting and blackening ofsuckers.
- Vascular discolouration (pale yellow to dark brown or bluish black) is concentrated near the centre of the pseudostem, becoming less apparent on the periphery.
- Greyish brown bacterial ooze is seen when the pseudostem of affected plant is cut transversely.
- A firm brown dry rot is found within fruits of infected plants (characteristic symptom).

Etiology

Scientific Classification

Kingdom: Prokaryotae

Division: Gracilicutes

Class: Proteobacteria

Family: Pseudomonadaceae

Genus: *Ralstonia*

Species: *solanacearum*

- Pathogen:*Pseudomonas solanacearum / Burkholderia solanacearum*
- Moko disease is caused by race 2 of *Ralstonia solanacearum* which infects *Musa* and *Heliconia*.
- It is rod shaped, gram negative bacterium with one polar flagellum.

Disease Cycle

- The pathogen survives in crop residues left in the field.
- Primary infection occurs when the bacteria enter the roots and rhizomes via wounds, caused by nematodes and farm implements.
- Secondary infections from plant to plants are caused by insects that transmit the bacterial oozes form infected male flowers to healthy flowers.

Management

- Eradicate infected plant.
- Expose soil to direct sunlight.

- Use of clean planting material.
- Fallowing and crop rotation is advisable.
- Disinfestation of tools with formaldehyde diluted with water in 1:3 ratio.
- Crop rotation (3 years rotation with sugarcane or rice) & providing good drainage.
- Biocontrol with *Pseudomonas fluorescens.*

7. Bunchy Top

Diagnostic Symptoms

- Subsequent leaving show the same symptoms and are dwarfed.
- Dark broken bands of green tissues on the veins, leaves and petioles.
- Plants are extremely stunted.
- Leaves are reduced in size marginal chlorosis and curling.
- Leaves upright and become brittle.
- Many leaves are crowded at the top. Branches size will very small. If infected earlier no bunch will be produced.

Transmission

- Pathogen:*Banana bunchy top virus.*
- The virus is not transmitted by mechanical inoculation.
- The chief insect vector is *Pentalonia nigronervosa.*
- The aphid acquires the virus in a feeding period of 24 hours and transmits the virus to healthy plants to an infection feeding period of one and a half years.
- The aphid retains the virus for a period of 13 days and the incubation period for the virus is a few hours to two days.

Perpetuation

- The virus perpetuates through infected plant suckers.

Management

- Adaptation of strict quarantine measures
- Eradication of all infected suckers .
- Certified viru free suckers should be for planting.

Model Question Paper

A. Objective Type Questions

- Check plants regularly for symptoms, such as dark green streaks on the leaves.
- Use insecticidal soap or an approved pesticide to kill aphids. Soak the plant thoroughly, including underneath the petiole.
- Destroy infected plants immediately, including any suckers that may be infected.

a. Choose the correct answer from the following:

1. The main cause of the panama wilt disease in India

(a)	*Phytophthora parasitica*	(b)	*Phytophthora palmivora*
(c)	*Pseudomonas solanacearum*	(d)	*Fusarium oxysporum* f.sp *cubense*

2. *Pseudomonas solanacearum* pv *citri* is a type of bacteria

(a)	Gram -ve	b.	Gram -ve
(c)	Both	(d)	None

3. Optimum temperature for development of Panama wilt of banana

(a)	20-30^{0}C	(b)	30-40^{0}C.
(c)	10-15^{0}C	(d)	40-45^{0}C.

4. The fungicide Thiophanate methyl@ 0.1% used for control disease of banana

(a)	Moko	(b)	Cigatoka
(c)	Grey blight	(d)	Bunchy top

5.The color of colony of*Pseudomonas solanacearum* is

(a)	Yellow	(b)	White
(c)	Green	(d)	Pink

6. Small spindle shaped spots on foliage with greyish centre and yellowish halo running parallel to veins is characteristic symptoms of

(a)	Powdery mildew	(b)	Gummosis
(c)	Grey blight	(d)	Sigatoka

7. Moko disease is caused by which race of *Ralstonia solanacearum*

(a)	Race-1	(b)	Race-2
(c)	Race-3	(d)	Race-4

8. *Ralstonia solanacearum* belongs to the division

(a)	Gracilicutes	(b)	Firmicutes
(c)	Tenericutes	(d)	Mandicutes

9. The fungus *Cercospora musae* survives on dry infected leaves of banana on the field soil and primary infection takes place through in the infected plant debris.

(a)	Zoospore	(b)	Aplanospore
(c)	Ascospore	(d)	Basidiospore

10. The pathogen *Fusarium oxysporum* f.sp *cubense* perennates in soil through its

(a)	Oospores	(b)	Cleistothecium
(c)	Chlaymydospore	(d)	Apothecium

(15).A firm brown dry rot is found within fruits of infected plants is characteristic symptom of

(a)	Powdery mildew	(b)	Gummosis
(c)	Grey blight	(d)	Moko

16. The chief insect vector of bunchy top of banana is
   (a) *Pentalonia nigronervosa* (b) *Bemesia tabaci*
   (c) *Aphid gossypii* (d) *Acaria cajani*

Answer

S.No	Answer	S.No.	Answer
1	b. *Phytophthora palmivora*	6	d. Sigatoka
2	a. Gram -ve	7	b. Race-2
3	a. 20-30^{0}C	8	a. Gracilicutes
4	b. Cigatoka	9	c. Ascospore
5	b. White	10	c. Chlaymydospore

b.State whether the following statements are True or False.

1. Insects that spread the bacterial oozes from sick male flowers to healthy flowers are the source of secondary infections in banana bacteiral wilt.
2. The gram-negative, rod-shaped bacterium *Pseudomonas solanacearum* has a single polar flagellum.
3. Panama wilt in bananas is caused by an soilborne fungus in nature
4. *Fusarium oxysporum* f. sp. cubense generates chlamydospores as well as micro and macro conidia.
5. One of the main signs of the banana wilt disease is sudden wilting of the plant or individual leaves.
6. The hallmark sign of banana sigatoka disease is little spindle-shaped spots on foliage with a greyish center and a yellowish halo that run parallel to veins.
7. *Mycosphaerella musicola* yields perithecia with ostioles that range in color from dark brown to black.
8. Primary infection always takes place through wounded roots in panama wilt disease of banana.
9. Mechanical inoculation is not a means of transmission for the Banana bunchy top virus.
10. Plant suckers that are infected are the source of the Banana bunchy top virus.

Answer

S.No	Answer	S.No.	Answer
1.	True	6.	True
2.	True	7.	True

3.	True	8.	True
4.	True	9.	True
5.	True	10.	True

B. Descriptive Questions

a. Long Answer Questions

1. Explain in detail the occurrence, importance, symptoms, etiology, disease cycle and management of sigatoka diseae of banana.
2. Write down the disease cycle ad epidemiology of the following diseases of banana
   (i) Panama wilt (ii) Bacterial wilt
   (iii) Sigatoka
3. Briefly describe the diagnostic symptoms and integrated management of any two of the following banana diseases:
   (i) Bunchy top (ii) Sigatoka
   (iii) Panama wilt
4. Describe the panama wilt disease of banana in the following headings:
   (i) Pathogen (ii) Symptoms
   (ii) Disease cycle (iv) Management
5. Explain the bacterial wilt of banana in the following headings:
   (i) Symptoms (ii) Disease cycle
   (iii) Epidemiology (iv) Management
6. Explain in detail symptoms, transmission, perpetuation and management of bunchy top of banana

b. Short answer Questions

1. Write down the disease cycle of sigatoka disease of banana.
2. Write down the etiology of panama wilt of banana.
3. Which pathogen causes bacterial wilt of banana? Give its systematic position.
4. Write down the integrated management practices of panama wilt.
5. Describe the causal organisms and transmission of banana bunchy top.
6. Differentiate between bacterial wilt and panama wilt of banana.

38

Major Diseases of Apple Crop & Management

Sl No	Name of Diseases	Causal Organism
1	Scab	*Venturia inaequalis*(Cooke) Wint
2	Powdery mildew	*Podosphaera leucotricha*(Ell. And Ev.)
3	Fire blight	*Erwinia amylovora* (Burrill) Winslow et al
4	Crown gall	*Agrobacterium tumefaciens* (E.F.Sm.&Towns)Conn

1. Scab

Diagnostic Symptoms

- Symptoms generally noticed on leaves and fruits.
- Affected leaves become twisted or puckered and have black, circular spots on their upper surface.
- On the under surface of leaves, the spots are velvety and may coalesce to cover the whole leaf surface. Severely affected leaves may turn yellow and drop.
- Scab can also infect flower stems and cause flowers to drop.
- Lesions later become sunken and brown and may have spores around their margins.
- Infected fruit become distorted and may crack, allowing entry of secondary organisms.

Etiology

Scientific Classification

Kingdom: Fungi

Division: Ascomycota

Class: Loculoascomycetes

Order: Pleosporales

Family: Venturiaceae

Genus: *Venturia*

Species: *inaequalis*

- Pathogen: *Ventruia inaequalis* (*Spilocaea pomi* Fr.)
- Pathogen has got two distinct stages- the imperfect stage(*Spilocaea pomi*)

stage on living plant parts and perfect (*Ventruia inaequalis*) on fallen dead leaves.

- Fungus produces ascocarps in a stroma in over wintered leavesin orchard floor.
- Pseudothecia are dark brown to black and spherical 90-150μm inn diameter, having short beak and a distinct ostiole having single celled bristles at the apex.
- Asci: are cylindrical, fasciculate short stipulate produced in large number in the pseudotheicum which vary in size from 55-75x6-12 μm.
- Ascospores: measure 11-15x5-7 μm which are yellowish green to unequally bicelled, with upper cell smaller and wider than the lower one.
- Conidia are produced on conidiophores whicharise from short erect, closely septate brown mycelium and are non septate or have a single septum. Conidia are produced on leaf surface and fruit lesions throughput the spring and summer and measure 12-12x6-9μm.
- Fungus is bipolar heterothallic.
- There are five races of the pathogen .Out of these two have been reported to occur in India.

Disease cycle

- Pathogen survives through perithecia in the soil debris.
- Soil inoculums is the source of primary infection.
- Secondary infections occurs by conidia through rain splash or wind.

Management

- Use resistant or tolerant varieties.
- Collect and burn of fallen leaves, shoots and fruits.
- Ensure pruning method that allow for more air circulation.
- Avoid getting foliage wet when watering.
- Resistant varieties: Ashwini, Ambika, Angeles, American pride, Surabhi.

2. Powdery Mildew

Diagnostic Symptoms

- Symptoms produced on young shoots, leaves, blossoms, and fruit.
- White or grey patches appears on the leaves.
- Leaf fold, twist and cracking may takes place.
- Buds get infected and may be killed.
- Both leaves and twigs may be covered with white powdery mass.
- Defoliation of leaves and deformation of fruits takes place.

Etiology

Scientific Classification

Kingdom: Fungi

Division: Eumycota

Phyllum: Ascomycota

Class: Ascomycetes

Order: Erysiphales

Family: Erysiphaceae

Genus: *Oidium*

Species: *farinosum*

- Pathogen: *Podosphaera leucotricha* (*Oidium farinosum*)
- Obligate parasite
- Pathogen consists of weft of white hyphal threads and draw nutrition through th haustoria.
- Conidia:produced in chain on conidiophores,hyaline clear, without color, measure 20-38 × 12 µm, and contain distinct fibrosin bodies.
- Cleistothecia: ascocarp of pathogen, globose, black, partly embedded in mycelium.Appendages are long and stiff or short and tortuous.
- Asci: sac-like, enclosed in fruiting bodies ascocarps, contains a single ascus with eight ascospores, each of which is elliptical and measures 22-36 x 12-15 µm. Ascocarps are densely grouped together, measure 75-96 µm in diameter and have apical and basal appendages.

Disease Cycle

- Pathogen survives in the form of resting mycelium or encapsulated haustoria in the in dormant terminal and lateral shoot buds and in blossom buds produced and infected the previous growing season.
- Abundant sporulation from overwintering shoots and secondary lesions on young foliage leads to a rapid build up of inoculum.
- Secondary spread occurs through wing borne conidia.
- Cleistotheica are produced on heavily infected shoot and leaves in mid summer but they are not regarded as an important inoculums source because the ascoposres do not germinate readily.

Management

- Plant resistant varieties.
- Plant crops with sufficient spacing to allow for good ventilation.
- Remove infect leaves when the first spots appear.
- Donot touch healthy plants after touching infected plants.
- Fertilize with a balanced nutrient supply.

- Remove plant residue after harvest.
- Monitor fields regularly to assess the incidence of a disease or pest.

3. Fire blight

Erwinia amylovora pv.***pyri***

Diagnostic Symptoms

- Term 'fire blight' sums up the symptoms as if scorched by fire. Primary infection occurs on flowers causing 'blossom blight', followed by the more serious 'shoot blight'.
- Infected blossoms become brown, later turning black and giving a burnt appearance.
- Flower become water soaked and brownish to black in color,may fall or remain hanging.
- Brwon black blotches appear on the leaves,blackening advances and leaves curl and hang.
- Watersprouts and terminal twigs wilt, die-back extends down the water sprout and twigs.
- Canker appear on branches which may girdle the banch and trunk to cause collar rot.
- Infected fruit becomes water soaked and brown to black in color.Brown ooze may be seen on the surface of infected parts under humid conditions.

Etiology

Scientific Classification

Kingdom: Bacteria

Phyllum: Protobacteria

Class: Gammaproteobacteria

Order: Enterobacteriales

Family: Enterobacteriaceae

Genus: *Erwinia*

Species: *Amylovora* pv.*pyri*

- The pathogen is of historical importance. It has the distinction of being the first bacterium of plants shown to be a pathogen.
- The bacterium can be a plant pathogen was given by T.J.Burrill in 1878.
- *Erwinia amyovora* is also the first pathogen to be associated with an inscet vector.
- Rod shaped, filamentous bacterium.
- Gram negative.
- Size 1.1-3.0 µm x 0.5-1.2 µm.

- Motile by many peritrichous flagella, aerobic or facultative anaerobic, non sporulating.
- Colonies of virulent isolates are small, round and white with typical glistening shine.
- Bacterium requires nicotininc acid as a growth factor.

Disease Cycle

- Pathogen survives primarily in cankers, blighted twigs and bud of symptomless twigs and serve as primary inoculums for infection next spring.
- During spring and early summer, infected parts reactivate and produce bacterial ooze consisting of millions of bacterial cells and easily transported to blossoms by insects such as flies, ants, and beetles.
- Bacteria can then be spread very efficiently from blossom to blossom by honey bees causes blossom blight.
- Once blossoms are infected, the bacteria can quickly spread and cause secondary infection into shoots and branches, which can be spread by rain, insects, and contaminated pruning tools.Secondary infections may continue to occur throughout the growing season.
- Secondary infections multiply the disease and finally result in infections of the tree trunk and the root stock.

Management

- Grow resistant varieties, if available.
- Monitor orchards regularly for signs of the disease.
- Prune out infected branches and burn them, preferably by the end of the winter.
- Donot apply excess nitrogen on the trees.
- Shave off the cankers and treat with Bordeaux paste.
- Regulate soil fertility to reduce succulence of new growth.

4. Crown Gall

Agrobacterium tumefaciens

Diagnostic Symptoms

- Small tumours or galls develop in the wound just below the soil surface. They enlarge rapidly with outer tissues becoming brown or black.
- Galls can also appear on roots, stem, branches, petioles and leaf veins.
- In early stages, tumours are more or leass spherical, white colored and quite soft.Later, the outer tissue becomes dark brown.
- Gall tissue is disorganized growth with an enlarged cambium and irregular vascular tissue. Movement of water and nutrients is cleverly impaired by galls.

- Movement of water and nutrients is severly impaired by galls.
- Attacked plants may remain stunted producing small chlorotic leaves. Main roots grow poorly and their yiled is reduced (Figure-4).

Etiology

Scientific Classification

Kingdom: Bacteria

Division: Gracilicutes

Class: Proteobacteria

Order: Rhizobiales

Family: Rhizobeaceae

Genus:*Agrobacterium*

Species: *tumefaciens*

- Shape: Rod shaped.
- Staining: Gram negative.
- Size: 0.6-1.0 μm x 1.5-3.0 μm.
- Nature: Aerobic.
- Donot produce endospores.
- Flagella: 2-4, peritrichous flagella.
- Plasmids: Pathogen contain one to several large plasmids composed of circular double stranded DNA.
- Ti-Plasmid: Plasmid carries the genes for gall or tumour induction and is called tumour inducing plasmid (Ti-Plasmid).
- Most important characteristic feature of pathogen is its ability to transfer part of Ti-plasmid (T-DNA) into host plant chromosome and transform normal plant cells to tumor cells.
- This has two principal physiological effects; it disrupt the metabolism of growth hormones in the tissues, brining about oncogenesis and it causes the synthesis of specific chemicals called opines. These opines can be utilized only by a bacteria that contain an appropriate Ti-plasmid and make it a 'genetic parasite'.
- High phytohormone level results from expression of three oncogenes transferred stably into the plant genome from *A. tumefaciens*: iaaM, iaaH and ipt.
- iaaM and iaaH oncogenes direct auxin biosynthesis and ipt oncogene cause cytokinin production

Disease Cycle

- Pathogen is a soil inhabitant and perennates in infested soil where it can live as a saprophyte for several years.

- When susceptible plants are raised at such sites, the pathogen reaches the crown region through the freshly developed wounds due to cultural practices, grafting or by insects.
- Once bacteria reaches inside the tissue, produce an enzyme that cuts the Ti-plasmid at specific sites leleasing the segment of T-DNA form the Ri or Ti plasmid is integrated into a plant cell chromosome at various site and expressed.
- T-DNA expression results in overproduction of plant hormones and usually the synthesis of opines. Uncontrolled synthesis of hormones stimulates plant cells to divide, enlarge and form a tumor.
- With the increase in number and size of tumor cells,tumours exert pressure on surrounding and underlying normal tissues, which may be distorted xylem vessels and ultimately impaired the water supply to upper parts of plants .
- Due to invasion of tumours by secondary microbes decay of peripheral cells takes place.
- The bacterium from these galls may escape to contaminate the surrounding soil of healthy roots.
- The pathogen is then disseminated to new planting sites and healthy plants by splashing rain, irrigation water, tools, wind, insects and plant parts used for propagation.

Management

- Grow resistant varieties
- Avoid wounding the crown and control root cheming insects.
- Regulate soil fertility to reduce succulence of new growth.
- Deep seeds or root systems of nursery seedlings or root stocks upto crown region in *A. tumefaciens* strain K84 suspension.

Model Question Paper

- Clean tools, containers, and work surfaces with a disinfectant.
- Use high-quality plants.
- Avoid injuring plants, especially around the soil level.
- Steam the soil at 140°F for 30 minutes.
- Solarize the soil at 160°F for 30 minutes or 140°F for 1 hour
- Remove galls when they are small.
- Remove stunted trees.
- Remove severely affected trees less than 4 years old.
- Remove trunks with large galls.
- Plant in sites with no history of crown gall
- Maximize air drainage.

- Grow crops of potatoes or other vegetables to help eliminate the bacteria from the soil
- Grass the area over for one or more years

A. Objective Type Questions

a. Choose the correct answer from the following:

1. *Ventruia inaequalis* belongs to
   (a) Pleosporales (b) Peronosporales
   (c) Melanconiales (d) Moniliales
2. Scab of apple is caused by
   (a) *Venturia inaequalis* (b) *Rhizoctonia bataticola*
   (c) *Sphacelotheca sorghi* (d) *Sphacelotheca cruenta*
3. Fire blight of apple is caused by
   (a) *Erwinia amylovora* (b) *Uncinula necator*
   (c) *Sphacelotheca sorghi* (d) *Sphacelotheca cruenta*
4. Powdery mildew of apple is caused by
   (a) *Podosphaera leucotricha* (b) *Oidium mangiferae*
   (c) *Sphacelotheca sorghi* (d) *Sphacelotheca cruenta*
5. The pathogen *Ventruia inaequalis* produces
   (a) Oospores (b) Cleistothecium
   (c) Pseudotheicum (d) Apothecium
6. *Ventruia inaequalis* represents the anamorphic state of ascomycetous fungus
   (a) *Spilocaea pomi* (b) *Oidium mangiferae*
   (c) *Sphacelotheca sorghi* (d) *Sphacelotheca cruenta*
7. The fruiting body cleistothecia formed in which fungus
   (a) *Podosphaera leucotricha* (b) *Puccinia graminis*
   (c) *Sphacelotheca sorghi* (d) *Sphacelotheca cruenta*
8. White superficial powdery fungal growth appear on leaves, stalks of panicles, flowers and young fruits of grape is characteristic symptoms of
   (a) Powdery mildew (b) Bud rot
   (c) Grey blight (d) None
9. The pathogen *Podosphaera leucotricha is*
   (a) Biotroph (b) Hemibiotroph
   (c) Necrotroph (d) None
10. *Podosphaera leucotricha* belongs to order.
   (a) Erysiphales (b) Peronosporales
   (c) Melanconiales (d) Moniliale

Answer

S. No	Answer	S. No	Answer
1	a. Pleosporales	6	a. *Spilocaea pomi*
2	a. *Venturia inaequalis*	7	a. *Podosphaera leucotricha*
3	a. *Erwinia amylovora*	8	a. Powdery mildew
4	a. *Podosphaera leucotricha*	9	a. Biotroph
5	c. Pseudotheicum	10	a. Erysiphales

c. State whether the following statements are True or False.

1. The imperfect stage (*Spilocaea pomi*) survive on living apple plant parts and perfect (*Ventruia inaequalis*) on fallen dead leaves.
2. *Venturia inaequalis* ia an obligate parasite.
3. *Erwinia amylovora* is a Gram +ve bacteria.
4. The fungus *Venturia inaequalis* is bipolar heterothallic.
5. The ascocarp of pathogen*Oidium farinosum* is known as cleistotheica.
6. White floury patches symptom appear on both side of the leaves of apple in powdery mildew disease.
7. *Erwinia amylovora* pv.*pyri* has the distinction of being the first bacterium of plants shown to be a pathogen.
8. Plasmid carries the genes for gall or tumour induction and is called tumour inducing plasmid (Ti-Plasmid).
9. *Agrobacterium tumefaciens* as an endospore forming bacteria.
10. The telemorphic stage of *Spilocaea pomi* is *Venturia inaequalis.*

Answer

Sl. No	Answer	Sl. No	Answer
1	True	6	False
2	False	7	True
3	False	8	True
4	True	9	False
5	False	10	True

Descriptive Question

(a) Long Answer Questions

1. Name two bacterial diseases of apple with their causal organism. Describe etiology, disease cycle and management of any one disease.
2. Explain in detail the occurrence, importance, symptoms, etiology, disease cycle and management of scab of apple.
3. Briefly describe the diagnostic symptoms, epidemiology & management of any two of the following apple diseases:

   (i) Fire blight (ii) Crown gall

4. Describe the powdery mildew of apple in the following headings:
   (i) Pathogen (ii) Symptoms
   (iii) Disease cycle (iv) Disease management
5. Describe in detail the most distinguishing symptoms, causal organism, disease cycle and management of crown gall of apple.

(b) Short Answer Questions

1. Write down the etiology of powdery mildew of apple.
2. Which fungal pathogen causes crown gall of apple? Give its systematic position.
3. Write down the management practices of scab of apple.
4. Describe the causal organisms and mode of penetration of firelight of apple.
5. Why the Fire blight disease of apple is historically important ?

39

Major Diseases of Grape Crop Management

S.No.	Name of Disease	Causal organism
1	Downy mildew	*Plasmopara viticola*
2	Powdery mildew	*Uncinula necator*
3	Anthracnose	*Gloeosporium ampelophagum* (*Elsinoe ampelina)*

1. Downy Mildew

Diagnostic Symptoms

- Symptoms appear on all aerial and tender parts of the vine.
- Irregular, yellowish, translucent spots on the upper surface of the leaves.
- Correspondingly on the lower surface, dirty white, powdery growth of fungus appears.
- Affected leaves become, yellow and brown and gets dried due to necrosis
- Premature defoliation.
- Dwarfing of tender shoots.
- Infected leaves, shoots and tendrils are covered by whitish growth of the fungus.
- White growth of fungus on berries which subsequently becomes leathery and shrivels. Infected berries turn hard, bluish green and then brown.
- Later infection of berries results in soft rot symptoms. Normally, the fully grown or maturing berries do not contact fresh infection as stomata turn non-functional.
- No cracking of the skin of the berries .

Etiology

Scientific Classification

Kingdom: Fungi

Division: Eumycota

Class: Oomycetes

Order: Peronosporales

Family: Peronosporaceae

Genus: *Plasmopara*

Species: *viticola*

- Pathogen:*Plasmopara viticola.*
- The pathogen is biotroph.
- Mycelium is intercellular with spherical haustoria, coenocytic, thin walled and hyaline.
- Sporangiophores arise from hyphae in the sub stomatal spaces.
- It branched at right angle to the main axis and at regular intervals. Secondary branches arise from lower branches.
- The sporangia are thin walled, oval or lemon shaped.
- The sporangia germinate either by forming zoospores, or behave as conidia and germinate by germ tube.
- The zoospores are egg shaped, biflagellate and 7 – 9 micron meter. The oospores are thick walled.
- The oospores are thick walled with a rough epispore and are produced mostly in tissues adjacent to the midrib.

Disease Cycle

- Pathogen overwinters mainly as oospores in the debris in the soil, serve as the primary inoculum.
- They germinate and form numerous biflagellate zoospores.
- Primary infection is by zoospores transported by wind or water to the wet lower leaves near the soil line.
- They encyst and form a germ tube that penetrates through stomata or the epidermis.
- Infection becomes systemic and symptom appear after 10-15 days of infection.
- Secondary infection takes place by the motile zoospores by splashing rain.
- As the close of the season, oospores are formed, which fainally reach the soil.

Management

- Deep ploughing of fields during summer .
- Make sure soils are well drained.
- Field sanitation.
- Remove weed and plant residues from the field.
- Keep tools and equipment clean.
- Apply a balanced fertilizer.
- Destroy the alternate host plants as well as crop debris .
- Vine should be kept high above ground to allow circulation of air by proper spacing.

- Pruning (April- May & September –October) and burning of infected twigs.
- Grow resistant varieties like Amber Queen, Cardinal, Champa, Champion, Dogridge and Red Sultana.

2. Powdery Mildew: *Uncinula necator*

Diagnostic Symptoms

- Powdery growth mostly on the upper surface of the leaves.
- Malformation and discolouration of affected leaves.
- Discolouration of stem to dark brown.
- Floral infection results in shedding of flowers and poor fruit set.
- Early berry infection results in shedding of affected berries.
- Powdery growth is visible on older berries and the infection results in the cracking of skin of the berries.

Etiology

Scientific Classification

Kingdom: Fungi

Division: Eumycota

Class: Leotiomycetes

Order: Erysiphales

Family: Erysiphaceae

Genus: *Uncinula*

Species: *necator*

- The fungus is oidium type.
- White growth consists of mycelium, conidiophores and conidia.
- Mycelium is external, septate and hyaline.
- Conidiophores are short and arise from external mycelium.
- Conidia are produced in chain. They are single celled, hyaline and barrel shaped.
- When the mating types are present cleistothecia can form on all infected tissues during later part of the growing season.
- Mature cleistothecia are black, depressed, contains 4-8 asci.
- 4-6 oval ascospores occur within each ascus.

Disease Cycle

- Pathogen overwinters as mycelium in the dormant buds and as cleistotheica in the cracks in the bark.Besides, the collateral hosts and volunteer plants provide conidia as the primary mycelium.
- Infected buds give rise to systematically infected shoots.
- Cleistothecia, when wet, discharge the ascospores, which infect the leaves.

- Ascospores and conidia germinate and push haustoria into the epidermal cells from the appresorium.
- Mycelium grows externally and forms conidia in 7 to 10 days after infection.
- Secondary infections can occur through spores produced, released when conditions are favorable for growth of the fungus in spring and cause new infections.

Management

- Deep ploughing of fields during summer .
- Use tolerant varieties if available.
- Keep a wide distance between vines to allow for good air circulation.
- Alternativley chose pruning practices that favor an open canopy.
- Monitor the field regularly for symptom of the disease.
- Use nitrogen fertilizers with caution to avoid excessive vegetative growth.
- Field sanitation.
- Destroy the alternate host plants as well as crop debris.
- Adopt ecological engineering by growing the attractant, repellent, and trap crops around the field bunds.
- Grow resistant varieties like ChholthRed,Skibba Red, Chholth white.

3. Anthracnose

Diagnostic Symptoms

- Symptoms visible on leaves, stem, tendrils and berries.
- Young shoots and fruits are more susceptible than leaves.
- Circular, greyish black spots or red spots with yellow halo appear.
- Later the centre of the spot becomes grey, sunken and fall off resulting in a symptom called 'shot hole'.
- Black, sunken lesions appear on young shoots.
- Cankerous lesions on older shoots. Girdling and death of shoots occur.
- Infection on the stalk of bunches and berries result in the shedding of bunches and berries respectively.
- Sunken spots with ashy grey centre and dark margin on fruits (Birds eye symptom).
- Mummification and shedding of berries.

Etiology

Scientific Classification

Kingdom: Fungi

Division: Eumycota

Phyllum: Deuteromycotina

Class: Coelomycetes

Order: Melancoliales

Family: Melanconiaceae

Genus: *Gloeosporium*

Species: *ampelophagum*

- Pathogen: *Gloeosporium ampelophagum* (*Elsinoe ampelina)*
- Perfect stage: *Elsinoe ampelina*
- Hyphae: well developed, septate and branched.They accumulate in subcuticular region of infected plant and develop acervuli.
- Acervuli: single celled, colorless, delicate and small conidiophores are formed in large number in each acervulus.
- Conidia: produce on tips of conidiophores, unicellular, hyaline, slightly narrower in the middle than at the ends, measure 5-6 x2.5-3.5µm.

Disease Cycle

- Pathogen survives in infected plant debris and act as a source of inoculums.
- Primary infection is through spores produced by disease vines.
- Secondary infection by wind borne conidia.

Management

- Choose sites with suitable sun exposure and good air circulation.
- Removal of infected twigs.
- Use of resistant varieties such as– Bangalore Blue, Beauty Seedless, Bharat Early, Golden queen, Large white.
- Make sure to keep a wide distance between vines.
- Remove any wild grapes near the vineyard.
- Monitor the vines and remove fruit or plant parts that show sign of the disease.
- Prune the vines in early winter during dormancy.
- Remove plant residues from the vineyard plow field and burry.

Model Questions

A.Objective Type Questions

a. Choose the correct answer from the following

1. *Colletotrichum gloeosporioides* belongs to order.

   (a) Albuginales (b) Peronosporales

   (c) Melanconiales (d) Moniliales

2. Downy mildew of grape is caused by

   (a) *Plasmopara viticola* (b) *Rhizoctonia bataticola*

   (c) *Sphacelotheca sorghi* (d) *Sphacelotheca cruenta*

3. Anthracnose of grape is caused by
   (a) *Gloeosporium ampelophagum* (b) *Uncinula necator*
   (c) *Sphacelotheca sorghi* (d) *Sphacelotheca cruenta*
4. Powdery mildew of grape is caused by
   (a) *Uncinula necator* (b) *Oidium mangiferae*
   (c) *Sphacelotheca sorghi* (d) *Sphacelotheca cruenta*
5. Acervuli of *Colletotrichum gloeosporioides* may posses sterile hair like structures called
   (a) Oospores. (b) Cleistothecium
   (c) Setae (d) Apothecium
6. *Gloeosporium ampelophagum*represents the anamorphic state of ascomycetous fungus
   (a) *Elsinoe alpelina* (b) *Oidium mangiferae*
   (c) *Sphacelotheca sorghi* (d) *Sphacelotheca cruenta*
7. Hyphae of *Colletotrichum gloeosporioides*well developed, septate and branched, subcuticular region of infected plant and develop
   (a) Acervuli (b) Cleistothecium
   (c) Setae (d) Apothecium
8. Optimum epidemiological condition for development of powdery mildew of grape
   (a) Optimum temperature- 15-30^{0}C, Relative humidity-60-90 %.
   (b) Optimum temperature- 15-40^{0}C, Relative humidity-50-70 %.
   (c) Optimum temperature- 15-50^{0}C, Relative humidity-40-60 %.
   (d) Optimum temperature- 15-20^{0}C, Relative humidity-40-50 %.
9. White superficial powdery fungal growth appear on leaves, stalks of panicles, flowers and young fruits of grape is characteristic symptoms of
   (a) Powdery mildew (b) Bud rot
   (c) Grey blight (d) None
10. The pathogen *Plasmopara viticola is*
   (a) Biotroph (b) Hemibiotroph
   (c) Necrotroph (d) None

Answer

S. No	Answer	S. No	Answer
1	c. Melanconiales	6	a. *Elsinoe alpelina*
2	a. *Plasmopara viticola*	7	a. Acervuli
3	a. *Gloeosporium ampelophagum*	8	a. Optimum temperature- 15-30^{0}C, Relative humidity-60-90%.
4	a. *Uncinula necator*	9	a. Powdery mildew
5	c. Setae	10	a. Biotroph

b. State whether the following statements are *True* or *False:*

1. Rain splash or wind-driven rainwater transmission of conidia is the primary cause of secondary anthracnose infection.
2. *Colletotrichum gloeosporioides* hyphae collect beneath the host cuticle and give rise to acervuli, which are fruiting bodies.
3. At a mature stage, fruits infected with anthracnose convey the pathogen into storage, where it causes significant loss during transportation, storage, and marketing.
4. Airborne conidia produced by *Uncinula necator* can cause the powdery mildew disease to spread secondary.
5. A white growth of *Plasmopara viticola* on grape berries that later shrivels and turns leathery.
6. Foliar spray of 0.02% Streptomycin followed by Copper oxychloride (0.3%) regulates bacterial blight of mango.
7. To combat grape powdery mildew, apply foliar sprays containing either 0.1% or 0.3% wettable sulfur or karathane.
8. The sporangia of Plasmopara viticola are spherical, lemon-shaped, and have thin walls.
9. *Plasmopara viticola* produces egg-shaped biflagellate zoospores.
10. Most *Plasmopara viticola* oospores develop in tissues next to grape leaf midribs.

Answer

S. No	Answer	S. No	Answer
1	True	6	True
2	True	7	True
3	True	8	True
4	True	9	True
5	True	10	True

B. Descriptive Questions

a. Long Answer Questions

1. Describe in detail about symptoms, etiology, disease cycle and management of powdery mildew of grape.
2. Describe in detail the most distinguishing symptoms, causal organism, disease cycle and management of anthracnose of grape.
3. Describe in detail the most distinguishing symptoms, causal organism, epidemiology, disease cycle and management of downy mildew of grape.

4. Briefly describe the diagnostic symptom and management of powdery mildew and anthracnose of grape.
5. Explain the downy mildew of grape in the following headings:

(i)	Pathogen	(ii)	Symptoms
(ii)	Disease cycle	(iv)	Disease management

b. Short answer questions

1. Which fungal pathogen cause anthracnose disease of grape? Give its systematic position.
2. Elaborate disease cycle of Powdery mildew of grape in pictorial form .
3. Describe in brief about managment of downy mildew of grape.
4. How would you identify the powdery mildew of grape.
5. Write the epidemiological condition for development of powdery mildew of grape.

40

Major Diseases of Coconut Crop & Management

S. No	Disease	Causal Organism
1	Bud rot	*Phytophthora palmivora*
2	Ganoderma wilt	*Ganoderma lucidum*
3	Stem bleeding	*Ceratocystis paradoxa/ Ophiostoma paradoxa*
4	Grey blight	*Pestalotiopsis palmarum*

1) Bud rot

Diagnostic Symptoms

- Severe on young palms.
- Yellowish green discoloration of the heart leaf or crown leaf.
- The basal tissues of the leaf rot quickly and can be easily separated from the crown.
- Spindle withers and droop down.
- Older leaves develop irregular, water soaked spots which are sunken in nature.
- The leaves and sheath in the central spindle fall off leaving an outer whorl of green leaves.
- The withered central shoot can be pulled out very easily from the crown.
- The central crown may rot and in few months the tree may wilt.
- Young nuts fail to mature and fall.

Etiology

- Pathogen: *Phytophthora palmivora.*
- The fungus produces intercellur, non septate, hyaline mycelium.
- Sporangiophores are hyaline and simple or branched occasionally.
- The sporangiophores are hyaline, thin walled, pear shaped with a prominent papillae.
- Sporangia releases reniform, biflagellate zoospores upon germination.
- The fungus also produces thick walled, spherical oospores.
- In addition, thick walled, yellowish brown chlamydospores are also produced.

Disease cycle

- Primary infection occurs through dormant mycelium, oospores or chlamydospores carried over summer months on the host debris.
- With the onset of monsoon rains, the fungus becomes active producing cottony mycelium that infects tender host tissue.
- Secondary infection occurs through sporangia with numerous zoospores which spread rapidly in the rain water or sporangia spread through wind and insects also.

Management

- Cutting and burning of badly infected palms.
- If the disease is detected early remove the infected portions and protect with Bordeaux paste (Tree surgery).
- Give prophylactic spray with 1% Bordeaux mixture to all the healthy plams in the vicinity of diseases one and also before onset of monsoon rains.

2) Ganoderma Wilt

Diagnostic Symptoms

- Called as Basal stem rot /Thanjavur wilt / Bole rot.
- The most usual symptoms are yellowing, withering and drooping of the outer fronds which remain hanging around the trunk for several months before shedding.
- The younger leaves remain green for sometime and later turn yellowish brown.
- The new fronds produced become successively smaller and yellowish in colour which do not unfold properly.
- The wilting plants also show bleeding patches near the base of the trunk.
- A brown gummy liquid oozes out from the cracks in the tree which slowly result in the death of outer tissues.
- As the infection advances, fresh bleeding patches appear above the old once, up to 3-5 meters height. The decay of the basal portion occurs slowly and tree succumbs to the diseases in 2-3 years.

Etiology

- Pathogen: *Ganoderma lucidum.*
- The fungus produces a semi circular basidiocarp (bracket), which is attached to the tree with a stalk.
- The bracket is very big about 10-12 cm diameter and woody.
- The upper surface is tough, shining, light to dark brown or almost black with concentric furrows.
- The lower surface is white and soft with numerous minute pores.

- These pores represent the opening of the hymenial tubes, which are lined with basidia and basidio-spores.
- Basidiospores are oval, brown and thick walled.

Disease Cycle

- The fungus is soil-borne and survives in the soil for long time.
- The primary infection is through basidiospores in the soil, which attack roots.
- The irrigation water and rain water also help in the spread of the fungus.

Management

- Remove and burn severely infected trees which are beyond recovery.
- Isolate the diseased trees by digging a trench all around to check further spread. Irrigate the palms at least once in a fortnight during summer months.
- Apply heavy doses of farm yard manure or compost for green manure at 50 Kg/tree/year along with 5 kg of neem cake.
- Drench the soil near the tree with 40 litres of 1 per cent Bordeaux mixture at quarterly interval for thrice a year and repeat after 2-3 years.

Model Question Paper

Objective Type Questions

A.Choose the correct answer from the following:

1. *Phytophthora palmivora* belongs to order.
   (a) Agonomycetales (b) Peronosporales
   (c) Peronosporales (d) Moniliales
2. Bud rot of coconut is caused by
   (a) *Phytophthora palmivora* (b) *Rhizoctonia bataticola*
   (c) *Sphacelotheca sorghi* (d) *Sphacelotheca cruenta*
3. Ganoderma wilt of coconut is caused by:
   (a) *Cercospora arachidicola* (b) *Ganoderma lucidum*
   (c) *Sphacelotheca sorghi* (d) *Sphacelotheca cruenta*
4. *Phytophthora palmivora* produces thick walled sexual spores
   (a) Oospore (b) Zygospore
   (c) Pycnidium (d) Apothecium
5. Secondary infection of bud rot of coconut is by.
   (a) Soil borne conidia (b) Air borne sporangia
   (c) Externally seed borne conidia (d) Internally seed borne conidia
6. Optimum epidemiological condition for development of bud rot of coconut
   (a) Low temperature – 18-20^{0}C, Hidh relative humidity- above 90 percent

(b) Low temperature – 15-18^{0}C, High relative humidity- below 90 percent
(c) Low temperature – 10-12^{0}C, High relative humidity- above 70 percent
(d) Low temperature – 8-10^{0}C, High relative humidity- below 70 percent

7. Coconut plants show bleeding patches near the base of the trunk is a characteristic symptoms of
   (a) Thanjavur wilt (b) Bud rot
   (c) Grey blight (d) None
8. Primary infection of Thanjavur wiltis through which spores in the soil, which attack roots
   (a) Basidiospores (b) Ascospores
   (c) Oospores (d) Teliospores

Answer

S. No	Answer	S. No	Answer
1	c. Peronosporales	6	a. Low temperature – 18-20^0 C, High relative humidity- above 90 percent
2	a. *Phytophthora palmivora*	7	a. Thanjavur wilt
3	b. *Ganoderma lucidum*	8	a. Basidiospores
4	a. Acervuli	9	--
5	b. Air borne sporangia	10	--

b. State whether the following statements are *True* or *False:*

1. The central shoot of a coconut plant with bud rot readily breaks off with a modest tug.
2. The mycelium of the fungus *Phytophthora palmivora* is intercellur, non septate, and hyaline.
3. Upon germination, the sparangia of *Phytophthora palmivora* produce biflagellate, reniform zoospores.
4. *Phytophthora palmivora* produces thick walled, yellowish brown chlamydospores that cause coconut bud rot.
5. In agricultural wastes in the soil, *Phytophthora palmivora*, which causes coconut bud rot, lives as oospores and chamydospores.
6. The primary means of transmission for bub rot infections are airborne sporangia and zoospores.
7. In the later phases of infection, the pathogen *Ganoderma lucidium* forms fruiting bodies (Bracket) along the side of the basal trunk.
8. The main way that *Ganoderma lucidium* is transmitted is by soil-borne basidiospores that target roots.
9. Ganoderma wilt can be decreased by soaking the soil next to the coconut tree three times a year with a 1% Bordeaux mixture applied at quarterly intervals.

10. Coconut root wilt disease is caused by a complex of factors , primarily fungal and bacterial pathogens.

Answer

S. No	Answer	S. No	Answer
1.	True	6.	True
2.	True	7.	True
3.	True	8.	True
4.	True	9.	True
5.	True	10.	True

B. Descriptive questions

a. Long answer questions

1. Which are the important fungal diseases of coconut ? Describe bud rot under heads, diagnostic symptoms, causal organism, etiology, disease cycle, epidemiology and disease management.
2. Describe in detail the causal organism, symptoms, disease-cycle, epidemiology and management of bud rot of coconut.
3. Describe Ganoderma wilt of coconut under heads, diagnostic symptoms, causal organism, disease cycle and disease management.
4. Describe in detail the symptoms, causal organism, etiology, disease-cycle, epidemiology and management of Ganoderma wilt of coconut.
5. Write short notes on

a. Disese cycle of ganoderma wilt of coconut.

b. Management of bud rot of coconut.

c. Etiology of *Phytophthora palmivora* cusing bud rot of coconut.

b.Short answer questions

1. Which fungal pathogen cause bud rot of coconut? Give its systematic position.
2. Elaborate disease cycle of Ganoderma wilt of coconut in pictorial form .
3. Describe in brief about mangment of Ganoderma wilt of coconut.
4. How would you identify the Ganoderma wilt of coconut disease in the field.
5. Write the epidemiological condition for development of bud rot of coconut.
6. Write down the systemic position, etiology of pathogen *Ganoderma lucidum* causing ganoderma wilt of coconut.

41

Major Diseases of Tea Crop & Management

S. No	Name of Disease	Causal organism
1	Blister blight	*Exobasidium vexans*
2	Red rust	*Cephaleurus mycoidea*
3	Black rot	*Corticium invisum* and *C. theae*
4	Black root	*Rosellina areuata*

1) Blister Blight

Diagnostic Symptoms

- Initially oily, yellowish, translucent spots appear on the tender leaf and turn to deep red shiny blisters.
- The circular spot gradually enlarges to 3 to 13mm diameter, bulged on the under surface of the leaf with a concave trough like depression on the upper surface.
- Leaves become curled and distorted.
- First flush of 2-3 young leaves are attacked and the young shoots and buds are killed.
- Mature leaf is not affected.
- In nursery infection, seedlings are stunted with many thin stems instead of a single stalk.
- Repeated attacks cause death of seedlings.
- Badly affected nurseries will have to be abandoned.
- Succulent leaves and green shoots of newly pruned tea are most susceptible.
- Basidiospores cause secondary infection.

Etiology

- Pathogen:*Exobasidium vexans*
- The mycelium is confined to the blistered areas on the leaves.
- They are septae and collect in bundles below the lower epidermis.
- Later by rupturing the epidermis continuous layer of vertical hyphae are projected on the surface of spot.
- The fungus produces two kind of spores viz.,the conidia and basidiospores.
- The conidia are most abundant, borne singly at the tip of long stalks.

- Basidia are formed on the surface in large number but never form a continuous hymenium.

Disease Cycle

- The fungus completes its life cycle in 11-28 days and several generations of spores are produced in a season.
- It produces conidia and basidiospores in the same blister.
- Spores are air borne.
- The basidiospores germinate on tea leaves by forming appresorium from which infection takes place through an infection hypha.
- It is a polycyclic diseases and the secondary infections are repeated till plants are available in the field.
- The perpetuation of the fungus appears to be form the pre existing infected bushes.

Management

- Remove infected leaves and shoots, and prune shade trees to allow sunlight to reach the tea plants.
- Growing resistant cultivars: This is another option to control the disease.
- Direct sunlight: The basidiospores of the pathogen are sensitive to direct sunlight, so you can plant rows further apart to allow sunlight to reach young leaves.
- Apply bio formulations of *Pseudomonas fluorescens* Pf1 or antagonistic agents like *Trichoderma harzianum, Gliocladium virens, and Bacillus subtilis.*
- Use sodium bicarbonate (baking soda) or salicylic acid.
- Seedlings should be protected in nursery by weekly sprays of COC @0.3%.

Model Question Paper

A.Objective Type Questions

a. Choose the correct answer from the following:

1. *Exobasidium vexans*belongs to order.
   (a) Exobasidiales (b) Peronosporales
   (c) Melanconiales (d) Moniliales
2. Blister blight is caused by
   (a) *Hemileia vastatrix* (b) *Rhizoctonia bataticola*
   (c) *Exobasidium vexans* (d) *Sphacelotheca cruenta*
3. Blister blight of tea is a type of disease
   (a) Monocyclic (b) Polycyclic
   (c) Polyetic (d) None

4. The fungus *Exobasidium vexans* completes its life cycle in how many days:

(a) 11-28 (b) 28-35

(c) 10-15 (d) None

(5) Blister blight disease is

(a) Soil borne (b) Seed borne

(c) Air borne (d) All

Answer

S. No	Answer	S. No	Answer
1.	a.Exobasidiales	4.	a.11-28
2.	c.*Exobasidium vexans*	5.	c.Air borne
3.	b.Polycyclic	6.	---

b. State whether the following statements are True or False

1. Mature leaf of tea is mostly not affected with blister blight.
2. Basidiospores of *Hemiliea vastatrix* cause secondary infection in tea plant.
3. Succulent leaves and green shoots of newly pruned tea are most susceptible to blister blight of tea.
4. The fungus*Hemiliea vastatrix*produces two kind of spores viz.,the conidia and basidiospores.
5. Relative humidity greater than 83% for 7 to 10 days favours blister blight disease.
6. Temperature above 35°C inhibit the blister blight disease.
7. The fungus*Hemiliea vastatrix*produces conidia and basidiospores in the same blister of tea plant.
8. *Hemiliea vastatrix*spores are air borne in nature.
9. Blister blight of tea is a polycyclic diseases and the secondary infections are repeated till plants are available in the field
10. The perpetuation of the fungus *Hemiliea vastatrix*appears to be from the pre existing infected bushes.

Answer

S. No	Answer	S. No	Answer
1	True	6	True
2	True	7	True
3	True	8	True
4	True	9	True
5	True	10	True

B. Descriptive Questions

a. Long answer questions

1. Explain in detail about symptoms, etiology, disease cycle and management of blister blight of tea.

2. Describe the blister blight of tea in the following headings:
   (i) Pathogen (ii) Symptoms
   (iii) Disease cycle (iv) Disease management
3. Describe in detail the most distinguishing symptoms, causal organism, disease cycle, epidemiology and management of blister blight of tea

b.Short answer questions

1. Why the blister blight of tea is historically important in India ?
2. Write down the etiology of blister blight of tea.
3. Which fungal pathogen causes blister blight of tea? Give its systematic position.
4. Write down the management practices of blister blight of tea.
5. Describe the causal organisms and diagnostic symptoms of blister blight of tea.

42

Major Diseases of Coffee Crop Management

S. No	Name of Disease	Causal organism
1	Rust	*Hemileia vastatrix*
2	Black rot or Koleroga	*Cortieium salminicolor*
3	Die back or Anthranose	*Collectorichum coffeanum*
4	Damping off / Collar rot	*Rhizocotonia solani*
5	Berry blotch	*Cercospora coffeicola*

1. Rust

Diagnostic Symptoms

- The disease is restricted to leaves although sometimes it can seen on tender shoots and berries.
- The rust pustules appear as small spots, 1-2 mm in diameter.
- In early stages these spots are yellowish becoming orange in color with increase in size.
- On the upper surface, opposite the spots on the lower surface, the color is often brownish.
- In severe attack leaves may dry and wither.
- The berries remain small and fail to ripen.
- Extensive defoliation weakens the trees and results in poor yield, severe die back of twigs, and death of trees.

Etiology

- Pathogen: *Hemileia vastatrix*
- The mycelium is intercellular and sends haustoria into the cells.
- The mycelium sends out erumpent stalks through stomata which bear the uredospores.
- The uredospores are reniform or orange segment like in shape.
- The convex side of the spores are echinulated and the lower side is smooth and measure 26 – 40 x 20 – 30 micron meter.
- The telial stage succeeds the uredial stage in the later stage.
- The pycnial and aecial stages of *H. vastatrix* has not been seen and it is still not confirm whether the fungus is autoecious or heterocious

Disease Cycle

- The fungus survives primarily as mycelium, uredia and uredospores on infected leaves.
- The uredospores are easily disseminated by wind, rain and possibly insects.
- Under favourable conditions the spores germinate and enter the host through stomata.
- The mycelium grow intercellularly sending haustoria into the cells.
- Repeated crops of uredospores are produced during the growing season and, as a result, several secondary infection cycles take place if environmental conditions are favourable.
- Young leaves are more susceptible than older leaves.

Management

- Fallen leaves should be composted or destroyed.
- Grow resistant varieties.
- Spray the bushes once with Bordeaux mixture @0.5% or Copper oxy chloride @0.25% and subsequently 2-3 times during monsoon.

Model Question Paper

A. Objective Type Questions

a. Choose the correct answer from the following:

1. *Hemileia vastatrix* belongs to order.
   (a) Pucciniales (b) Peronosporales
   (c) Melanconiales (d) Moniliales
2. Rust of coffee is caused by
   (a) *Hemileia vastatrix* (b) *Rhizoctonia bataticola*
   (c) *Sphacelotheca sorghi* (d) *Sphacelotheca cruenta*
3. The uredospores of *Hemileia vastatrix* are like
   (a) Orange segment (b) Reniform
   (c) Both (d) None
4. The uredospores are easily disseminated by
   (a) Wind (b) Rain
   (c) Insects (d) All
5. Several secondary infection cycles of rust are produced during the growing season as a result of
   (a) Uredospore (b) Teliospore
   (c) Basidiospore (d) Aeciospore

Answer

S. No	Answer	S. No	Answer
1	a. Pucciniales	4	d. All
2	a. *Hemileia vastatrix*	5	a. Uredospore
3	c. Both	6	

b.State whether the following statements are True or False

1. *Hemileia vastatrix* mycelium produces erumpent stalks through stomata that hold uredospores.
2. *Hemileia vastatrix* uredospores have an orange segment-like or reniform shape.
3. The telial stage follows the uredial stage in the later phases of *Hemileia vastatrix* life cycle.
4. There are no reports on the pycnial or aecial stages in *H. vastatrix*.
5. The fungus *Hemileia vastatrix* mostly survives on diseased leaves as uredospores, uredia, and mycelium.
6. The uredospores of coffee rust are easily dispersed by wind, rain, and maybe insects.
7. A rust infection causes repeated crops of uredospores to be produced during the coffee growing season.
8. Young leaves are more susceptible to rust disease than older leaves of coffee plants.
9. Uredospores of *Hemileia vastatrix* germinate only in free moisture..

Answer

S. No	Answer	S. No	Answer
1	True	6	True
2	True	7	True
3	True	8	True
4	True	9	True
5	True		

B. Descriptive Questions

a. Long answer Questions

1. Explain in detail about symptoms, etiology, disease cycle and management of rust of coffee.
2. Describe the rust of coffee in the following headings:

   i. Pathogen (ii) Symptoms

   iii Disease cycle (iv) Disease management
3. Describe in detail the most distinguishing symptoms, causal organism, disease cycle, epidemiology and management of rust of coffee.

b. Short answer questions

1. Write down the etiology of rust of coffee.
2. Which fungal pathogen causes rust of coffee? Give its systematic position.
3. Write down the management practices of rust of coffee.
4. Describe the causal organisms and diagnostic symptoms of rust of coffee.
5. Why the rust of coffee is historically important ?

43

Major Diseases of Mango Crop & Management

S. No	Name of Disease	Causal organism
1	Anthracnose	*Colletotrichum gloeosporioides* Ston, Spaull and Schrenk
2	Malformation	*Fusarium mangiferae* Wollenweber & Reinking
3	Bacterial blight	*Xanthomonas campestris* pv. *mangiferae*
4	Powdery mildew	*Oidium mangiferae* Berthet
5	Stem end rot	*Botrydiplodia theobromae*
6	Red rust	*Cephaleurus mycoides*

Anthracnose

Colletotrichum gloeosporioides (Glomerella cingulata)

Diagnostic Symptoms

- Disease produces symptoms like leaf spot, blossom blight, withertip, twig blight and fruit rot .
- Symptoms on leaves show as gray to brown spots with dark margins and a yellow halo. The spots later enlarge and coaleasce to form sizable necrotic areas.
- Tender shoots and foliage are easily affected which ultimately cause die back young branches.
- Blossom blight may vary in severity from slight to a heavy infection of the panicles .
- Black spots develop on panicles as well as on fruits. Severe infection destroys the entire inflorescence resulting in no setting of fruits.
- Young infected fruits develop black spots, shrivel and drop off.
- Fruits infected at mature stage carry the pathogen into storage and cause considerable loss during storage, transit and marketing.

Etiology

Scientific Classification

Kingdom: Fungi

Division: Eumycota

Sub-**division:** Deuteromycotina

Class: Coelomycetes

Order: Melanconiales

Family: Melanconiaceae

Genus: *Colletotrichum*

Species: *gloeosporioides*

- *Colletotrichum gloeosporioides* represents the anamorphic state of ascomycetous fungus *Glomerella cingulata*(Telemorphic state).
- Mycelium:septate and slightly dark coloured.
- Hyphae: accumulate below the host cuticle and develop fruiting bodies called acervuli.
- Acervuli: consists of numerous, closely packed, single celled, colorless conidiophores that produce conidia on their tips.It may also posses sterile hair like structures called 'setae'.
- Conidia: broadly oval to oblong with obtuse ends, non septate, some times posses oil globules, measure 12-16x4-6μm.
- Setae: common on twigs but not on fruits.

Disease Cycle

- Pathogen survives in plant debris like twigs and leaves which fall on the ground .
- Primary infection caused by condia developed on disesed plant parts on ground, are disseminated by air, brought on to the susceptible part of plant, germinate in favourable conditions.
- Secondary infection is mainly due to transfer of conidia through rain splash or wind driven rain water.

Management

- Plant resistant varieties and use healthy seedlings. Leave sufficient space between plants.
- Prune trees yearly to enhance ventilation.
- Implement good drainage methods.
- Diseased leaves, twigs, gall midge infected leaves and fruits, should be collected and burnt.
- Cover the fruits on tree, 15 days prior to harvest with news or brown paper bags.
- Foliar spray of *Pseudomonas fluorescens* @ 5g/liter of water at 21 days interval commencing from october on flower branches.5-7 sprays one to be given on flowers and bunches.
- Store fruits in a well ventilated environment.

2. Powdery Mildew

Oidium mangiferae (Acrosporum mangiferae)

Diagnostic Symptoms

- Pathogen parasitizes young tissues of all parts of the inflorescence, leaves and fruits.
- Initially white superficial powdery fungal growth appearon leaves, stalks of panicles, flowers and young fruits.
- Young leaves are attacked on both the sides but it is more evident on the upper surface. Often these patches coalesce and occupy larger areas turning into purplish brown in colour
- Affected flowers and fruits drop pre-maturely reducing the crop load considerably or might even prevent the fruit set.

Etiology

Scientific Classification

Kingdom: Fungi

Division: Eumycota

Class: Leotiomycetes

Order: Erysiphales

Family: Erysiphaceae

Genus: *Oidium*

Species: *mangifeare*

- Pathogen is an ectoparasitic biotroph and oogamous type.
- Hyphae are superficial, septate, hyaline, and ramify over the host surface forming white dense coating..
- Conidiophores short, hyaline and conidia single celled –barrel shaped, produced in chain.

Disease Cycle

- Pathogen overwinters in dormant bud or on old leaves.
- Resting spores are source of primary infection.
- Secondary spread of disease can occur thorugh air borne conidia produced in these infections.
- Spores blown wind from infected areas readily adhere to hairy, unopened flowers near tip of the inflorescence and germinate in five to seven hours and causes infection.

Epidemiology

- Optimum temperature- 15-30^{0}C.
- Relative humidity-60-90 %.

- Rains or mists accompanied by cooler nights during flowering.
- Most susceptible stage: Full bloom and fruit set at pea size stage.

Management

- Use resistant/tolerant varieties likeNeelum, Zardalu, Banglora, Torapari-khurd and Janardhan pasand.
- Spraying at full bloom needs to be avoided.
- Grow mango on dry and well ventilated areas.
- Intercrop with other non host tree species.
- Prune diseased leaves and malformed panicles harbouring the pathogen to reduce primary inoculums load.

3. Mango malformation

Fusarium moniliforme* var. *subglutinans

Diagnostic Symptoms

- Two types of symptoms are generally produced; floral malformation and vegetative malformation.
- **Vegetative malformation:**
    - Pronounced in young seedlings.
    - Affected seedlings develop vegetative growths which are abnormal growth, swollen and have very short internodes, giving it a bunch like an appearance on the shoot apex.
    - Seedling remain stunted and eventually die.
- **Floral malformation:**
    - Flower buds are transformed into vegetative buds and a large number of small leaves and stems, which are characterized by appreciably reduced internodes and give an appearance of witches" broom.
    - Flower buds seldom open and remain dull green.

Etiology

- Physiological disorder, eriophyid mites, fungal infection, virus, herbicides and other toxic compounds aids in the production of pathogen.
- Deficiency of iron, zinc and copper can also cause the malformation.

Disease Cycle

- Disease is mainly spread via infected plant material.
- Wound by mango bud mite lead to facilitate primary infection.
- Secondary infectionis either by air borne conidia or by conidia carried by eriophid mite *Aceria mangiferae.*

Epidemiology

- Optimm temperature- 10-15^{0}C.
- More disease in young than in old palnts.

Integrated Management:

- Use of disease free planting material.
- Monitor orchard regularly for any signs of deformed plants parts.
- Plough the field before planting to destroy existing weeds in the field.
- Deep summer ploughing.
- Irrigate the orchards as and when required
- Provide proper shade, irrigation & drainage.
- Floral malformed panicles/ vegetative malformed shoots should be pruned and burnt.
- Application of NAA (200 ppm) in the first week of October (Before bud differentiation stages) followed by deblossoming in the late December or January
- Spray zinc, boron and copper before bloom and after fruit harvesting..
- Avoid mechanical injury to the mango trees during field work.
- Protect them from strong winds and heavy rains with wind breaks.
- Disinfect working tools and equipments.

Model Question Paper

A.Objective Type Questions

a. Choose the correct answer from the following

1. *Colletotrichum gloeosporioides* belongs to order.

(b) Agonomycetales (b) Peronosporales
(c) Melanconiales (d) Moniliales

2. Anthracnose of mango is caused by

(b) *Colletotrichum gloeosporioides* (b) *Rhizoctonia bataticola*
(c) *Sphacelotheca sorghi* (d) *Sphacelotheca cruenta*

3. Powdery mildew of mango is caused by:

(b) *Cercospora arachidicola* (b) *Oidium mangiferae*
(c) *Sphacelotheca sorghi* (d) *Sphacelotheca cruenta*

4. Acervuli of *Colletotrichum gloeosporioides* may posses sterile hair like structures called

(b) Oospores. (b) Cleistothecium
(c) Setae (d) Apothecium

5. *Colletotrichum gloeosporioides* represents the anamorphic state of ascomycetous fungus

(b) *Glomerella cingulata* (b) *Oidium mangiferae*
(c) *Sphacelotheca sorghi* (d) *Sphacelotheca cruent*

6. Hyphae of *Colletotrichum gloeosporioides* accumulate below the host cuticle and develop fruiting bodies called
   (b) Acervuli (b) Cleistothecium
   (c) Setae (d) Apothecium
7. Optimum epidemiological condition for development of powdery mildew of mango
   (b) Optimum temperature- 15-30^{0}C, Relative humidity-60-90 %.
   (b) Optimum temperature- 15-40^{0}C, Relative humidity-50-70 %.
   (c) Optimum temperature- 15-50^{0}C, Relative humidity-40-60 %.
   (d) Optimum temperature- 15-20^{0}C, Relative humidity-40-50 %.
8. White superficial powdery fungal growth appear on leaves, stalks of panicles, flowers and young fruits of mango is characteristic symptoms of
   (b) Powdery mildew (b) Bud rot
   (c) Grey blight (d) None
9. *Oidium mangiferae is*
   (b) Biotroph (b) Hemibiotroph
   (c) Necrotroph (d) None
10. *Oidium mangiferae* belongs to order.
   (b) Erysiphales (b) Peronosporales
   (c) Melanconiales (d) Moniliales

Answer

S. No	Answer	S. No	Answer
1	c. Melanconiales	6	a. Acervuli
2	a. *Colletotrichum gloeosporioides*	7	a. Optimum temperature- 15-30^{0}C, Relative humidity-60-90 %
3	b. *Oidium mangiferae*	8	a. Powdery mildew
4	c. Setae	9	a. Biotroph
5	a. *Glomerella cingulata*	10	a. Erysiphales

b State whether the following statements are *True* or *False:*

1. Rain splash or wind-driven rainwater transmission of conidia is the primary cause of secondary anthracnose infection.
2. *Colletotrichum gloeosporioides* hyphae aggregate beneath the host cuticle and give rise to acervuli, which are fruiting bodies.
3. At a mature stage, fruits infected with anthracnose carry the virus into storage, where it causes significant loss during transportation, storage, and marketing.
4. Airborne conidia caused by these diseases can cause the secondary spread of powdery mildew disease.
5. The eriophid mite *Aceria mangiferae* or airborne conidia are the two main sources of secondary infection causing malformations in mangos.

6. *Xanthomonas mangiferae*- indicae belong to family Psudomonadaceae.
7. Yellowish color of the colonies of *Xanthomonas mangiferae- indicae* is due to production of characteristic water in soluble, non-diffusible pigment.
8. Foliar spray of 0.02% Streptomycin followed by Copper oxychloride (0.3%) manages bacterial blight of mango.
9. Foliar sprays with wettable sulphur@ 0.3% orKarathane@ 0.1% control powdery mildew of mango.
10. Deficiency of iron, zinc and copper can also cause the malformation.

Answer

S. No	Answer	S. No	Answer
1	True	6	True
2	True	7	True
3	True	8	True
4	True	9	True
5	True	10	True

B. Descriptive questions

a. Long answer questions

1. Describe in detail about symptoms, etiology, disease cycle and management of powdery mildew of mango.
2. Describe in detail the most distinguishing symptoms, causal organism, disease cycle and management of anthracnose of mango.
3. Describe in detail the most distinguishing symptoms, causal organism, epidemiology, disease cycle and management of bacterial blight of mango.
4. Briefly describe the diagnostic symptom and management of powdery mildew and anthracnose disease of mango.
5. Explain the mango malformation in the following headings:

   (i) Pathogen (ii) Symptoms

   (ii) Disease cycle (iv) Disease management

b. Short answer questions

1) Which bacterial pathogen cause bacterial blight of mango? Give its systematic position.
2) Elaborate disease cycle of powdery mildew of mango in pictorial form .
3) Describe in brief about managment of mango malformation.
4) How would you identify the powdery mildew of mango.
5) Write the epidemiological condition for development of powdery mildew of mango.
6) Write down the systemic position etiology of pathogen *Colletotrichum gloeosporioides causing* anthracnose of mango.

44

Major Diseases of Potato Crop & Management

S. No	Name of Disease	Causal organism
1	Early blight	*Alternaria solani*
2	Late blight	*Phytophthora infestans*
3	Black scurf	*Rhizoctonia solani*
4	Common scab	*Streptomyces scabies*
5	Viral diseases	
6	Mild mosaic/Interveinal mosaic	Potato virus X
7	Severe mosaic	Potato virus Y
8	Leaf roll	Potato leaf roll virus

1. Early blight

Diagnostic Symptoms

- Brown-black necrotic spot-angular, oval shape characterized by concentric rings .
- The concentric ridges produce a target board effect, the most characteristic symptom of the disease.
- Several spot coalesce & spread all over the leaf.
- In severe attacks leaves shrivel and fall down.
- Infection and rotting of tubers has also been reported (Fig.1).

Etiology

Scientific Classification

Kingdom: Fungi

Phylum: Ascomycota

Class: Dothideomycetes

Subclass: Pleosporomycetidae

Order: Pleosporales

Family: Pleosporaceae

Genus: *Alternaria*

Species: *solani*

- Pathogen; *Alternaria solani.*
- Hyphae are light brown or olivaceous which become dark coloured with age.
- The hyphae are branched, septate and inter and intra cellular.
- The coniophores emerge through the stomata or between the epidermal cells.
- The conidia are club shaped with a long beak which is often half the long of the whole conidium.
- The lower part of the conidium is brown while the neck is colorless.
- The body of the conidium is divided by 5 – 10 transverse septa and there may or may not be a few longitudinal septa.

Disease cycle

- The conidia and the mycelium in the soil or in the debris of the affected plants can remain viable for more than 17 months.
- These conidia or the new conidia found on the overwintered mycelium bring about the primary infection of the succeeding potato crop.
- The conidia formed on the spots developed due to primary infection are disseminated by wind to long distances.
- The conidia from the affected plant may also be disseminated to the adjoining plants by rain and insects.
- Infection occurs through the stomata but direct penetration may also takes place.

Management

- Summer deep ploughing.
- Soil solarization during summer.
- Field sanitation, rogueing.
- Avoid water logged conditions in the field.
- Follow crop rotation.
- Apply manures and fertilizers as per soil test recommendations.
- Use resistant/tolerant varieties.
- Use healthy, certified and weed seed free tubers.
- Removal and destruction of infected plant debris should be done because the spores lying in the soil are the primary source of infection.

2) Late blight

Diagnostic Symptoms

- Initially starts from leaf tips or margins and spread inward.
- Small faded green patches on upper surface of leaf which turn into brown spots.

- Downy growth of the pathogen on subsequent lower surface.
- Progressive defoliation and collapse of plants under favourable conditions.
- Water soaked stripes on stem which becomes necrotic.
- Purplish brown spots appear on skin of tubers.
- On cutting, the affected tubers show rusty brown necrosis spreading from surface to the centre.

Etiology

Scientific Classification

Phylum: Heterokontophyta

Class: Oomycota

Order: Peronosporales

Family: Peronosporaceae

Genus: *Phytophthora*

Species: *infestans*

- Pathogen – *Phytophthora infestans*
- The mycelium is endophytic, coenocytic and hyaline which are inter cellular with double club shaped haustoria type.
- Sporangiophores are hyaline, branched intermediate and thick walled.
- Sporangia are thin walled, hyaline, oval or pear shaped with a definite papilla at the apex.
- The sporangium may act as a conidium and germinate directly to form a germ tube.
- Zoospores are biflagellate possess fine hairs while the other does not.
- The sexual reproduction in *Phytophthora* is not very common.

Disease cycle

- Primary infection is through use of infected tubers.
- Mycelium spreads into shoots produced from infected tubers and reaches the aerial parts of the plant.
- Sporangiophore emerges through stomata on stem and leaves and produce sporangia, which are spread by rain to wet potato leaves or stems and cause disease.
- Large number of asexual generation in a growing season kills the foliage rapidly.
- The zoospores found in the soil germinate, penetrate through lentils or wounds into the tubers and send intercellular mycelium and haustoria into the cells and cause infection.

Management

- Regulatory measures.
- Select healthy tubers for planting.
- Delayed harvesting.
- High ridging to about 10-15cm height reduces tuber infection.
- Apply potasic fertilizer.
- Grow resistant varieties such as Kufri Jyothi, Kufri Badshah, Kufri Jeevan, Kufri Sherpa, etc.

3. Black scurf

Diagnostic Symptoms

- Black speck, black speck scab, russet scab on tubers.
- At the time of sprouting dark brown colour appear on the eyes.
- Affected xylem tissue causes to wilting of plants.
- Infected tuber contains russeting of the skin.
- Hard dry rot with browning on internal tissue.
- Spongy mass appear on the infected tuber.

Etiology

- Pathogen: *Rhizoctonia solani.*
- Perfect stage-*Thanetophorus cucumeris.*
- The mycelium is hyaline when young and brown at maturity.
- Hyphae are septate and branched with a characteristic constriction at their junction with the main hyphae.
- The branches arise at a right angle to main axis. Sclerotia are black.
- A basidium bears four sterimata each with a basidiospore at the end.
- The basidiospores are hyaline, elliptical to obovate and thin walled.
- They are capable of forming secondary basidiospores.

Disease cycle

- The fungus is capable of leading a saprophytic life on the organic material and can remain viable in the soil for several years.
- The sclerotia on the seed tubers is the principal source of infection of the subsequent crop raised with these tubers.
- On return of favourable conditions the mycelium present in the soil may develop producing new hypae.
- The new hypha infects fresh tubers, germinating buds and stem.
- Under adverse condtion, the mycelium turns into sclerotia and serves as means for survival.

Management

- Disease free seed tubers alone should be planted.
- Soak seed tubers in a solution of trisodium phosphate (90 g/l of water) one day before sowing. The tubers should be thoroughly rinsed and dried in shade.
- Well sporulated tubrs planted shallow.
- The disease severity is reduced in the land is left fallow for 2 years.

4) Viral diseases

(a) Mild mosaic/Interveinal mosaic

Diagnostic Symptoms

- Often referred as latent potato mosaic.
- Light yellow mottling with slight crinklingon potato plants.
- Interveinal necrosis of top foliage.
- Stunting of diseases plants.
- Leaves may appear slightly rugose where strains of PV Y combines.

Spread

- Potato virus X (PV X)
- Spreads mechanically through rubbing of leaves, contact of infected plants, seed cutting knives, farm implements.
- Root clubbing of healthy and diseased plants in field.

Management

- Disease free seed tubers for planting.
- Rouging of diseased plants.

(b) Severe mosaic –

Diagnostic Symptoms

- Also called potato leaf drop streak.
- Chlorotic streaks on leaves which become necrotic.
- Necrosis of leaf veins and leaf drop streak.
- Interveinal necrosis and stem/petiole necrosis.
- Plant remain stunted in growth.
- Rugosity and twisting of the leaves occurs in combination with PV X and PV A.

Survival and spread

- Pathogen: Potato virus Y (PVY)
- Infected tubers
- Spread by aphids, *Myzus persicae* and *Aphis gossypii*

Management

- Disease free seed tubers for planting
- Rouging of diseased plants
- Aphid control

(c) Leaf roll

- Diagnostic Symptoms
- Upward rolling of leaves, which have a stiff leathery texture.
- Plants stunted and have a stiff upright growth.
- Phloem necrosis of tubers in some varieties.

Spread

- Pathogen: *Potato leaf roll virus.*
- Infected seed tubers or by aphids.

Management of viral diseases

- Disease free seed tubers for planting.
- Use healthy seed, hot and cold weather cultivation, green manuring, irrigation, fertilizer application.
- Plant early bulking and/or maturing cultivars to help seed production programme in areas having short aphid-free periods so that the seed crop may escape the population pressure of aphid vectors.

Model Question Paper

Objective Type Questions

a. Choose the correct answer from the following:

1. *Alternaria solani* belongs to order

(b)	Pleosporales	(b)	Peronosporales
(c)	Hyphomycetales	(d)	Moniliales

2. Early blight of potato is caused by:

(b)	*Alternaria solani*	(b)	*Rhizoctonia bataticola*
(c)	*Sphacelotheca sorghi*	(d)	*Sphacelotheca cruenta*

3. The Late blight of potato is caused by

(b)	*Phytophthora infestans*	(b)	*Sphacelotheca cruenta*
(c)	*Rhizoctonia solani*	(d)	*Rhizoctonia bataticola*

4. The most characteristic symptom of the potato disease like brown-black necrotic spot-angular, oval shape characterized by concentric rings produce a target board effect

(b)	Alternaria blight	(b)	Angular leaf spot
(c)	Anthracnose	(d)	Late blight

5. The body of the conidia of *Alternaria solani* is divided by
   (b) Horizontal septa (b) Vertical septa
   (c) Both (d) None
6. *Phytophthora infestans* belongs to order.
   (b) Pleosporales (b) Peronosporales
   (c) Hyphomycetales (d) Moniliales
7. Dutch rule is followed for which disease of potato
   (b) Alternaria blight *(b)* Late blight
   (c) Black scurf (d) None
8. The optimum temperature for germination of the sporangia by zoospores of *Phytophthora infestans* is.
   (b) 12-13^{0}C (b) 22-24^{0}C
   (c) 25-30 ^{0}C (d) 30-35^{0}C
9. The optimum temperature for germination of the sporangia by germ tube of *Phytophthora infestans* is.
   (b) 12 ^{0}C (b) 24^{0}C
   (c) 30 ^{0}C (d) 35^{0}C
10. The perfect stage of *Rhizoctonia solani* is
   (b) *Ascochyta rabiei* (b) *Rhizoctonia bataticola*
   (c) *Thanetophorus cucumeris* (d) *Sphacelotheca cruenta*

Answer

S. No	Answer	S. No	Answer
1	a. Pleosporales	6	b.Peronosporales
2	a. *Alternaria solani*	7	a. Alternaria blight
3	a. *Phytophthora infestans*	8	a. 12-13^{0}C
4	a. Alternaria blight	9	a. 12 ^{0}C
5	c. Both	10	d *Thanetophorus cucumeris*

(b) State whether the following statements are *True* or *False*

1. *Alternaria solani* conidia are fashioned like clubs and have a large beak that is frequently half the length of the conidium.
2. The conidia or the newly discovered conidia on the *Alternaria solani* overwintered mycelium cause the initial infection of the potato crop that follows.
3. The *Alternaria solani* conidia that grow on the patches that appear in early potato blight as a result of primary infection are carried great distances by wind.
4. *Phytophthora infestans* is an endophytic, coenocytic, and hyaline mycelium that is intercellular and has haustoria that resemble double clubs.

5. *Sporangia of Phytophthora infestans* are thin walled, hyaline, oval or pear shaped with a distinct papilla at the tip.
6. *Phytophthora infestans* zoospores are biflagellate and have fine hairs.
7. Rugosity and twisting of the potato leaves occurs in combination with PV X and PV A.
8. The mycelium of *Rhizoctonia solani* turns into sclerotia and serves as means for survival in potato crop under adverse condtion.
9. The black scurf disease in potato develops more in sandy soil than in clay.
10. Severe mosaic in potato is spread by *Myzus persicae* and *Aphis gossypii.*

Answer

S. No	Answer	S. No	Answer
1	True	6	True
2	True	7	True
3	True	8	True
4	True	9	True
5	True	10	True

Descriptive Questions

a. Long answer questions

1. Describe in detail the symptoms, etiology, disease cycle and management of Late blight of potato .
2. Give the diagrammatic representation of disease cycle and epidemiology late blight and black scurf disease of potato.
3. Briefly describe the diagnostic symptoms and management of any two of the following viral diseases of potato

   (i). Mosaic (ii) Leaf curl
4. Illustrate the Black scurf of potato in the following headings:

   (i) Pathogen (ii) Symptoms

   (ii) Disease cycle (iv) Disease management
5. Describe in detail the most distinguishing symptoms, causal organism, disease cycle and management of early blight of potato .

b. Short answer questions

1. Compare between symptom of late blight and early blight of potato .
2. Give the disease cycle and epidemiological factor for development of late blight of potato.
3. Write an integrated management scheduele of late blight of potato. .
4. Explain the disease cycle and epidemiological factor of black scurf of potato.
5. Write systemic position and etiology of pathogen causing early blight of potato.

45

Major Diseases of Tomato Crop & Management

S. No	Name of Disease	Causal organism
	Damping off	*Pythium aphanidermatum*
	Early blight	*Alternaria solani*
	Late blight	*Phytophthora infestans*
	Fusarium wilt	*Fusarium oxysporum* f. sp. *lycopersici*
	Bacterial wilt	*Burkholderia solanacearum*
	Mosaic	Tomato mosaic virus (TMV)
	Leaf curl	Tomato leaf curl virus (TLCV)

1. Damping off

Diagnostic Symptoms

- Disease occurs in two stages, i.e. the pre-emergence and the post-emergence phase.
- In the pre-emergence phase the seedlings are killed just before they reach the soil surface.
- The young radical and the plumule are killed and there is complete rotting of the seedlings.
- The post-emergence phase is characterized by the infection of the young, juvenile tissues of the collar at the ground level.
- The infected tissues become soft and water soaked. The seedlings topple over or collapse.

Etiology

- Pathogen; *Pythium aphanidermatum*
- The fungus has a characteristic mycelium, which coenocytic, hyaline, freely branching and thick hyphae.
- It reproduces both asexually and sexually.
- The asexual reproduction is through the formation of sporangia.
- The sporangium germinates by giving rise to a tubular structure which ends in a vesicle.
- The zoospores are kidney shaped with two flagella.

- The zoospores germinate to produce a germ tube which gives rise to a characteristic coenocytic mycelium.
- The sexual reproduction of the fungus is characterized by the formation of oogonium, antheridium and oospore.
- Oospore is aplerotic, i.e does not fill the oogonial cell completely.
- The oospores may germinate when conditions are favourable, giving rise to germ tube and mycelium.

Disease cycle

- The fungus is mainly soil borne.
- The oospores persist in soil, resisting adverse conditions.
- When there is sufficient moisture, they germinate and produce the mycelium which infect the host plant.
- The mycelia stage of the fungus is capable of infection the host plant and multiplying very rapidly.
- The zoospores formed in the vesicles of sporangia are commonly released in soil water to swim about and spread from place to place.
- When sexual reproduction starts, resulting in the formation of oospores which help to survive adverse environmental conditions.
- Thus the fungus is capable of living for many years in soil, completing its life cycle both saprophytically and as a facultative parasite.

Management

- Used raised seed bed.
- Provide light, but frequent irrigation.
- Provide better drainage system.
- Seed treatment with fungal culture *Trichoderma viride* (4 g/kg of seed)

2. Early Blight

Diagnostic Symptoms

- The disease occurring on the foliage at any stage of the growth.
- Early blight is first observed on the plants as small, black lesions mostly on the older foliage.
- Spots enlarge, and by the time they are one-fourth inch in diameter or larger, concentric rings can be seen in the center of the diseased area.
- Tissue surrounding the spots may turn yellow. If high temperature and humidity occur at this time, much of the foliage is killed.
- Lesions on the stems are similar to those on leaves, sometimes girdling the plant if they occur near the soil line.
- The fungus also infects the fruit, generally through the calyx or stem attachment.
- Lesions attain considerable size, usually involving nearly the entire fruit; concentric rings are also present on the fruit.

Etiology

Scientific Classification

Kingdom: Fungi

Phylum: Ascomycota

Class: Dothideomycetes

Subclass: Pleosporomycetidae

Order: Pleosporales

Family: Pleosporaceae

Genus: *Alternaria*

Species: *solani*

- Pathogen ;*Alternaria solani*
- Mycelium is septate, branched, light brown which become darker with age.
- Conidiophores are dark colored.
- Conidia are beaked, muriform, dark colored and borne singly.

Disease cycle

- The disease spreads through infected seeds.
- It may survive in seed bed soils and field, with plant refuse.
- Pathogen already present in soil and plant debris causes primary infection on next crop by wind, rain, irrigation and water splash on the lower leaves.
- Secondary spread occurs through conidia developed on primary spots.
- Infected seed also produce diseased seedlings and serves as primary source.
- Infection takes place through stomata and wounds.

Epidemiology

- Optimum temperature-20-30^{0}C
- Dry warm weather alternating with the intermittent rains.
- Reduction in plant vigour and senescence.

Management

- Use resistant or tolerant cultivars.
- Change the nursery beds location every season,h eradicate weeds and volunteer tomato plants, fertilize properly.
- Avoid planting overlapping crops in adjacent area.

Late blight

Diagnostic Symptoms

- At first water soaked spots appear usually at the edge of the lower leaves.
- In moist weather the spot enlarge and form brown blightened areas with indefinite borders.

- A zone of whitish downy fungal growth can be observed on lesions on the undersurface of leaves. Death of leaves occurs rapidly.
- Petioles and stems are affected with lesions and may die in a short time.
- Purplish brown spots appear on skin of tubers.
- Symptoms on fruits are expressed in the form of dark, greasey spots, which can enlage to occupy the entire fruit. Complete rotting of fruits can be observed.

Etiology

- Pathogen: *Phytophthora infestans.*
- Mycelium is endophytic, coenocytic and hyaline which are inter cellular with double club shaped haustoria type.
- Sporangiophores are hyaline, branched intermediate and thick walled.
- Sporangia are thin walled, hyaline, oval or pear shaped with a definite papilla at the apex.
- Sporangium may act as a conidium and germinate directly to form a germ tube.
- Zoospores are biflagellate possess fine hairs while the other does not.
- Sexual reproduction in *Phytophthora* is not very common.

Disease cycle

- Infected seed and the infected soil may serve as a source of primary infection.
- Infected seeds are mainly responsible for persistence of the disease from crop to crop.
- Secondary infection is caused by air borne sporangia.

Management

- Select healthy tubers for planting.
- Delayed harvesting.
- High ridging to about 10-15cm height reduces tuber infection.
- Apply potasic fertilizer.

3. Fusarium Wilt

Diagnostic Symptoms

- The first symptom of the disease is clearing of the veinlets and chlorosis of the leaves.
- The younger leaves may die in succession and the entire may wilt and die in a course of few days. Soon the petiole and the leaves droop and wilt.
- In young plants, symptom consists of clearing of veinlet and dropping of petioles. In field, yellowing of the lower leaves first and affected leaflets wilt and die.

- The symptoms continue in subsequent leaves. At later stage, browning of vascular system occurs.
- Plants become stunted and die.

Etiology

- Pathogen ;*Fusarium oxysporum f. sp. lycopersici.*
- Mycelium is septate and hyaline.
- They produce macro and micro conidia.
- Micro conidia are one celled, hyaline, ovoid to ellipsoid.
- Chlamydospores are formed in older mycelium.
- Three races of pathogen, Races 1,2, and 3, have been identified.

Disease cycle

- The fungus is seed borne and soil borne.
- The fungus survives in the soil as chlamydospores or as saprophytically growing mycelium in infected crop debris for more than 10 years.
- One of the chief methods of its distribution is by seedlings raised in infected soil.
- Transfer of conidia through surface drainage water and agricultural implements also help in distribution of the pathogen from field to field.

Epidemiology

- Optimum temperature-28^{0}C.
- Other conditions which predispose the plant to wilt are short day length, low light intensity, low nitrogen and phosphorus and high potassium.

Management

- Summer ploughing.
- The affected plants should be removed and destroyed.
- Seed treatment with *Trichoderma viride* 1% WP @ 9 g/kg seed·
- Root zone application: Mix thoroughly 2.5 kg of the *T. viride* 1% WP in 150 kg of compost or farmyard manure and apply this mixture in the field after sowing/ transplanting of crops.
- Crop rotation with a non-host crop such as cereals.

5. Bacterial Wilt

Diagnostic Symptoms

- Characteristic symptoms of bacterial wilt are the rapid and complete wilting of normal grown up plants.
- Lower leaves may drop before wilting. Pathogen is mostly confined to vascular region; in advantage cases, it may invade the cortex and pith and cause yellowbrown discolouration of tissues.

- Infected plant parts when cut and immersed in clear water, a white streak of bacterial ooze is seen coming out from cut ends.

Etiology

- Pathogen; *Burkholderia solanacearum.*
- The bacterium is gram negative, rod shaped often occurs in pairs, motile with 1 – 4 flagella..

Disease cycle

- The bacterium is soil borne, survives for 3 years in fallow and for a unlimited period in cultivated land..
- Infection always occurs through wounds caused by transplanting, cultural operation or nematode infections.
- The organism first moves into the large xylem vessels.
- When a single lateral bundle is invaded drooping of leaves is common but if all bundles are invaded the plant wilt.
- The chilli, egg plant, grount nut, potato and tobacco are alternative hosts of the pathogen which help it to survive between tomato crops.

Management

- Avoid damage to seedling while transplanting.
- Apply bleaching powder @ 10kg/ha.
- Crop rotations, viz., cowpea-maize-cabbage, okra-cowpea-maize, maize-cowpea-maize and finger millet-egg plant are reported effective in reducing bacterial wilt of tomato.
- Rotate with non-host crops, particularly with paddy.
- Use seedlings from pathogen free seed beds.
- Restriction of irrigation water flowing from affected field to healthy field.

6. Mosaic

DiagnosticSymptom

- The disease is characterized by light and day green mottling on the leaves often accompanied by wilting of young leaves in sunny days when plants first become infected.
- The leaflets of affected leaves are usually distorted, puckered and smaller than normal.
- Sometimes the leaflets become indented resulting in "fern leaf" symptoms.
- The affected plant appears stunted, pale green and spindly.
- The virus is spread by contact with clothes, hand of working labour, touching of infected plants with healthy ones, plant debris and implements (Fig.5).

Etiology

- Pathogen Tomato mosaic virus (TMV)
- Virus particles are rod shaped, not enveloped, usually straight with helical construction and measure 300x 8nm.
- The single stranded RNA constitutes about 5 percent of the particle weight.
- Thermal inactivation point is 85 - 90°C.

Mode of spread and survival

- The virus is seed borne and upto 94% of seeds may contain the virus.
- The virus infection occurs during transplanting. It is readily transmissible.
- Many solanaceous plants are susceptible to Tomato mosaic virus.
- The virus is spread easily by man and implements in cultural operations or by sap or by animals and by leaf contact.
- The virus particles get entry in the cells after contact and induce symptoms.

Management

- Seeds from disease free healthy plants should be selected for sowing.
- Soaking of the seeds in a solution of Trisodium Phosphate (90 g/litre of water) a day before sowing helps to reduce the disease incidence.
- In the nursery all the infected plants should be removed carefully and destroyed.
- Seedlings with infected with the viral disease should not be used for transplanting.
- Crop rotation with crops other than tobacco, potato, chilli, capsicum, brinjal, etc. should be undertaken.

7. Leaf curl

Diagnostic Symptoms

- Leaf curl disease is characterized by severe stunting of the plants with downward rolling and crinkling of the leaves. The newly emerging leaves exhibit slight yellow colouration and later they also show curling symptoms.
- Older leaves become leathery and brittle. The nodes and internodes are significantly reduced in size.
- The infected plants look pale and produce more lateral branches giving a bushy appearance. The infected plants remain stunted.

Etiology

- Pathogen: Tomato leaf curl virus (ToLCV)
- The particles of the virus have a geminate structure, that is, they are made up of two incomplete icosahedra(isometric particles).
- It is a gemini virus.
- The nucleic acid is a circular DNA molecule.
- The virus particles are 80nm in diameter.

Disease cycle

- It is neither seed nor sap transmissible.
- But seeds from fresh fruits having infection may have the virus on the seed coat.
- The virus is transmitted by white fly, *Bemisia tabaci* and grafting.
- Even a single viruliferous insect is able to transmit the virus.

Management

- Raising nursery in protected condition (with net of sufficient mesh size to prevent the entry of vector, whitefly.
- Seeds from disease free healthy plants should be selected for sowing.
- In the nursery all the infected plants should be removed carefully and destroyed. Seedlings infected with the viral disease should not be used for transplanting.
- Keep yellow sticky traps @ 12/ha to monitor the white fly.
- Raise barrier crops-cereals around the field.
- Removal of weed host. Protected nursery in net house or green house.

Model Question Paper

Objective Type Questions

A.Choose the correct answer from the following:

1. The sexual reproduction of the *Pythium aphanidermatum* is characterized by the formation of

(a) Oogonium (b) Antheridium
(c) Oospore (d) All

2. Damping off disease of tomato is caused by:

(a) *Pythium aphanidermatum* (b) *Sphacelotheca cruenta*
(c) *Rhizoctonia bataticola* (d) *Rhizoctonia solani*

3. Late blight of tomato is caused by

(a) *Phytophthora infestans* (b) *Sphacelotheca cruenta*
(c) *Rhizoctonia bataticola* (d) *Rhizoctonia solani*

4. *Fusarium oxysporum* f. sp. *lycopersici belongs to order.*

(a) Pleosporales (b) Peronosporales
(c) Hyphomycetales (d) Moniliales

5. Bacterial wilt of tomato is caused by

(a) *Burkholderia solanacearum* (b) *Xanthomonas solanacearum*
(c) *Erwinia carotovora* (d) *Bacilus subtilis*

6. Late blight of tomato is caused by

(a) *Phytophthora infestans* (b) *Sphacelotheca cruenta*
(c) *Rhizoctonia bataticola* (d) *Rhizoctonia solani*

7. The most characteristic symptom of the tomato disease like brown-black necrotic spot-angular, oval shape characterized by concentric rings produce a target board effect
   (a) Alternaria blight (b) Angular leaf spot
   (c) Anthracnose (d) None
8. The body of the conidia of *Alternaria solani* is divided by
   (a) Horizontal septa (b) Vertical septa
   (c) Both (d) None
9. *Phytophthora infestans* belongs to order
   (a) Pleosporales (b) Peronosporales
   (c) Hyphomycetales (d) Moniliales
10. Alternaria solanibelongs to order
   (a) Pleosporales (b) Peronosporales
   (c) Hyphomycetales (d) Moniliales

(a) Answer

S. No	Answer	S. No	Answer
1.	d. All	6.	a. *Phytophthora infestans*
2.	a. *Pythium aphanidermatum*	7.	a. Alternaria blight
3.	a. *Phytophthora infestans*	8.	c. Both
4.	c. Hyphomycetales	9.	b. Peronosporales
5.	a. *Burkholderia solanacearum*	10.	a. Pleosporales

(b) State whether the following statements are *True* or *False*.

1. The seedlings of tomato infected with *Pythium aphanidermatum* topple over or collapse.
2. The zoospores of *Pythium aphanidermatum* are kidney shaped with two flagella.
3. Oospore is *Pythium aphanidermatum* is aplerotic.
4. Damping off disease of tomato is soil borne in nature.
5. Conidia of *Alternaria solani* are beaked, muriform, dark colored and borne singly.
6. Alternaria blight disease spread through infected seed of tomato .
7. A zone of whitish downy fungal growth can be observed on lesions on the undersurface of leaves of tomato is characteristic symptoms of late blight disease of tomato.
8. The mycelium *Phytophthora infestans* is endophytic, coenocytic and hyaline which are inter cellular with double club shaped haustoria type.
9. Sporangia of *Phytophthora infestans* are thin walled, hyaline, oval or pear shaped with a definite papilla at the apex.

10. Zoospores of *Phytlphthora infestans* are biflagellate possess fine hairs .

Answer

S. No	Answer	S. No	Answer
1	True	6	True
2	True	7	True
3	True	8	True
4	True	9	True
5	True	10	True

Descriptive Questions

a. Long answer questions

1. Describe in detail the symptoms, etiology, disease cycle, epidemiology and management of damping off of tomato.
2. Describe in detail the symptoms, etiology, disease cycle and management of Fusarium wilt of tomato.
3. Describe in detail the symptoms, etiology, disease cycle and management of Late blight of tomato.
4. Give the diagrammatic representation of disease cycle and epidemiology late blight and early blight disease of tomato.
5. Describe in detail the most distinguishing symptoms, causal organism, disease cycle and management of early blight of tomato .

b. Short answer questions

1. Compare between symptom of late blight and early blight of tomato .
2. Compare between symptom of Fusarium wilt and Bacterial wilt of tomato.
3. Give the disease cycle and epidemiological factor for development of late blight of tomato.
4. Write an integrated management schedule of bacterial wilt of tomato. .
5. Explain the disease cycle and epidemiological factor of Late blight of tomato.

46

Major Diseases of Brinjal Crop & Management

S. No	Disease	Causal organism
1	Damping off	*Phythium aphanidermatum*
2	Phomopsis fruit rot or blight	*Phomopsis vexans*
3	Sclerotinia blight	*Sclerotinia sclerotiorum*
3	Collar rot	*Sclerotium rolfsii*
4	Bacterial wilt	*Pseudomonas solanacearum*
5	Little leaf	*Mycoplasma*

1) Phomopsis fruit rot or Blight

Diagnostic Symptoms

- Plants are attacked at all stages of growth.
- On leaves, circular to irregular, clearly defined grayish brown spots having light centers appear. The diseased leaves become yellowish in colour and may drop off.
- Several black pycnidiacan be seen on older spots.
- Lesions on stem are dark brown, round to oval and have grayish centers where pycnidia develop.
- At the base of the stem, the fungus causes characteristic constrictions leading to canker development and toppling of plants.
- On fruits, small pale sunken spots appear which on enlargement cover entire fruit surface. These spots become watery leading to soft rot phase of the disease. A large number of dot like pycnidia also develop on such spots.
- Infection of fruit through calyx leads to development of dry rot and fruits appear black and mummified (Fig.1).

Etiology

Scientific Classification

Kingdom: Fungi

Division: Ascomycota

Class: Sordariomycetes

Order: Diaporthales

Family: Valsaceae

Genus: *Phomopsis*

Species: *vexans*

- Pathogen: *Phomopsis vexans*
- Perfect stage: *Diaporthe vexans*
- Mycelium of the pathogen is septate and hyaline becoming dark with age.
- Pycnidia are submerged and later becoming erumpent with a prominent ostiole.
- Pycnidia with or without beak are found in the affected tissue .They are globose or irregular.
- Conidiophores in the pycnidium are hyaline, simple or branched.
- Conidia are hyaline, one celled and sub cylindrical.
- Conidia are produced on simple to branched conidiophores and are of 2 types: alpha conidia, which are sub cylindrical and beta conidia, which are filiform and curved. Role of beta-conidia in the epidemiology of the disease is not very clear.
- Ascospores are hyaline, narrowly ellipsoid to bluntly fusoid with one septum.
- Perfect stage produces perithecia in which asci with 8 hyaline, bicelled, ellipsoid-fusoid ascospores are produced which are usually constricted at septum.

Disease Cycle

- Pathogen is seed borne and also survives in plant debris as mycelium and pycnidia.
- Pathogen is locally disseminated as water borne pycniospores.
- Spore exude out through the ostiole and can easily be thrown away rain drops.They can also disseminated tools and inscets.
- Inoculum present in the form of mycelium and pycnidia results in the diseased seedlings.

Management

- Use of disease free seed.
- Deep summer ploughings.
- Practicing crop rotation.
- Adopting good field sanitation.
- Destruction of infected plant material.
- Seeds obtained from disease plants should be used for planting.
- Hot water treatment of seed at 50 ^{0}C for 30 minutes.

2. Sclerotina Blgiht

***Sclerotinia sclerotiorum* (Lib.) De Bary.**

Diagnostic Symptom

- Infection may occur at any part of the foliage, mainly the stem or branches.
- A dry and discoloured spot develops at the point of infection.
- It gradually girdles the entire stem and also progresses up and down.
- As a result of tissue necrosis, the portion of the plant beyond the point of infection wilts.
- Damage to brinjal is mainly through partial or complete wilting of the plant.
- Occasionally, the fungus may attack the seedlings in nursery (Fig.1).

Etiology

Scientific Classification

Kingdom: Fungi

Division: Ascomycota

Class: Leotiomycetes

Order: Helotiales

Family: Sclerotiniaceae

Genus: *Sclerotinia*

Species: *sclerotiorum*

- Mycelium is hyaline,much branched, septate hyphae which are inter and intra cellular and invades all tissues of affected host portion.
- Pathogen donot produce true conidia or macroconidia.
- Microsclerotia (Spermatia) produced on sclerotia on the discs of over mature apothecia and in culture, as the food supply declines.
- Spermatia germinate very sparsely in water or culture media and apparently do not serve as source of infection or dissemination of the pathogen.
- They spermatize the ascogenous cells formed beneath the rind of the sclerotium and thus initiate apothecial development.
- Sclerotia formed on host surface are usually loaf shaped or globose while those formed in the pith of the stem are elongated.
- On germination these scleortia give rise to several stipes which develop the funnel shaped cup at the tip.
- Ascospores are discharged in abundance from these cup.
- Ascospores are hyaline, 1 celled and ovate, always 8 in each ascus which has an apical pore through which spore discharge occurs with violence.
- Pathogen always disseminated by means of ascospores in the absence of asexual reproduction .

Disease Cycle

- Sclerotia survive in soil or in plant debris.
- Pathogen always disseminated by means of ascospores in the absence of asexual reproduction which are the most common structures of infection.

Management

- Immediately after harvest, the affected plants and debris should be collected and burnt.
- Deep summer ploughing should be given in such a way that surface soil is buried deep.
- Rotation of cropping pattern with crops like beet root, onion, maize, paddy and gingelly eliminate the fungal inoculum in the field.

Model Question Paper

Objective Type Questions

A. Choose the correct answer from the following

1. Phomopsis fruit rot disease of brinjal is caused by:

(a) *Phomopsis vexans* (b) *Sphacelotheca cruenta*
(c) *Rhizoctonia bataticola* (d) *Rhizoctonia solani*

2. Sclerotina blight of brinjal is caused by

(a) *Sclerotinia sclerotiorum* (b) *Sphacelotheca cruenta*
(c) *Rhizoctonia solani* (d) *Rhizoctonia bataticola*

3. *Sclerotinia sclerotiorum belongs to order*

(a) Pleosporales (b) Peronosporales
(c) Helotiales (d) Moniliales

4. Which disease of brinjal is characterized by the development of many dot-like pycnidia on sunken spots that arise and cover the entire fruit surface?

(a) Alternaria blight (b) Angular leaf spot
(c) Anthracnose (d) Phomopsis fruit rot

5. *Phomopsis vexans* belongs to order.

(a) Pleosporales (b) Diaporthales
(c) Hyphomycetales (d) Moniliales

6. The perfect stage of fungus *Phomopsis vexans*

(a) *Diaporthe vexans* (b) *Sphacelotheca cruenta*
(c) *Rhizoctonia bataticola* (d) *Rhizoctonia solani*

7. The optimum temperature for development of Phomopsis fruit rot in brinjal is

(a) 12^0C (b) >18 ^{0}C
(c) 15 ^{0}C (d) 10 ^{0}C

8. The optimum temperature for germination of the sporangia by germ tube of *Phytophthora infestans* is.
   (a) 12 ^{0}C (b) 24^{0}C.
   (c) 30 ^{0}C (d) 35^{0}C
9. Microsclerotia formed on sclerotia in culture and on the overmature apothecia when the food supply is reduced
   (a) *Sclerotinia sclerotiorum* (b) *Rhizoctonia bataticola*
   (c) *Thanetophorus cucumeris* (d) *Sphacelotheca cruenta*
10. *Sclerotinia sclerotiorum* survives in soil as
   (a) Sclerotia (b) Chlaymydospore
   (c) Oospore (d) Teliospore

(a) Answer

S. No	Answer	S. No	Answer
1	a. *Phomopsis vexans*	6	a. *Diaporthe vexans*
2	a. *Sclerotinia sclerotiorum*	7	c. 26 ^{0}C
3	c. Helotiales	8	*a.* >18^{0}C
4	d. Phomopsis fruit rot	9	a. *Sclerotinia sclerotiorum*
5	b. Diaporthales	10	a. Sclerotia

(b) State whether the following statements are *True* or *False*

1. Several black pycnidia can be seen on older spots present on brinjal leaf and fruit infected with Phomopsis fruit rot.
2. Brinjals afflicted with *Phomopsis vexans* have pycnidia, either with or without a beak.
3. *Diaporthe vexans* produces perithecia in brinjal in which asci with 8 hyaline, bicelled, ellipsoid-fusoid ascospores are produced .
4. The pathogen *Diaporthe vexans* is seed borne and also survives in brinjal debris as mycelium and pycnidia.
5. *Diaporthe vexans* is locally disseminated as water borne pycniospores.
6. *Sclerotinia sclerotiorum* produces microsclerotia on the discs of over mature apothecia and in culture, as the food supply declines.
7. Ascospores of *Sclerotinia sclerotiorum* are hyaline, 1 celled and ovate, always 8 in each ascus which has an apical pore through which spore discharge occurs with violence.
8. Sclerotia of *Sclerotinia sclerotiorum* survive in soil or in brinjal plant debris.
9. *Sclerotinia sclerotiorum* always disseminated by means of ascospores in the absence of asexual reproduction which are the most common structures of infection.
10. Brinjal sclerotinia blight is controlled by foliar spraying with Mancozeb 75% WP @ 0.25 percent.

Answer

S. No	Answer	S. No	Answer
1	True	6	True
2	True	7	True
3	True	8	True
4	True	9	True
5	True	10	True

Descriptive Questions

a. Long answer questions

1. Describe in detail the symptoms, etiology, disease cycle, epidemiology and management of Phomopsis fruit rot of brinjal .
2. Describe in detail the symptoms, etiology, disease cycle and management of Sclerotina blight of brinjal .
3. Briefly describe the diagnostic symptoms and management of any two of the following fungal diseases of brinjal
   (i) Sclerotina blight (ii) Phomopsis fruit rot
4. Illustrate the Phomopsis fruit rot of brinjal in the following headings:
   (i) Pathogen (ii) Symptoms
   (ii) Disease cycle (iv) Disease management
5. Describe in detail the most distinguishing symptoms, causal organism, disease cycle and management of Sclerotina blight of brinjal.

b. Short answer questions

1. Give the disease cycle and epidemiological factor for development of Sclerotina blight of brinjal.
2. Write an integrated management schedule of Sclerotina blight of brinjal.
3. Explain the disease cycle and epidemiological factor of Phomopsis fruit rot of brinjal.
4. Write systemic position and etiology of pathogen causing Phomopsis fruit rot of brinjal.
5. Write systemic position and etiology of pathogen Sclerotina blight of brinjal

47

Major Diseases of Chilli Crop & Management

S. No	Disease	Pathogen
1	Anthrac nose & Fruit Rot	*Colletotrichum capsici*
2	Fusarium wilt	*Fusarium oxysporum f.sp.capsici*
3	Powdery mildew	*Leveillula taurica*
4	Viral diseases	
a	Mosaic	Chilli mosaic virus
b	Leaf curl	Tobacco leaf curl virus

1. Anthracnose or Fruit Rot and Die Back

Diagnostic Symptoms

The disease occurs in two forms: the die back and the ripe -rot

Die back symptoms

- Small, circular to irregular, brownish black scattered spots appear on leaves.
- Severely infected leaves defoliate.
- Infection of growing tips leads to necrosis of branches from tip backwards.
- Necrotic tissues appear grayish white with black dot like acervuli in the center.
- Shedding of flowers due to the infection at pedicel and tips of branches.

Fruit symptoms

- Ripe fruits are more liable for attack than the green ones.
- Small, circular, yellowish to pinkish sunken spots appear on fruits.
- Spots increase along fruit length attaining elliptical shape.
- Severe infection result in the shrivelling and drying of fruits.
- Such fruits become white or greyish in colour and lose their pungency.
- On the surface of the lesions minute black dot like fruiting bodies called 'acervuli'.
- Develop in concentric rings and fruits appear straw coloured.
- The affected fruits may fall off subsequently.

Etiology

Scientific Classification

Kingdom: Fungi

Division: Eumycota

Sub-division: Deuteromycotina

Class: Coelomycetes

Order: Melanconiales

Family: Melanconiaceae

Genus: *Colletotrichum*

Species: *capsici*

- Pathogen: *Colletotrichum capsici*
- The mycelium is septate and inter and intra cellular.
- Conidia in mass appear pinkish.
- Conidia are falcate, fusiform with acute apices and narrow truncated base.
- Acervuli formed are rounded and elongated.
- Setae are brown 1-5 septate, rigid, hardly swollen at the base, slightly tapered towards the paler acute apex.

Disease cycle

- The pathogen survives in the plant debris, which include infected twigs, stem and fruits.
- The secondary spread take place through the wind borne conidia produced in the infected left out plants.
- The conidia germinate within four hours in water and an appresoria is developed for further growth of the fungus between the plant tissues.
- Infected seeds or badly infected fruits also carry the pathogen and serves as source of primary inoculums.

Management

- Collect and destroy all infected plant parts.
- Collect seeds only from fruits without infection.
- Removal and destruction of solanaceous weed hosts and infected plant debris.
- Resistant varieties: Perennial, Bengal Green, S20-1, H-4, H-6, Lorai, etc.

2. Fusarium wilt

Diagnostic Symptoms

- Characterised by wilting of the plant and upward and inward rolling of the leaves. The leaves turn yellow and die.

- Commonly appear localised areas of the field where a high percentage of the plants wilt and die, although scattered wilted plants may also occur.
- Disease symptoms are characterised by an initial slight yellowing of the foliage and wilting of the upper leaves that progress in a few days into a permanent wilt with the leaves still attached.

Etiology

Scientific Classification

Kingdom: Fungi

Division: Eumycota

Sub-**division:** Deuteromycotina

Class: Hyphomycetes

Order: Moniliales

Family: Dematiaceae

Genus: *Fusarium*

Species: *solani*

- Pathogen:*Fusarium oxysporum f.sp.capsici.*
- Mycelium is grayish white.
- Microconidia are formed singly, hyaline and cylindrical.
- Macro conidia are cylindrical to falcate.
- Chlamydospores are globose to oval and rough walled. discoloured, particularly in the lower stem and roots.

Disease cycle

- The fungus is mainly soil borne and survive in soil mainly as chlaymydospores formed by the hyphal and conidial cells.
- Infection is always through injured roots.
- Deep wounds to expose the xylem help in easy infection.
- After entry into the root, the fungus multiply extensively in vascular tissues, interrupting the transport of nutrients to the above ground plant parts, where the mycelium is formed in abundance to partially or completely plug the vessels, interrupting the transport of nutrients to the above ground plant parts, thus stunting the plant growth.
- Continued dislocation of flow of plant nutrients leads to plant wilt.

Epidemiology

- Optimum soil temperature-20-30^{0}C
- Hot and dry summers followed by rains sees to favor the disease.

Management

- Use of wilt resistant varieties.
- Seed treatment with 4g *Trichoderma viride* formulatiois effective.
- Mix 2kg *T.viride* formulation with 50kg FYM, sprinkle water and cover with a thin polythene sheet. When mycelia growth is visible on the heap after 15 days, apply the mixture in rows of chilli in an area of one acre.

(3) Viral diseases

a. Mosaic Viruses

Diagnostic Symptoms

- Light green and dark green patches on the leaves.
- These may be sunken or raised (puckering).
- The leaves are generally reduced in size and somewhat become filamentous like.
- Diseased plants produce less flowers and fruits.

Causal Agent

- The virus is sap transmissible and can be easily transmitted through contact during cultivation.
- Natural transmission of virus through aphid vectors such as *Myzus persicae*, *Aphis gossypii*.

Management

- Grow resistant/tolerant varieties.
- Use healthy, certified and weed free seeds.
- Avoid planting overlapping crops in adjacent area.
- Every infected plant in the nursery needs to be carefully taken out and destroyed.
- Ecological engineering of chilli with growing intercrops such as cowpea, maize, coriander etc.
- Rotate the chilli crop with a non-host cereal crop, cucurbit, or cruciferous vegetable.

b. Leaf curl

Dianostic Symptoms

- Leaves curl towards midrib and become deformed.
- Stunted plant growth due to shortened internodes and leaves greatly reduced in size.
- Flower buds abscise before attaining full size and anthers do not contain pollen grains.
- The virus is generally transmitted by whitefly. So control measures of whitefly in this regard would be helpful (Fig.1).

Causal agent

- The tobacco leaf curl virus causes the leaf curl disease.
- The virus is not sap transmitted and is also not seed borne.
- In nature, it is transmitted by the white fly (*Bemesia tabaci*).

Management of viral diseases

- Selection of healthy and disease - free seed.
- Nursery beds should be covered with nylon net or straw to protect the seedlings from viral infection.
- Raise 2-3 rows of maize or sorghum as border crop to restrict the spread of aphid vectors.
- Collect and destroy infected virus plants as soon as they are noticed.

Model Question Paper

Objective Type Questions

a. Choose the correct answer from the following:

1. Anthrac nose disease of chilli is caused by:
   (a) *Colletotrichum capsici* (b) *Sphacelotheca cruenta*
   (c) *Rhizoctonia bataticola* (d) *Rhizoctonia solani*
2. Fusarium wilt of chilli is caused by
   (a) *Fusarium oxysporum f.sp.capsici* (b) *Sphacelotheca cruenta*
   (c) *Rhizoctonia solani* (d) *Rhizoctonia solani*
3. *Fusarium oxysporum f.sp.capsici* belongs to order.
   (a) Pleosporales (b) Peronosporales
   (c) Helotiales (d) Moniliales
4. Minute black dot like fruiting bodies called 'acervuli' produced in which disease of chilli
   (a) Alternaria blight (b) Angular leaf spot
   (c) Anthracnose (d) Leaf spot
5. *Colletotrichum capsici* belongs to order.
   (a) Pleosporales (b) Diaporthales
   (c) Hyphomycetales (d) Melanconiales
6. The optimum temperature for development of Anthracnose in chilli is
   (a) 12^0C (b) 22^0C.
   c. 28 ^{0}C (d) 30^0 C
7. Microsclerotia formed on sclerotia in culture and on the overmature apothecia when the food supply is reduced
   (a) *Sclerotinia sclerotiorum* (b) *Rhizoctonia bataticola*

(c) *Thanetophorus cucumeris* (d) *Sphacelotheca cruenta*

8. *Fusarium oxysporum f.sp.capsici* survives in soil as
   (a) Sclerotia (b) Chlaymydospore
   (c) Oospore (d) Teliospore

9. The nature of *Fusarium oxysporum f.sp.capsici is*
   (a) Soil borne (b) Seed borne
   (c) Air borne (d) Vector borne

10. Leaf curl of chilli is transmitted by.
   (a) Aphid (b) White fly
   (c) Hopper (d) Termite

Answer

S. No	Answer	S. No	Answer
1	a. *Colletotrichum capsici*	6	c.28 ^{0}C
2	a. *Fusarium oxysporum f.sp.capsici*	7	a. *Sclerotinia sclerotiorum*
3	d. Moniliales	8	b. Chlaymydospore
4	c.Anthracnose	9	a. Soil borne
5	d. Melanconiales	10	b. White fly

b State whether the following statements are *True* or *False*

1. Anthracnose disease occurs in two forms: the die back and the ripe –rot.
2. Minute black dot like fruiting bodies called 'acervuli' develop on the chilli fruit infected with *Colletotrichum capsici* .
3. Tobacco leaf curl virus is generally transmitted by whitefly.
4. Leaf curl disease is caused by the tobacco leaf curl virus.
5. Setae of *Colletotrichum capsici*are brown 1-5 septate, rigid, hardly swollen at the base, slightly tapered towards the paler acute apex.
6. The secondary spread of anthracnose take place through the wind borne conidia produced in the infected left out chilli plants.
7. The *Fusarium oxysporum f.sp.capsici*is mainly soil borne and survive in soil mainly as chlaymydospores .
8. *Fusarium oxysporum f.sp.capsici* belongs to family dematiaceae
9. Tobacco leaf curl virus is not sap transmitted and is also not seed borne.
10. Natural transmission of Chilli mosaic virus through aphid vectors

Answer

S. No	Answer	S. No	Answer
1	True	6	True
2	True	7	True
3	True	8	True
4	True	9	True
5	True	10	True

Descriptive Questions

a. Long answer questions

1. Explain the symptoms and management of the following fungal diseases that affect chillies
   (i) Anthracnose (ii) Wilt
2. Use the following headings to illustrate the anthracnose of chilli
   (i) Pathogen (ii) Symptoms
   (iii) Disease cycle (iv) Management of the disease
3. Give a thorough description of the diagnostic symptoms, causes, disease cycle, epidemiology, and management of Phomopsis fruit rot of brinjal .
4. Clearly describe the symptoms, causes, development of the disease, and managment for chilli anthracnose.
5. Give a thorough description of the most distinctive symptoms, the etiological agent, the progression of the disease, and how to manage Fusarium wilt of chilli.

b. Short answer questions

1. Describe the epidemiological factors and disease cycle that lead to the development of Fusarium wilt of chilli.
2. Develop an integrated management plan for chilli Fusarium wilt
3. Describe the systemic position and etiology of the organism that causes chilli anthracnose.
4. Describe the systemic position and etiology of the organism that causes chilli Fusarial wilt
5. Explain about the cultural management procedures that need to be followed to control fungal disease in chilli.

48

Major Diseases of Rose Crop & Management

S.No.	Name of Diseases	Causal Organism
1	Powdery mildew	*Sphaerothecapannosa var. rosae*
2	Die-back	*Diplodia rosarum*
3	Black spot	*Diplocarpan rosae*
4	Rust	*Phragmidum mucronatum*

1) Powdery mildew

Diagnostic Symptoms

- A white, powdery fungal growth on the upper, lower or both leaves and shoots.
- There may be discolouration of the affected parts of the leaf, and heavily infected young leaves can be curled and distorted.
- Mildew growth may also be found on the stems, flower stalks, calyces and petals.
- Heavily infected flower buds frequently fail to open properly.
- Mildew growth on stems and flower stalks is usually thicker and more mat-like than that on the leaves.
- Mildew growth on all parts may turn browner as it ages.

Etiology

Scientific Classification

Kingdom: Fungi

Division: Eumycota

Class: Leotiomycetes

Order: Erysiphales

Family: Erysiphaceae

Genus: *Sphaerotheca*

Species: *pannosa*

- Pathogen: *Sphaerotheca pannosa var. rosae/Sphaerotheca pannosa.*
- Obligate parasite: Pathogen must have a living host complete its life cycle, which can be as short as 72to 96 hours in favouable conditions.

- Mycelium: white, septate, ectophytic and sends globose haustoria into the epidermal cells of the host.
- Conidiophores: are short and erect.
- Conidia: one celled, oblong, minutely verrucose with many large fat globules .
- Cleistothecia: formed towards the end of the season on the leaves, petals, stems and thornswith simple myceloid appendages.
- Ascus: contains eight ascospores

Disease cycle

- Pathogen survives as mycelium in dormant buds and shoots or as cleistothecia.
- They discharge their spores in the spring.
- Secondary infection occurs through wind borne conidia throughout year.
- However, when it grows it can survive in buds, emerging with the appearance of the flowers.

Epidemiology

- Hot and warm days with cool nights.
- Water on leaf surface prevents the spores from germinating.

Management

- Collection and burning of fallen leaves.
- Prune the infected plants and dispose of them properly.
- Provide plants with adequate nutrients and water to maintain their immune defences.
- Keep the soil well watered and mulched to prevent moisture loss and to cover up overwiterinng spores.
- Maintain proper spacing to provide good air circulation and prune them regularly to prevent overcrowding.
- Use fan to provide adequate ventilation during humid nights.
- Resistant varieties: Ashwini, Ambika, Angeles, American pride, Surabhi.

2. Die-back

Diagnostic Symptoms

- Browning and dieback of the tips of young shoots in spring.
- Browning and dieback of a pruning stub, which then progresses further down the branch.
- Dieback of twigs, branches, main stems or even the whole plant at any time of year.
- Fungal structures, such as tiny black fruiting bodies, are sometimes visible on the affected parts of the plant
- In some instances there may also be associated root decay.

Etiology

Scientific Classification

Kingdom: Fungi

Division: Ascomycetes

Class: Dothideomycetes

Order: Botriosphaeriales

Family: Botriosphaeiaceae

Genus:*Diplodia*

Species: *rosarum*

- Pathogen*: Diplodia rosarum/Botryodiplodia theobromae.*
- Pathogen produces round, black pycnidia which bear spores.
- Pycnidia: round, black pycnidia which bear spores.
- Pycnidiospores: dark coloured and two celled.
- Perithecia: immersed in the host tissue and are surrounded by a pseudostroma.
- Ascospores: ellipsoidal or fusoid, hyaline, two celled with the septum in or near the middle.

Disease Cycle

- Pathogen survives inplant debris.
- Spores from pycnidia are considered as primary source of infection and spread.
- Scondary infection takes place thorugh spores spread through air current or blowing rain.

Management

- Plant roses in well-prepared soil, making sure that the roots are well spread out.
- Do not plant roses into soil that has grown roses previously, without taking remedial action. Change the soil to a depth of at least 30cm and a width of at least 60cm.
- Use resistant / tolerant varieties: Blue moon, Red gold, Summer queen, etc.
- Pruning should be done so that lesions on the young shoots will be eliminated and apply chaubatia paste in the pruned area.
- Avoid soils that are prone to either drought or waterlogging.
- Feed plants in spring with a proprietary rose fertiliser and mulch the soil to prevent water loss.

C. Black spot

Diagnostic Symptoms

- Dark brown tar coloured spots with fringed borders.
- Spots coalesce forming large patches.

- Infected leaves turn brown and defoliate.
- Fungus may also attack stems and flowers of rose bushes.
- On stems, infected areas are blackened with blistered appearance dotted withpustules.

Etiology

Scientific Classification

Kingdom: Fungi

Division: Ascomycota

Class: Leotiomycetes

Order: Helotiales

Family: Dermateaceae

Genus: *Diplocarpan*

Species: *rosae*

- Pathogen*: Diplocarpan rosae (Anamorph: Marssonina rosae)*
- Vegetative body of the fungus consists of two parts *viz*., the subcuticular mycelium and the internal mycelium.
- Fungus produces acervuli between the outer wall and cuticle of the epidermis.
- Fungus produces *Marssonina*-type conidia in acervuli and ascospores in tiny apothecia formed in old lesions.
- Conidiophore: short arise from a thin black stroma and give rise to successive crops of conidia.
- Conidia push up and rupture the cuticle.

Disease Cycle

- Pathogen survives as mycelia, ascospores and conidia in infected leaves and canes.
- Both kinds of spores can cause primary infections of leaves in the spring by direct penetration.
- The mycelium grows in the mesophyll, but within two weeks forms acervuli and conidia at the upper surface.
- Conidia are produced throughout the growing season and cause repeated infections during warm, wet weather.

Management

- Use resistant / tolerant varieties: Bebebune, Coronado,Grand opera, Sphinx.
- Affected parts should be collected and destroyed.
- Space roses far enough apart for good air circulation.
- Avoid overhead watering and keep foliage as dry as possible.
- Radiance - escape infection due to waxy surface.

Model Question Paper

a. Objective Type Questions

1. Powdery mildew of rose is caused by.
   (a) *Botrytis cinerea* (b) *Colletotrichum lindemuthianum*
   (c) *Sphaerotheca pannosa var. rosae* (d) *Sphacelotheca cruenta*
2. Black spot of rose is caused by
   (a) *Diplocarpan rosae* (b) *Colletotrichum lindemuthianum*
   (c) *Pythium aphanidermatum* (d) *Sphacelotheca cruenta*
3. The mycelium *of Diplocarpan rosae* forms structure at the upper surface within two weeks of infection
   (a) Chlamydospores (b) Pycnidia
   (c) Acervuli (d) Basidiospore
4. Foliar spray with sulfur fungicide control which disease of rose
   (a) Stem gall (b) Leaf spots
   (c) Powdery mildew (d) Fusarium wilt
5. The *Diplocarpan rosae* produces *Marssonina*-type conidia in acervuli and ascospores in tiny fruiting body called
   (a) Apothecia (b) Pycnidia
   (c) Acervuli (d) Perithecia
6. The telemorphic stage of *Diplocarpan rosae is*
   (a) *Marssonina rosae* (b) *Colletotrichum lindemuthianum*
   (c) *Pythium aphanidermatum* (d) *Sphacelotheca cruenta*
7. The optimum temperature for development of dieback of rose is
   (a) 8-12^{0}C (b) 15-20^{0}C.
   c. 16-26 ^{0}C (d) 30-32^{0} C
8. A fruiting body called cleistothecia formed towards the end of the season on the leaves, petals, stems and thorns of rose with simple myceloid appendages due to infection of
   (a) *Sphaerothecapannosa var. rosae* (b) *Colletotrichum lindemuthianum*
   (c) *Pythium aphanidermatum* (d) *Sphacelotheca cruenta*
9. *Sphaerotheca pannosa var. rosae is*
   (a) Obligate parasite (b) Facultative parasite
   (c) Obligate saprophyte (d) Facultative saprophyte

10. *Sphaerothecapannosa var. rosae* survives in soil as resting strucuture called
    (a) Apothecia (b) Pycnidia
    (c) Cleistothecia (d) Perithecia

(a) Answer

S. No	Answer	S. No	Answer
1	a. *Sphaerotheca pannosa* var. *rosae*	6	a.*Marssonina rosae*
2	a. *Diplocarpan rosae*	7	d. 30-32^0 C
3	c. Acervuli	8	a. *Sphaerothecapannosa var. rosae*
4	c. Powdery mildew	9	a. Obligate parasite
5	a. Apothecia	1o	c. Cleistothecia

b. State whether the following statements are True or False.

1. *Diplodia rosarum* produces round, black pycnidia which bear spores.
2. *Diplocarpan rosae* survives as mycelia, ascospores and conidia in infected leaves and canes.
3. *Diplocarpan rosae* produces *Marssonina*-type conidia in acervuli and ascospores in tiny apothecia formed in old lesions of rose.
4. The pathogen *Sphaerotheca pannosa* is an obligate parasite.
5. When rose leaves have powdery mildew disease, white, floury patches appear up on both sides of the leaves.
6. The fungus *Diplocarpan rosae*produces acervuli between the outer wall and cuticle of the epidermis.
7. Diplocarpan rosae infections and spread are thought to be mostly caused by pycnidia spores.
8. Scondary infection of die-backtakes place thorugh spores spread through air current or blowing rain.
9. *Sphaerotheca pannosa* ascus contains eight ascospores.
10. Foliar spray with combination fungicide viz., Captan + Hexaconazole WP @ 0.3% control powdery mildew of rose.

b Answer

S. No	Answer	S. No	Answer
1	True	6	True
2	True	7	True
3	True	8	True
4	True	9	True
5	True	10	True

B. Descriptive Questions

a. Long answer questions

1. Explain in detail about symptoms, etiology, disease cycle, epidemiology and management of powdery mildew of rose .
2. Briefly describe the diagnostic symptoms, etiology, disease cycle, epidemiology and management of dieback diseases of rose.
3. Describe the black spot of rose in the following headings:

   (i) Pathogen (ii) Symptoms

   (ii) Disease cycle (iv) Disease management
4. Describe in detail the most distinguishing symptoms, causal organism, disease cycle and integrated management of die back of rose.
5. Give the diagrammatic representation of the disease cycle of the following diseases of rose

   (i) Powdery mildew (ii) Die back

   (iii) Black spot

b. Short answer questions

1. Write down the etiology of pathogen causing powdery mildew of rose.
2. Which fungal pathogen causes powdery mildew of rose? Give its systematic position.
3. Write down the integrated management practices of black spot diseases of rose.
4. Suggest some chemicals for management of powdery mildew of rose.
5. Describe favourable condition for development of black spot and dieback disease of rose.

Index